Coatings on Food Packaging and Shelf Life

Coatings on Food Packaging and Shelf Life

Editors

Lili Ren
Liyan Wang

Basel • Beijing • Wuhan • Barcelona • Belgrade • Novi Sad • Cluj • Manchester

Editors

Lili Ren
Jilin University
Changchun, China

Liyan Wang
Jilin Agricultural University
Changchun, China

Editorial Office
MDPI
St. Alban-Anlage 66
4052 Basel, Switzerland

This is a reprint of articles from the Special Issue published online in the open access journal *Coatings* (ISSN 2079-6412) (available at: https://www.mdpi.com/journal/coatings/special_issues/coatings_shelf_life).

For citation purposes, cite each article independently as indicated on the article page online and as indicated below:

Lastname, A.A.; Lastname, B.B. Article Title. *Journal Name* **Year**, *Volume Number*, Page Range.

ISBN 978-3-0365-9178-0 (Hbk)
ISBN 978-3-0365-9179-7 (PDF)
doi.org/10.3390/books978-3-0365-9179-7

© 2023 by the authors. Articles in this book are Open Access and distributed under the Creative Commons Attribution (CC BY) license. The book as a whole is distributed by MDPI under the terms and conditions of the Creative Commons Attribution-NonCommercial-NoDerivs (CC BY-NC-ND) license.

Contents

Aoxue Hu and Yingming Mao
Effects of Pullulan-Based Coatings Incorporating ε-Polylysine and Glutathione on the Preservation of Cowpeas (*Vigna unguiculata* L.) Postharvest
Reprinted from: *Coatings* 2023, 13, 125, doi:10.3390/coatings13010125 1

Mohamed M. Gemail, Ibrahim Eid Elesawi, Muthana M. Jghef, Badr Alharthi, Woroud A. Alsanei, Chunli Chen, et al.
Influence of Wax and Silver Nanoparticles on Preservation Quality of Murcott Mandarin Fruit during Cold Storage and after Shelf-Life
Reprinted from: *Coatings* 2023, 13, 90, doi:10.3390/coatings13010090 9

Sarengaowa, Liying Wang, Yumeng Liu, Chunmiao Yang, Ke Feng and Wenzhong Hu
Screening of Essential Oils and Effect of a Chitosan-Based Edible Coating Containing Cinnamon Oil on the Quality and Microbial Safety of Fresh-Cut Potatoes
Reprinted from: *Coatings* 2022, 12, 1492, doi:10.3390/coatings12101492 27

Ignasius Radix A. P. Jati, Erni Setijawaty, Adrianus Rulianto Utomo and Laurensia Maria Y. D. Darmoatmodjo
The Application of *Aloe vera* Gel as Coating Agent to Maintain the Quality of Tomatoes during Storage
Reprinted from: *Coatings* 2022, 12, 1480, doi:10.3390/coatings12101480 43

Boran Hu, Lan Lin, Yujie Fang, Min Zhou and Xiaoyan Zhou
Application of Chitosan-Lignosulfonate Composite Coating Film in Grape Preservation and Study on the Difference in Metabolites in Fruit Wine
Reprinted from: *Coatings* 2022, 12, 494, doi:10.3390/coatings12040494 59

Hanaa S. Hassan, Mervat EL-Hefny, Ibrahim M. Ghoneim, Mina S. R. Abd El-Lahot, Mohammad Akrami, Asma A. Al-Huqail, et al.
Assessing the Use of *Aloe vera* Gel Alone and in Combination with Lemongrass Essential Oil as a Coating Material for Strawberry Fruits: HPLC and EDX Analyses
Reprinted from: *Coatings* 2022, 12, 489, doi:10.3390/coatings12040489 75

Xiaoguang Wu, Peiren Wang, Qiyao Xu, Bin Jiang, Liangyu Li, Lili Ren, et al.
Effects of *Pleurotus ostreatus* on Physicochemical Properties and Residual Nitrite of the Pork Sausage
Reprinted from: *Coatings* 2022, 12, 484, doi:10.3390/coatings12040484 99

Liangyu Li, Peiren Wang, Yanli Xu, Xiaoguang Wu and Xuejun Liu
Effect of Trehalose on the Physicochemical Properties of Freeze-Dried Powder of Royal Jelly of Northeastern Black Bee
Reprinted from: *Coatings* 2022, 12, 173, doi:10.3390/coatings12020173 113

Sherif Fathy El-Gioushy, Mohamed F. M. Abdelkader, Mohamed H. Mahmoud, Hanan M. Abou El Ghit, Mohammad Fikry, Asmaa M. E. Bahloul, et al.
The Effects of a Gum Arabic-Based Edible Coating on Guava Fruit Characteristics during Storage
Reprinted from: *Coatings* 2022, 12, 90, doi:10.3390/coatings12010090 129

Hoda A. Khalil, Mohamed F. M. Abdelkader, A. A. Lo'ay, Diaa O. El-Ansary, Fatma K. M. Shaaban, Samah O. Osman, et al.
The Combined Effect of Hot Water Treatment and Chitosan Coating on Mango (*Mangifera indica* L. cv. Kent) Fruits to Control Postharvest Deterioration and Increase Fruit Quality
Reprinted from: *Coatings* 2022, 12, 83, doi:10.3390/coatings12010083 145

Yuan Fu, Long Zhang, Mengdi Cong, Kang Wan, Guochuan Jiang, Siqi Dai, et al.
Application of *Auricularia cornea* as a Pork Fat Replacement in Cooked Sausage
Reprinted from: *Coatings* **2022**, *11*, 1432, doi:10.3390/coatings11111432 . . . . **161**

Communication

Effects of Pullulan-Based Coatings Incorporating ε-Polylysine and Glutathione on the Preservation of Cowpeas (*Vigna unguiculata* L.) Postharvest

Aoxue Hu [1,2,3] and Yingming Mao [1,2,4,5,*]

1. Jiangsu Key Laboratory of Marine Bioresources and Environment, Jiangsu Ocean University, 59 Cangwu Road, Lianyungang 222005, China
2. Co-Innovation Center of Jiangsu Marine Bio-Industry Technology, 59 Cangwu Road, Lianyungang 222005, China
3. School of Food Science and Engineering, Jiangsu Ocean University, 59 Cangwu Road, Lianyungang 222005, China
4. School of Environmental and Chemical Engineering, Jiangsu Ocean University, 59 Cangwu Road, Lianyungang 222005, China
5. Jiangsu Key Laboratory of Marine Biotechnology, 59 Cangwu Road, Lianyungang 222005, China
* Correspondence: maoym@jou.edu.cn

Abstract: Pullulan has a fine-coating-forming ability, ε-polylysine has an antibacterial activity, and glutathione has both a potent antioxidant activity and polyphenol-oxidase-inhibiting ability. This study explored the effects of pullulan-based coatings incorporating ε-polylysine and glutathione (1% pullulan + 0.2% ε-polylysine + 0.3% glutathione) on the preservation of cowpeas (*Vigna unguiculata* L.) during refrigerated storage. Pullulan-based coatings incorporating ε-polylysine and glutathione decreased the weight loss, decay and rust spot indices, respiratory rate and malondialdehyde by 49.01%, 60.38%, 91.09%, 69.09% and 49.23%, respectively, and increased soluble solid content by 34.21% compared with the control group after 15 days of refrigerated storage ($p < 0.05$). Results show that pullulan-based coatings incorporating ε-polylysine and glutathione treatment may be practical materials for the preservation of cowpeas during refrigerated storage.

Keywords: pullulan; ε-polylysine; glutathione

1. Introduction

Cowpea (*Vigna unguiculata* L.), an annual entwining herbaceous plant under the subfamily *Papilionidae*, is a common vegetable variety in China. This species is rich in vitamins, plant protein and mineral elements and has a high economic value [1]. Cowpeas are a highly seasonal vegetable that is usually harvested in summer in high temperatures and humidity. After harvest, this vegetable is very intolerant to storage because of its crisp tissue, high water content and strong respiration [2]. Generally, the storage period of cowpeas at room temperature is 3 days. In a short period, dehydration, wilting, fading, rust spots and even rotting occur, causing major economic losses [3,4]. Therefore, the postharvest preservation of cowpea is a major problem for the healthy development of the cowpea industry, requiring the study of postharvest preservation technologies of cowpea.

Edible coating preservation technology can control moisture and solute migration, internal gas exchange, respiration, and oxidative reaction rates and improve the appearance and surface characteristics of fruits and vegetables [5,6]. Moreover, edible coatings serve as the carriers of food additives, such as preservatives, antioxidants, colorants and flavors [7]. Pullulan is a homogeneous polysaccharide with a fine coating-forming ability [8] and ε-polylysine is a broad-spectrum preservative used in the food industry [9]. Glutathione is an antioxidant that can suppress polyphenol oxidase (PPO) activity and browning [10,11].

Citation: Hu, A.; Mao, Y. Effects of Pullulan-Based Coatings Incorporating ε-Polylysine and Glutathione on the Preservation of Cowpeas (*Vigna unguiculata* L.) Postharvest. *Coatings* **2023**, *13*, 125. https://doi.org/10.3390/coatings13010125

Academic Editor: Elena Torrieri

Received: 26 November 2022
Revised: 4 January 2023
Accepted: 5 January 2023
Published: 10 January 2023

Copyright: © 2023 by the authors. Licensee MDPI, Basel, Switzerland. This article is an open access article distributed under the terms and conditions of the Creative Commons Attribution (CC BY) license (https://creativecommons.org/licenses/by/4.0/).

Given their coating-forming ability and antibacterial and antioxidant activities, pullulan-based coatings incorporating ε-polylysine and glutathione may have antibacterial and antioxidant activities that can thus extend the shelf life of cowpea. Hence, further investigations into pullulan-based coatings incorporating ε-polylysine and glutathione are needed. Therefore, this study aims to investigate the preservative effects of pullulan-based coatings incorporating ε-polylysine and glutathione on cowpeas during refrigerated storage.

2. Methods and Materials

2.1. Materials

Cowpeas were obtained from a local agricultural product wholesale market in Haizhou, Lianyungang, China. Cowpeas with the same length, thickness and maturity and without mechanical damage, pests or diseases were selected, rinsed and drained naturally for use. Pullulan with a molecular weight of 3.1×10^5 Da was obtained from Pharmacopoeia, Japan. ε-polylysine was obtained from Zhejiang Tongfa Biotechnology Co., Ltd., Ningbo, China. Glutathione was obtained from Zhejiang Shenyou Biotechnology Co., Ltd., Huzhou, China. All the other chemicals were of reagent grade.

2.2. Treatment of Cowpeas

Based on our previous study, dipping solutions were prepared as follows: pullulan was dissolved in purified water to obtain a solution with a concentration of 1% (w/v) to which 0.2% (w/v) ε-polylysine and 0.3% (w/v) glutathione were added. The cowpeas were soaked in the dipping solutions for 15 min at 4 °C, rinsed, drained naturally, sealed in 25 cm × 38 cm × 0.03 cm polyethylene fresh-keeping bags and placed on racks at −4 °C for 15 days in cold storage. Cowpeas without any treatment were used as a control.

2.3. Weight Loss

The weight of the cowpeas was assayed during refrigerated storage. The percent weight loss was calculated by weighing the cowpeas every 3 days.

2.4. Decay Index

The decay index was divided into five grades as follows: grade 0, no decay; grade 1, decayed area < 10%; grade 2, decayed area of 10%–25%; grade 3, decayed area of 25%–50%; and grade 4, decayed area > 50%. Decay index (%) = 100 × $\sum$[(decay grade × number of cowpeas in the grade)]/[(number of the highest grade × total number of cowpeas)] [12].

2.5. Rust Spot Index

The rust spot index was divided into five grades as follows: grade 0, no rust spots; grade 1, rust spots of 1%–25%; grade 2, rust spots of 26%–50%; grade 3, rust spots of 51%–75%; and grade 4, rust spots of 76% to 100%. Rust spot index (%) =100 × $\sum$[(rust spot grade × number of cowpeas in the grade)]/[(number of the highest grade × total number of cowpeas)] [12].

2.6. Soluble Solids

The cowpeas were homogenized, filtered and centrifuged to yield a supernatant, in which soluble solids were measured using a refractometer (WYA-2WAJ, Shanghai Lichen Instrument Technology Co., Ltd., Shanghai, China).

2.7. Respiratory Rate

The cowpeas (approximately 1 kg) were tightly sealed in a 1000 mL glass container at 25 °C for 2 h. The carbon dioxide content of the headspace was determined using a Trace 2000 GC and a Thermo mass spectrometer [13].

2.8. Malondialdehyde (MDA)

The MDA content of the cowpeas was assayed using thiobarbituric acid reactive substances methods. Approximately 2 g of cowpea tissue sample was extracted with a trichloroacetic acid solution. Then, the absorbance was measured at 450, 530 and 600 nm [14].

2.9. Statistical Analysis

All experiments were conducted in sextuplicate. All data are expressed in mean ± standard deviation (SD). Origin 7.0 statistical analysis software was used for data collation and analysis. Significant difference analysis was performed using the paired sample t-test.

3. Results

3.1. Weight Loss

As shown in Figure 1, the weight loss rate of cowpeas during refrigerated storage continuously increased with time. At the beginning of the six-day refrigerated storage, the weight loss rate of each group increased rapidly and then increased slowly from the 9th day of refrigerated storage.

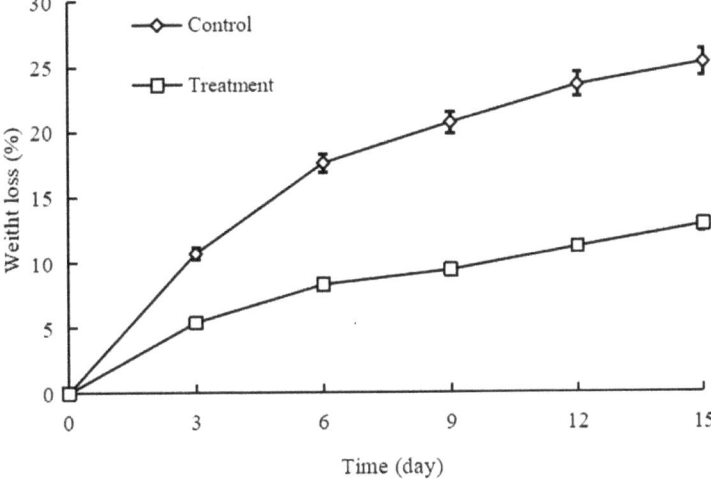

Figure 1. Effect of pullulan-based coatings on the weight loss of cowpeas during refrigerated storage. Bars represent the standard deviation ($n = 6$).

3.2. Decay Index

As shown in Figure 2, the cowpeas in the control group started to decay from the 3rd day of refrigerated storage. The decay index increased steadily with time during refrigerated storage. After the 6th day, the decay index increased sharply and reached 53.14% on the 15th day. However, the decay index of the cowpeas in the treatment group increased slowly during the entire storage period, and the decay index was only 21.04% on the 15th day, which is much lower than that of the control group ($p < 0.05$).

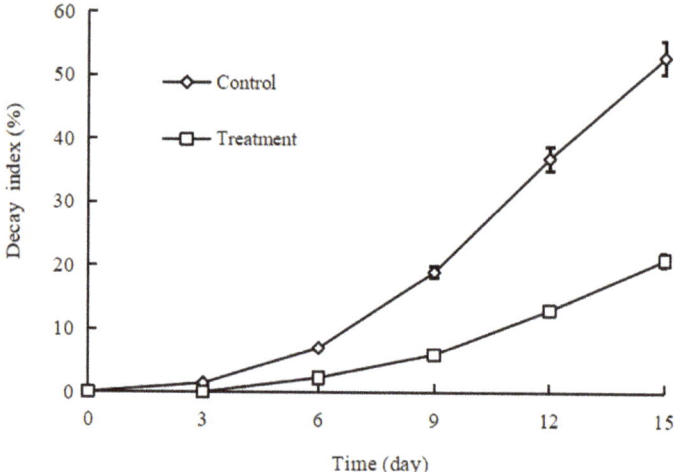

Figure 2. Effect of pullulan-based coatings on the decay index of cowpeas during refrigerated storage. Bars represent the standard deviation ($n = 6$).

3.3. Rust Spot Index

As shown in Figure 3, the cowpeas in the control group started to show rust spots from the 3rd day of refrigerated storage. The rust spot rate continued to increase with time during refrigerated storage. After the 6th day, the rust spot rate increased sharply, reaching 64.07% on the 15th day. However, the rust spot rate of the cowpeas in the treatment group increased slowly throughout the entire storage period, and the value was only 5.72% on the 15th day, which is much lower than that of the control group ($p < 0.05$).

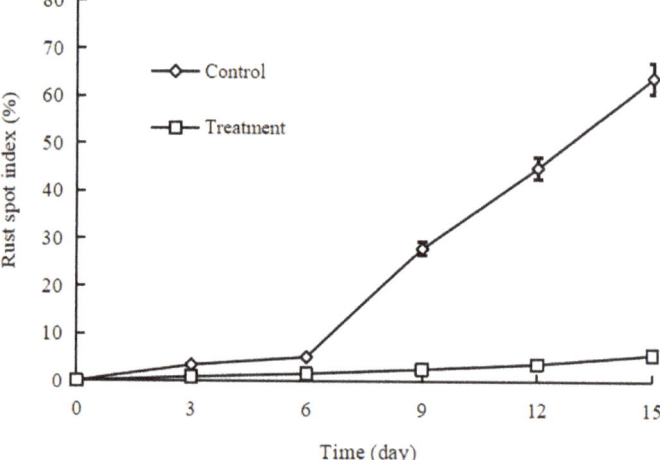

Figure 3. Effect of pullulan-based coatings on the rust spot index of cowpeas during refrigerated storage. Bars represent the standard deviation ($n = 6$).

3.4. Respiratory Rate

The peak respiratory intensity of the control group was observed on the 6th day, while that of the treatment group was delayed to the 9th day and is lower than that of the control group ($p < 0.05$, Figure 4). On the 15th day of storage, the respiration intensity of the cowpeas in the treatment group was 17.09 mg/kg·h, which is 69.09% lower than that of the control group

($p < 0.05$), indicating that treatment with pullulan-based coatings incorporating ε-polylysine and glutathione significantly inhibited the respiration intensity of the cowpeas and reduced the nutrient consumption during refrigerated storage ($p < 0.05$).

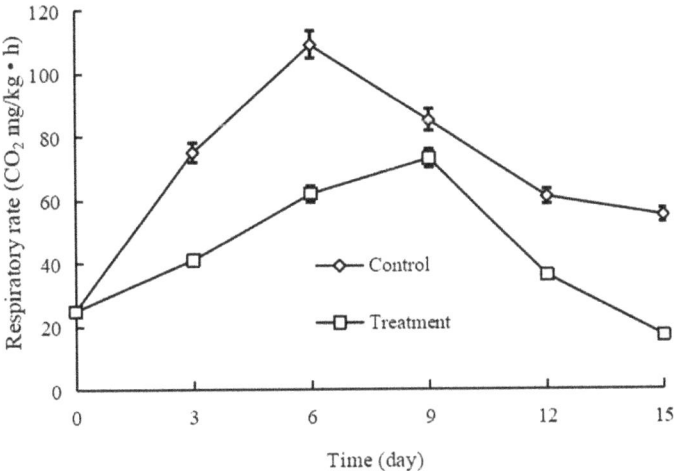

Figure 4. Effect of pullulan-based coatings on the respiratory rate of cowpeas during refrigerated storage. Bars represent the standard deviation ($n = 6$).

3.5. Soluble Solids

Soluble solid contents in the cowpeas in the control group continuously decreased with time during refrigerated storage. After the 6th day, the soluble solid contents decreased sharply, reaching 3.81% on the 15th day. However, the soluble solid content of the cowpeas in the treatment group decreased slowly throughout the entire storage period with a value of 5.14% on the 15th day, which is much higher than that of the control group (Figure 5, $p < 0.05$).

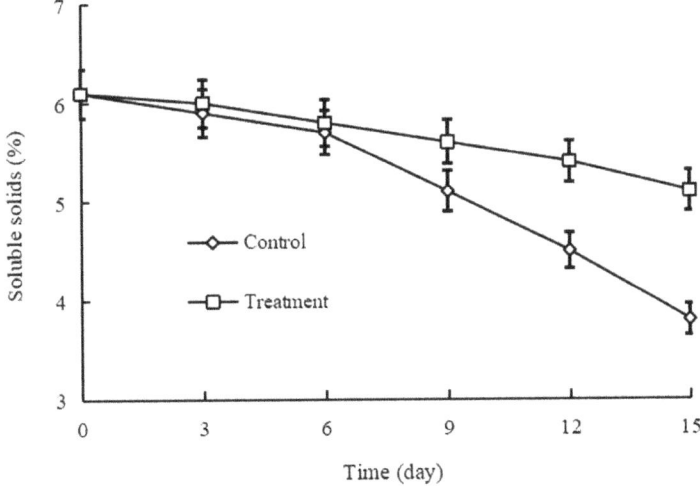

Figure 5. Effect of pullulan-based coatings on soluble solids in cowpeas during refrigerated storage. Bars represent the standard deviation ($n = 6$).

3.6. MDA

As shown in Figure 6, the MDA content of the control group increased slowly in the first 9 days of refrigerated storage and then increased sharply after 9 days of refrigerated storage, while the MDA content of the treatment group increased slowly during refrigerated storage and at lower levels than those of the control group ($p < 0.05$).

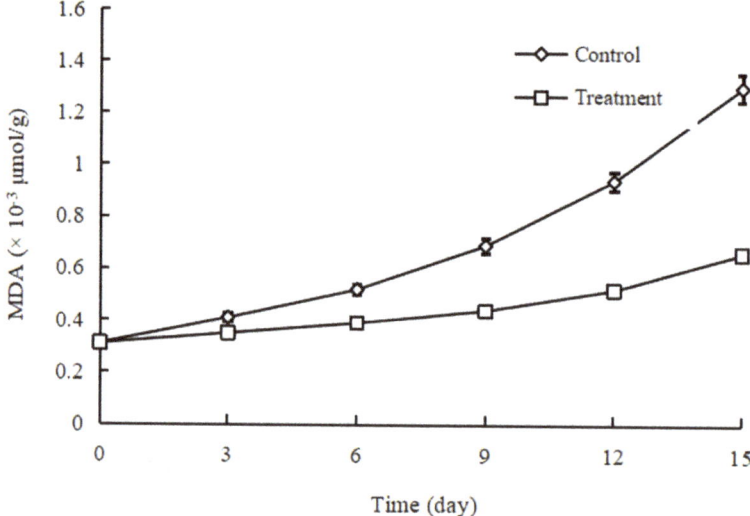

Figure 6. Effect of pullulan-based coatings on the MDA content of cowpeas during refrigerated storage. Bars represent the standard deviation ($n = 6$).

4. Discussion

The edible part of the cowpeas was fresh and had tender pods, and the pods maintained a high moisture content, which is an indication of the preservation technology. After refrigerated storage, the treatment with pullulan-based coatings incorporating ε-polylysine and glutathione significantly inhibited the increase in the weight loss rate of the cowpeas and effectively maintained a high moisture content ($p < 0.05$). Pullulan solution treatment could form a dense protective layer on the surface of the cowpeas, thus effectively reducing the respiration, transpiration and water loss of the pod [8].

Glutathione has potent antioxidant activity and can suppress PPO activity and browning and thus decrease the rust spot rate [10,11]. The preservation treatment formed wax-like coatings on the cowpea surface, thus reducing the mechanical loss during refrigerated storage. In addition, the coatings blocked the passage between the outside world and the fruit cells, preventing infection with aerobic microorganisms such as fungi, which effectively prevented spoilage of the cowpeas during refrigerated storage. Moreover, ε-polylysine has broad-spectrum antibacterial activity, suppressing the growth of spoilage bacteria and preventing the spoilage of the cowpeas during refrigerated storage [9].

The browning of cowpeas after harvesting is a common phenomenon, and it often occurs in injured areas, especially the stem. After the harvest, the color of the cowpeas changed from dark green to light green or even yellow, resulting in rust spots. Glutathione has potent antioxidant activity and can suppress the PPO activity and browning, thus decreasing the rust spot rate [10,11].

Cowpea is a respiratory climacteric vegetable. As shown in Figure 4, the respiration intensity of the cowpeas during the whole storage period increased first and then decreased. The treatment with pullulan-based coatings incorporating ε-polylysine and glutathione affected spontaneous air conditioning and preservation. The coatings formed on the surface of the cowpeas, delaying their physiological metabolism and remarkably reducing the respiration intensity of the cowpeas, thereby inhibiting the process of other metabolic activities, reducing the consumption of nutrients and delaying the senescence of the cowpeas.

Soluble solids closely contribute to the taste and flavor of cowpeas. After the harvest, some sugar was degraded to carbon dioxide and water respiration, thus decreasing the soluble solid content in the cowpeas. The coatings that formed on the surface of the cowpeas reduced the physiological metabolism during refrigerated storage and inhibited the consumption of organic matter [12]. The treatment with pullulan-based coatings incorporating ε-polylysine and glutathione maintained a higher soluble solid content during the refrigerated storage of the cowpeas and maintained a higher nutritional quality of the cowpeas than the control group ($p < 0.05$).

Environmental stress and tissue aging can decrease the ability of the tissue to scavenge active oxygen and induce the production of many free radicals that then produce MDA, and the MDA content can reflect the degree of membrane damage. The treatment with pullulan-based coatings incorporating ε-polylysine and glutathione inhibited the respiration of the cowpeas during refrigerated storage, thereby inhibiting other physiological and biochemical activities in the cowpea cells, delaying the senescence of the cowpeas and inhibiting the accumulation of MDA. The coating treatment also reduced the weight loss rate of the cowpeas during refrigerated storage, thereby reducing the increase in cell membrane permeability. This indicated that the coating treatment can inhibit the occurrence of membrane lipid peroxidation to a certain extent, thus preventing the destruction of the cell structure and inhibiting cell senescence and death. Glutathione has potent antioxidant activity and can suppress membrane lipid peroxidation and thus inhibit the accumulation of MDA [10,11].

In conclusion, the treatment with pullulan-based coatings incorporating ε-polylysine and glutathione effectively decreased the weight loss, decay and rust spot indices, the respiratory rate, and the MDA content, and increased the soluble solids during refrigerated storage compared with the control group. Therefore, the treatment with pullulan-based coatings incorporating ε-polylysine and glutathione might be a promising method for extending the shelf life of cowpeas.

Author Contributions: A.H.—Investigation, Software, Data curation and Writing Original draft preparation. Y.M.—Conceptualization, Supervision, Validation, Reviewing and Editing. All authors have read and agreed to the published version of the manuscript.

Funding: This research was supported by a project funded by the Priority Academic Program Development of Jiangsu Higher Education Institutions (PAPD).

Institutional Review Board Statement: Not applicable.

Informed Consent Statement: Not applicable.

Data Availability Statement: No new data were created or analyzed in this study. Data sharing is not applicable to this article.

Conflicts of Interest: The authors declare no conflict of interest.

References

1. Olabanji, I.O.; Ajayi, O.S.; Oluyemi, E.A.; Olawuni, I.J.; Olusesi, I.M. Nutraceuticals in different varieties of cowpeas. *Am. J. Food Sci. Technol.* **2018**, *6*, 68–75.
2. Baributsa, D.; Djibo, K.; Lowenberg-DeBoer, J.; Moussa, B.; Baoua, I. The fate of triple-layer plastic bags used for cowpea storage. *J. Stored. Prod. Res.* **2014**, *58*, 97–102. [CrossRef]
3. Baoua, I.B.; Amadou, L.; Lowenberg-DeBoer, J.D.; Murdock, L.L. Side by side comparison of Grain Pro and PICS bags for post-harvest preservation of cowpea grain in niger. *J. Stored. Prod. Res.* **2013**, *54*, 13–16. [CrossRef]
4. Murodock, L.L.; Baoua, I.B. On purdue improved cowpea storage (PICS) technology: Background, mode of action, future prospects. *J. Stored Prod. Res.* **2014**, *58*, 3–11. [CrossRef]
5. Rojas-Graü, M.A.; Tapia, M.S.; Martín-Belloso, O. Using polysaccharide-based edible coatings to maintain quality of fresh-cut Fuji apples. *LWT Food Sci. Technol.* **2008**, *41*, 139–147. [CrossRef]

6. Rojas-Graü, M.A.; Tapia, M.S.; Rodríguez, F.J.; Carmona, A.J.; Martin-Belloso, O. Alginate and gellan-based edible coatings as carriers of antibrowning agents applied on fresh-cut Fuji apples. *Food Hydrocolloid* **2007**, *21*, 118–127. [CrossRef]
7. Pranoto, Y.; Salokhe, V.; Rakshit, K.S. Physical and antibacterial properties of alginate-based edible film incorporated with garlic oil. *Food Res. Int.* **2005**, *38*, 267–272. [CrossRef]
8. Marvdashti, L.M.; Abdolshahi, A.; Hedayati, S.; Sharifi-Rad, M.; Riti, M.; Salehi, B.; Sharifi-Rad, J. Pullulan gum production from low-quality fig syrup using *Aureobasidium pullulans*. *Cell. Mol. Biol.* **2018**, *64*, 22–26. [CrossRef]
9. Hiraki, J.; Ichikawa, T.; Ninomiya, S.; Seki, H.; Uohama, K.; Seki, H.; Kimura, S.; Yanagimoto, Y.; Barnett, J.W. Use of ADME studies to confirm the safety of ε-Polylysine as a preservative in food. *Regul. Toxicol. Pharmacol.* **2003**, *37*, 328–340. [CrossRef]
10. Jiang, Y.M.; Fu, J.R. Inhibition of polyphenol oxidase and the browning control of litchi fruit by glutathione and citric acidic. *Food Chem.* **1998**, *62*, 49–52. [CrossRef]
11. Wu, S.J. Inhibition of enzymatic browning of the meat of *Clanis bilineata* (Lepidoptera) by glutathione. *Food Sci. Technol. Res.* **2013**, *19*, 347–352. [CrossRef]
12. Xing, S.J.; Zhang, Y.H.; Liu, K.H. Preservation Effect of Tea Polyphenol-incorporated Soy Protein Isolate Coating on Cowpea. *Food Ind.* **2015**, *36*, 192–195. (In Chinese)
13. Jongsri, P.; Wangsomboondee, T.; Rojsitthisak, P.; Seraypheap, K. Effect of molecular weights of chitosan coating on postharvest quality and physicochemical characteristics of mango fruit. *LWT Food Sci. Technol.* **2016**, *73*, 28–36. [CrossRef]
14. Zhang, F.P. Analysis about the effect of different temperature on preservation of *Vigna sinensis* during postharvest storage. *Food Res. Dev.* **2008**, *29*, 146–148. (In Chinese)

Disclaimer/Publisher's Note: The statements, opinions and data contained in all publications are solely those of the individual author(s) and contributor(s) and not of MDPI and/or the editor(s). MDPI and/or the editor(s) disclaim responsibility for any injury to people or property resulting from any ideas, methods, instructions or products referred to in the content.

Article

Influence of Wax and Silver Nanoparticles on Preservation Quality of Murcott Mandarin Fruit during Cold Storage and after Shelf-Life

Mohamed M. Gemail [1,†], Ibrahim Eid Elesawi [2,3,*,†], Muthana M. Jghef [4], Badr Alharthi [5], Woroud A. Alsanei [6], Chunli Chen [2], Sayed M. El-Hefnawi [1] and Mohamed M. Gad [1]

1. Horticulture Department, Faculty of Agriculture, Zagazig University, Zagazig 44511, Egypt
2. College of Life Science and Technology, Huazhong Agricultural University, Wuhan 430070, China
3. Agricultural Biochemistry Department, Faculty of Agriculture, Zagazig University, Zagazig 44511, Egypt
4. Department of Radiology, College of Medical Technology, Al-Kitab University, Kirkuk 36001, Iraq
5. Department of Biology, University College of Al Khurmah, Taif University, P.O. Box 11099, Taif 21944, Saudi Arabia
6. Department of Food and Nutrition, Faculty of Human Sciences and Design, King Abdulaziz University, Jeddah 21589, Saudi Arabia
* Correspondence: ibrahimeid@zu.edu.eg
† These authors contributed equally to this work.

Abstract: Citrus fruits are perishable and considered the most prominent and essential crops at the local and global levels. The world is focused on minimizing fruit postharvest losses, maintaining fruit quality, and prolonging its storability and marketability. Thus, this study was carried out throughout the two successive seasons of 2018 and 2019 on Murcott mandarin fruits, with the purpose of extending their storage period and shelf life by making a mixture of nanosilver and wax as a coating. The fruits were picked on the first of March, washed, and coated with the following treatments: 1000 ppm imazalil (IMZ as a control), wax, 50 ppm nanosilver, 100 ppm nanosilver, and finally, the combination of wax plus 100 ppm nanosilver, packaged in 0.005% perforated polyethylene (PPE), and stored at 5 ± 1 °C and 90%–95% relative humidity for four months. Samples of each treatment were randomly taken at monthly intervals to evaluate the tested treatments' effects on fruit quality during cold storage and 6 days of shelf life. The data proved that the combination of wax plus 100 ppm nanosilver packaged in 0.005% perforated polyethylene (PPE) was the most effective treatment for reducing discarded fruits, fresh weight loss, and catalase enzyme activity, as well as maintaining pulp firmness and vitamin C content and keeping a better taste panel index. Therefore, these coatings could be promising alternative materials for extending mandarin fruits' postharvest life and marketing period.

Keywords: mandarin; edible coating; AgNPs; postharvest; quality

1. Introduction

Citrus is considered the most prominent and important fruit crop at the local and global levels. Furthermore, in Egypt, citrus is the backbone of the fruit crop due to its most significant economic importance compared with other types of fruit, taking the first rank in the cultivated area as the first export crop. Moreover, it has been a large horticultural industry during the last few years, and the cultivated area has reached about 1941 Km² and produced 4,323,030 tons [1]. Furthermore, it is considered the most popular fruit in Egypt and has a high nutritional value with a rich content of vitamins, organic acids, pigments (carotenoids, flavonoids, anthocyanin, thiamine, riboflavin, niacin, etc.), sugars, fibers, essential and volatile oils, as well as mineral elements such as calcium, phosphorus, iron, sodium, and potassium [2–5].

Citrus fruits are deemed perishable and susceptible to a reduced quality after harvest due to decay and water loss during transpiration and respiration [6]. Considering that citrus fruits have natural wax on the cortex that gets eliminated through the fruit's prolonged washing process, accordingly, compensation is needed to avoid dehydration [7,8]. The Murcott mandarin is one of the most popular mandarin cultivars in Egypt, but it is also exposed to many losses after harvesting and during storage, leading to a shorter postharvest life [9].

In the past decade, the world has attempted to reduce the loss of crops postharvest and maintain the quality of the fruit during storage and marketing. Inferior handling affects postharvest quality, disease incidence, and sensitivity to a chilling injury and contributes to high postharvest and marketing chain losses and reduces the storage period. Thus, these losses can occur at all postharvest stages until consumption [10–13]. Currently, some researchers aim at reducing excessive chemical components in crop fertilization by inexpensively utilizing environmentally safe organic substances to improve plant quality. Consequently, postharvest practices control fruit damage by using safe, suitable, and efficient harvest, handling, and storage treatments to prolong postharvest shelf life [10,14].

Fruit coating is considered a practical technique to provide additional preservation versus physiological disorders after harvest like stem-end rind breakdown, chilling injury [15,16], and a prolonged storage period, improving fruit appearance and quality [12,17]. Furthermore, experience with papaya, mandarin, and plums shows that edible coatings have the effect of maintaining postharvest fruit quality [18–21]. Furthermore, guar gum can be commercially applied to coat fruit to extend its shelf life and preserve postharvest quality in mango and "Valencia" oranges [22,23]. Wax is considered the most remarkable postharvest implementation to limit unfavorable changes and elongate the shelf life of fruit. The important properties of wax coatings on citrus fruit are a good, lustrous, and appealing appearance, which continues during the marketing process, the reduction of fruit weight loss, and maintenance of fruit quality. Furthermore, wax is predicted to be beneficial as a transporter of fungicides [6,15,24,25].

Moreover, the wax implementation plays a paramount function in increasing fruit quality. Furthermore, the imazalil (IMZ) preserves fruit against green mold caused by *Penicillium digitatum* and a single application of IMZ in wax has controlled green mold well and inhibited sporulation, with differing impacts on many parameters of fruit quality [26]. Over the previous years, new technologies have been introduced to prolong fresh fruit shelf life, such as loading coating substances with nanoparticles, which has presented an innovative and safe fruit-defense mechanism that ensures minimum direct exposition and lower penetration of nanoparticles into the treated food products [27,28].

Nanotechnology has attracted attention in the last decade due to its vital applications in many fields such as medicine, pharmaceuticals, catalysis, materials, and energy [29]. Nanomaterials are used in sustainable agriculture as promising plant growth agents, fertilizers, and pesticides. Moreover, nanomaterials are used in the control of plant pests, including insects, fungi, and weeds [29,30]. Applications of NPs are used in agriculture for a more efficient and safe use of chemicals. Although there are slight effects of toxicity on seed germination and root growth of five higher plant species—radish, rape, lettuce, corn, and cucumber—silver nanoparticles aid seedling growth in wheat, when large amounts are used including alumina, magnetite (Fe_3O_4), zinc, and zinc oxide [30–34]. On the contrary, silver nanoparticles can stimulate wheat growth and production and their application in the soil has very promising growth-promoting effects on wheat growth and yield [33]. Many studies have used nanotechnology in the field of food production, where it is preferable to use biosynthetic nanoparticles [33,35].

The research community has the most interest in silver nanoparticles (AgNPs) due to their noteworthy properties in size and effective antibacterial activities [36,37]. Silver nanoparticles have been used as food additives and packaging materials to eliminate pathogens [27,38,39]. Additionally, edible coating formulations mixed with AgNPs can be applied as a palatable fruit coating to reduce the growth of microorganisms that cause postharvest diseases to increase their shelf life [40–42]. Furthermore, the treatment with AgNPs exhibited significantly decreased weight losses compared with uncoated orange fruits [43]. Adding AgNPs to the polyethylene significantly reduces weight loss, retards softening, prevents fruit corruption, reduces decay, maintains firmness, decreases the rapid reduction in citric acid and vitamin C contents, and increases total antioxidant activity in fruit [44–46]. The studies also showed an increase in fruit weight retention, rate of respiration, total sugars, total soluble solids, and total carotenoids through the storage period. In contrast, this increase was relatively minimal and significant in coated fruits compared with uncoated fruits. On the other hand, hardness and acidity are greatly reduced upon storage. Still, this reduction was low for coated fruits when AgNPs and carboxymethyl cellulose CMC-AgNP coatings were capable of retardation fruit ripening of mango and preserving fruit quality through a cold storage period [41,47–51].

Nevertheless, there are no issued data on the use of coating substances loaded with AgNPs for improving mandarin fruit behavior during storage, especially the fruit of the Murcott mandarin cultivar. Furthermore, the studies that concentrated on the attitude of this cultivar and its quality characteristics over cold storage and shelf life are few [52–56]. Therefore, this work aims to study the influence of coating with wax and different concentrations of AgNPs before packaging in perforated polyethylene on the postharvest storage behavior and quality attributes of a Murcott mandarin fruit cultivar during its cold storage and shelf-life period.

2. Materials and Methods

2.1. Fruit Material and Growth Conditions

This study was performed during two successive seasons, 2018 and 2019, on mature yellow mandarin (Citrus reticulata L. Osbeck) fruits cv. Murcott. Fruits (600 fruits, 150 fruits/treatment, 10 fruits/replicate) were selected from a private citrus orchard in Wadi El-Molak, Ismailia Region Governorate, Egypt. The trees were six years old, budded on Volkameriana lemon rootstock, grown in sandy loam soil 5 m apart, and received the standard horticultural practices adopted in the area. The fruits were collected on the first day of March in both seasons. The collected fruits were uniform, healthy, and as free of physiological disorders and visible pathological problems as possible. All fruits were carefully transferred to the postharvest laboratory in the Horticulture Department, Faculty of Agriculture, Zagazig University, and kept for 24 h at room temperature. After that, all fruits were washed completely with tap water and soap and quickly rinsed with water to remove soap residues. Then, the fruits were surface sterilized with a 0.5% solution of sodium ortho-phenylphenate (SOPP) at pH 11.8–12.1 and 32 °C for two minutes, then left to air dry before treatment. Then the fruits were dipped in the treatment solution for two minutes and then left to air dry. The treatments included coating with 1000 ppm IMZ (control), wax, 50 and, 100 ppm nanosilver, and finally, the combination of wax plus 100 ppm nanosilver. Coated mandarin fruits of all treatments were air dried and packaged in 0.005% perforated polyethylene (PPE) and stored at 5 ± 1 °C and 90%–95% relative humidity for four months. Samples of each treatment were randomly taken at monthly intervals to evaluate the effect of the tested treatment during cold storage and six days' shelf life. The silver nanoparticles were obtained from Ahmed Saad's lab [57], which used phenolic aqueous extracts from pomegranate and watermelon peels to convert silver nitrate to silver nanoparticles, which were described as stated in the report [57].

2.2. Fruit Quality Characteristics

2.2.1. Discarded Fruit

This parameter was calculated as the percentage of rejected fruits to the total fruits [58].

2.2.2. Fruit Weight Loss (FWL)

To estimate FWL at specific storage periods, the fruits of each replicate were separately weighed before and after treatment during the cold storage process and before and after six days of shelf life. *FWL* was calculated as a percentage of the initial weight according to [47,59] utilizing the following Equation (1):

$$FWL\ (\%) = ((Wi - Ws)/W1) \times 100. \tag{1}$$

where (*Wi*) is fruit weight at the initial period, and (*Ws*) is fruit weight at the sampling period.

2.2.3. Fruit Pulp Firmness (FPF)

Five fruits from each replicate were hand flaked and used to distinguish pulp firmness as g/cm^2 using a Push-Pull Dynamometer (Model FD 101) [58,60].

2.2.4. Juice Total Soluble Solid Percentage (TSS)

TSS percentage was determined using a hand refractometer as Brix°.

2.2.5. Juice Total Acidity Percentage (TA %)

TA percentage was estimated by titrating 0.1 N NaOH in the presence of phenolphthalein as an indicator, and the result was calculated as grams of citric acid per 100 mL of fruit juice using, according to A.O.A.C. [61].

2.2.6. Ascorbic Acid (Vitamin C) Content

The content of ascorbic acid was estimated by titration in 2,6 dichlorophenol-indophenol dye, estimated and expressed as milligrams per 100 mL of juice [62].

2.2.7. Juice Panel Test Index (PTI)

Five persons judged random fruit samples from each replicate to give PTI scores according to the following index: Excellent taste = 4; Very good taste = 3; Good taste = 2; Acceptable taste = 1, and Bad taste (unacceptable fruits) = 0 [63].

2.3. Enzymes and Antioxidant Determination

2.3.1. Preparation of the Extract

The mandarin peels were ground after drying in a vacuum oven (45 °C), and 10 g of each powder sample was taken and then homogenized in 100 mL of 50% ethanol (1:10, w/v) stirred for 3 h at room temperature. The samples were filtered to obtain the supernatant. Then the solvent was disposed of with a rotary evaporator [64].

2.3.2. DPPH Radical Scavenging Assay

The total antioxidant activities in the mandarin peel extracts with different treatments at a concentration of (500 µg/mL) were examined according to [65]. We mixed 100 µL of each solution with 1 µL of ethanolic DPPH in the microtiter plate wells and incubated it at room temperature in the dark for 30 min. A microtiter plate reader (BioTek Elx808, USA) was used to measure the absorbance at 517 nm b and then applied Equation (2).

$$Radical\ scavenging\ activity\ (\%) = \frac{(Abs.\ control - Abs.\ sample)}{(Abs.\ control)} \times 100 \tag{2}$$

2.3.3. Catalase Activity (CAT)

CAT activity was assayed by using Biodiagnostic, Kit No. CA 25 17, Egypt, according to the method described by [66,67]. The formed chromophore absorbance was inversely proportional to the amount of catalase in the experimented sample [68]. Briefly, we mixed 0.05 mL of the sample, 0.5 mL of phosphate buffer (pH = 7), and 0.1 mL of chromogen-inhibitor and incubated for one min at room temperature, added 0.50 mL H_2O_2 and 0.20 mL chromogen-inhibitor to the mixture then incubated for 10 min at 37 °C. The decrease in absorbance was recorded at 510 nm.

2.4. Statistical Analysis

Before running a one-way ANOVA, pretests were conducted. We tested the normality assumption on sample distributions and obtained p-values of 0.0001. for homogeneity; we used the Levene test with p-value = 0.01598. The triplicate data means were analyzed for statistical differences by one-way ANOVA at a confidence level of 95% [69], using Costat program version 6.4 (Costat 2008). The sample size was calculated using the following Equation (3)

$$n = \left(\frac{ZSD}{E}\right)^2 \qquad (3)$$

Means were compared with the least significant difference (LSD) as a post hoc test at a probability level of 5%.

3. Results

3.1. Discarded Fruit %

After harvest, citrus fruits are considered perishable and vulnerable to quality decline due to rot and water loss from transpiration and respiration [6]. The data referred to the influence of various postharvest treatments on discarded fruit percentage of Murcott mandarins, regardless of the cold storage period, as illustrated in (Table 1).

Table 1. Effect of postharvest treatments on discarded fruit percentage (DFP%) of Murcott mandarin fruits during 1, 2, 3, and 4 months of cold storage and after 6 days of shelf life during 2018 and 2019 seasons.

Treatments	Cold Storage Period (Month) (P)					6 Days Life (P)				
	The First Season (2018)									
	1	2	3	4	Mean	1	2	3	4	Mean
IMZ (Control)	0.00 e	3.32 cd	6.71 b	10.15 a	5.04 A	0.00 f	6.70 bc	10.03 a	10.20 a	6.73 A
Wax	0.00 e	0.00 e	3.40 cd	6.66 b	2.51 B	0.00 f	0.00 f	6.42 bc	6.83 b	3.31 B
not wax + 50 ppm nanosilver	0.00 e	0.00 e	3.34 cd	3.50 c	1.71 C	0.00 f	0.00 f	6.52 bc	6.62 c	3.20 B
not wax + 100 ppm nanosilver	0.00 e	0.00 e	3.28 d	3.39 cd	1.67 C	0.00 f	0.00 f	3.63 e	6.29 bc	2.01 C
wax + 100 ppm nanosilver	0.00 e	0.00 e	0.00 e	3.41 cd	0.85 D	0.00 f	0.00 f	0.00 f	4.40 d	1.66 D
Mean	0.00 D	0.66 C	3.35 B	5.42 A		0.00 D	1.34 C	5.32 B	6.87 A	
	The second season (2019)									
IMZ (Control)	0.00 e	3.17 d	6.85 b	10.16 a	5.04 A	0.00 d	4.44 c	10.03 a	10.23 a	6.18 A
Wax	0.00 e	0.00 e	3.44 cd	6.63 b	2.52 B	0.00 d	0.00 d	6.64 b	6.97 b	3.40 B
not wax + 50 ppm nanosilver	0.00 e	0.00 e	3.57 c	3.47 cd	1.76 C	0.00 d	0.00 d	6.49 b	6.42 b	3.23 B
not wax + 100 ppm nanosilver	0.00 e	0.00 e	3.27 d	3.46 cd	1.68 C	0.00 d	0.00 d	3.84 c	4.43 c	2.07 C
wax + 100 ppm nanosilver	0.00 e	0.00 e	0.00 e	3.64 c	0.91 D	0.00 d	0.00 d	0.00 d	6.67 b	1.67 C
Mean	0.00 D	0.63 C	3.43 B	5.47 A		0.00 D	0.89 C	5.40 B	6.94 A	

IMZ = imazalil, lowercase letters in the same column indicate significant difference, while uppercase letters in the rows and columns indicate significant difference between means by LSD at a 0.05 level.

This data indicated that, compared with other treatments, coating with wax plus 100 ppm nanosilver and packaging in perforated polyethylene (PPE) was the best treatment for reducing the percentage of discarded fruit during the two seasons of a study in 2018 (Table 1) and 2019 (Table 1). Moreover, coating wax and nano silver at 50 or 100 ppm reduced discarded fruit percentage relative to the control, which gave the highest ratio. Coating with nano silver at 50 or 100 ppm significantly affected the discarded fruit percentage.

The control treatment could be seen, regardless of shelf-life period, to have the highest discarded fruit percentage compared to the combination of wax mixed with 100 ppm nanosilver and packaged in PPE, which recorded the lowest percentage. Moreover, all other treatments were more effective in reducing the discarded fruit percentage relative to the control. Coating with wax and 50 ppm nanosilver and packaging in PPE treatments had a similar effect on discarded fruit percentage (Table 1).

Concerning the influence of the cold storage period on the discarded fruit percentage of Murcott mandarins, it was evident that after the second month of cold storage, the rate of discarded fruit increased as the cold storage period progressed, reaching the highest percentage after four months of cold storage (Table 1). The previous trend was typically repeated as the shelf-life period progressed in the two seasons of study.

The interaction between postharvest treatments and cold storage also affected the discarded fruit percentage. There was a significant increase in the discarded fruit percentage in the control treatment compared with other used treatments after the fourth month of cold Storage. On the other hand, coating treatments with nano silver—at both unmixed concentrations or at 100 ppm mixed with wax—were more effective in reducing the percentage of discarded fruit (similar in their effect) compared to using a wax coating alone or the control by the end of cold storage. This trend of results was nearly similar to that obtained with the interaction effect between postharvest treatments and shelf-life period, with one exception being that coating with nanosilver (100 ppm) alone before packaging in PPE had the greatest ability to reduce the percentage of discarded fruit relative to other treatments after the last period of shelf life in both seasons (Table 1).

3.2. Weight Loss %

Fruit weight loss is primarily connected to water loss, mainly because transpiration, which is responsible for 90% of overall weight reduction, initially originates from the peel [70–72]. The data in Table 2 demonstrate that during the 2018 and 2019 seasons, the percentage of weight lost generally rose with longer storage times in both the cold storage and shelf-life periods.

Table 2. Effect of postharvest treatments on fresh weight loss percentage (FWL%) of Murcott mandarin fruits during 1, 2, 3, and 4 months of cold storage and after 6 days of shelf life during the 2018 and 2019 seasons.

Treatments	Cold Storage Period (Month) (P)					6 Days Shelf Life (P)				
	The First Season (2018)									
	1	2	3	4	Mean	1	2	3	4	Mean
IMZ (Control)	1.80 e	2.17 cd	2.40 b	2.90 a	2.32 A	2.03 ef	2.30 cd	2.60 b	3.10 a	2.51 A
Wax	0.72 g	1.17 f	1.80 e	2.23 bc	1.48 C	1.15 hi	1.63 g	2.00 f	2.50 bc	1.82 B
not wax + 50 ppm nanosilver	0.87 g	1.33 f	2.13 cd	2.23 bc	1.64 B	1.13 i	1.50 g	2.27 cde	2.50 bc	1.85 B
not wax + 100 ppm nanosilver	0.77 g	1.27 f	2.00 d	2.03 d	1.52 C	1.09 i	1.43 g	2.10 def	2.57 b	1.80 BC
wax + 100 ppm nanosilver	0.75 g	1.30 f	1.70 e	2.03 d	1.44 C	1.03 i	1.40 gh	1.97 f	2.37 bc	1.69 C
Mean	0.98 D	1.45 C	2.01 B	2.29 A		1.29 D	1.65 C	2.19 B	2.61 A	
	The second season (2019)									
IMZ (Control)	1.73 g	2.33 bc	2.50 b	2.90 a	2.37 A	2.07 de	2.53 b	2.23 cd	3.03 a	2.47 A
Wax	0.80 i	1.30 h	1.77 g	2.30 cd	1.54 C	1.078 g	1.63 f	2.03 e	2.47 b	1.80 B
not wax + 50 ppm nanosilver	0.87 i	1.20 h	2.23 cde	2.27 cde	1.64 B	1.13 g	1.47 f	2.23 cd	2.53 b	1.84 B
not wax + 100 ppm nanosilver	0.82 i	1.27 h	2.03 f	2.13 def	1.56 BC	1.13 g	1.47 f	2.10 cde	2.53 b	1.81 B
wax + 100 ppm nanosilver	0.91 i	1.33 h	1.73 g	2.10 ef	1.52 C	1.07 g	1.47 f	2.03 e	2.27 c	1.71 C
Mean	0.00 D	0.63 C	3.43 B	5.47 A		0.00 D	0.89 C	5.40 B	6.94 A	

IMZ = imazalil, lowercase letters in the same column indicate significant difference, while uppercase letters in the rows and columns indicate significant difference between means by LSD at a 0.05 level.

Furthermore, the results showed that all applications significantly decreased weight loss compared to the control during both the cold storage period (Table 2) and days of shelf life (Table 2), and the applied treatments nanosilver (100 ppm) and wax with nanosilver

(100 ppm) with packaging in perforated polyethylene (PPE) were more effective at reducing weight loss.

Furthermore, the most pronounced effect in reducing the weight loss percentage was recorded by combining a wax coating and 100 ppm nanosilver, and packaging the sample in PPE. Moreover, coating with wax and nanosilver at either 50 or 100 ppm reduced the fresh weight loss percentage of mandarins relative to the control.

In this respect, the data in Table 2 indicates the effect of the postharvest treatments, cold storage period, and their interaction with the fresh weight loss percentage of Murcott mandarins. This data showed that after the first month of cold storage (Table 2), all applied treatments were capable of reducing weight loss compared with the control. The differences among these treatments were not big enough to be significant except for "coated with 100 ppm nanosilver"—alone or mixed with wax and packaged in PPE. The weight loss percentage for all applied treatments tended to increase significantly with the advancement of cold storage.

Similar results were nearly found when discussing the interaction effect between used postharvest treatments and shelf-life (Table 2) duration, except for the combination consisting of wax plus 100 ppm nanosilver and packaging in PPE, which was able to record the lowest weight loss percentage as compared with other treatments after the last period of shelf life, especially in the second season (Table 2).

3.3. Pulp Firmness

The strength and fruit hardness of coated mandarins were significantly improved. In comparison, uncoated fruit undergoes tissue suppleness with time while being stored [53,73]. For both coated and uncoated fruits, there is no discernible change in fruit firmness during the first few days of low-temperature storage; rather, variations emerge over time [74].

The effect of postharvest treatments during the cold storage period on pulp firmness during the 2018 and 2019 seasons is displayed in Table 3. The data revealed that the combination consisting of coating with wax plus 100 ppm nanosilver and packaging in perforated polyethylene (PPE) was the most effective treatment for reducing the loss of pulp firmness, and it gave the greatest value of firmness as compared with other used treatments. In addition, all applied treatments recorded a higher firmness value than the control treatment, which showed the lowest value. This trend was stable in the two seasons of study (Table 3).

During the shelf-life period, it could be noticed that coating with wax mixed with 100 ppm nanosilver and packaging in PPE had a more elevated value for pulp firmness. Moreover, other treatments also caused a higher firmness value relative to the control treatment in both seasons (Table 3). On the contrary, mandarins treated with imazalil and packaged in PPE had the lowest firmness value.

3.4. Vitamin C Content

Due to acid consumption as respiration substrates, ascorbic acid degrades over time when stored [75]. Nevertheless, the data shown in Table 4 indicated that the highest content of vitamin C in the juice of Murcott mandarin was obtained by coating with either wax, 50 ppm nanosilver, or the combination of wax and 100 ppm nanosilver treatments with packaging in perforated polyethylene (PPE) during cold storage period effect in the two seasons of study (Table 4). The highest impact was demonstrated by the last treatment (the combination of wax and 100 ppm nanosilver), which maintained the highest content of vitamin C. In contrast, both seasons found the least vitamin C content in the control treatment (imazalil followed by packaging in PPE).

The same trend was noticed throughout the shelf-life periods (Table 4) consistently during the two seasons of study. All used treatments recorded higher vitamin C content than the control. Moreover, samples coated in all treatments with wax and others before packaging in PPE were similar in their vitamin C content in the 2019 season (Table 4).

Table 3. Effect of some postharvest applied treatments on pulp firmness (g/cm^2) of Murcott mandarin fruits during 1, 2, 3, and 4 months of cold storage and after 6 days of shelf life during 2018 and 2019 seasons.

Treatments	Cold Storage Period (Month) (P)					6 Days Shelf Life (P)				
	The First Season (2018)									
	1	2	3	4	Mean	1	2	3	4	Mean
IMZ (Control)	180.00 b	165.00 c	141.67 e	115.00 g	150.42 C	165.00 bc	145.00 ef	137.33 f	106.67 i	138.50 C
Wax	190.00 a	173.00 b	145.00 de	125.00 f	158.25 B	170.00 ab	160.33 bc	144.33 ef	113.33 hi	148.75 AB
not wax + 50 ppm nanosilver	193.33 a	175.00 b	144.67 de	120.00 fg	158.25 B	170.00 ab	165.00 bc	142.33 ef	110.00 i	146.83 B
not wax + 100 ppm nanosilver	190.00 a	173.67 b	150.00 d	127.00 f	160.17 B	165.00 bc	155.00 d	145.00 ef	120.00 gh	146.25 B
wax + 100 ppm nanosilver	190.00 a	175.00 b	163.33 c	148.33 de	169.17 A	175.33 a	160.00 cd	145.67 e	123.33 g	151.08 A
Mean	188.67 A	172.33 B	148.93 C	127.07 D		169.07 A	158.47 B	142.93 C	114.67 D	
	The second season (2019)									
IMZ (Control)	180.00 b	160.00 e	141.00 g	113.33 i	148.58 C	160.00 cd	141.67 ef	137.67 f	103.33 h	135.67 C
Wax	195.00 a	173.00 cd	147.67 f	128.33 h	161.00 B	175.00 a	165.67 bc	145.33 e	115.00 g	150.25 A
not wax + 50 ppm nanosilver	192.33 a	171.67 d	146.67 f	123.33 h	158.50 B	172.67 ab	156.67 d	141.33 ef	113.33 g	146.00 B
not wax + 100 ppm nanosilver	191.00 a	167.67 d	151.67 f	126.67 h	159.25 B	169.33 ab	158.33 d	148.33 e	116.67 g	148.17 AB
wax + 100 ppm nanosilver	190.67 a	177.33 bc	167.00 d	148.00 f	170.75 A	172.33 ab	161.00 cd	147.67 e	118.33 g	149.83 A
Mean	189.80 A	169.93 B	150.80 C	127.93 D		169.87 A	156.67 A	144.07 C	113.33 B	

IMZ = imazalil, lowercase letters in the same column indicate significant difference, while uppercase letters in the rows and columns indicate significant difference between means by LSD at a 0.05 level.

Table 4. Effect of some postharvest applied treatments on vitamin C content (mg/100 mL juice) of Murcott mandarin fruits during 1, 2, 3, and 4 months of cold storage and after 6 days of shelf life during 2018 and 2019 seasons.

Treatments	Cold Storage Period (Month) (P)					6 Days Shelf Life (P)				
	The First Season (2018)									
	1	2	3	4	Mean	1	2	3	4	Mean
IMZ (Control)	51.27 bc	50.00 d	47.53 f	43.17 h	47.99 C	50 bcd	47.76 gh	44.93 kl	41.50 n	46.05 C
Wax	51.63 b	50.67 bcd	47.93 ef	45.00 g	48.81 AB	50.60 ab	48.90 ef	45.77 jk	42.00 n	46.82 B
not wax + 50 ppm nanosilver	52.90 a	50.50 bcd	48.23 ef	45.20 g	49.21 A	50.40 abc	48.17 fg	46.03 ij	43.67 m	47.07 B
not wax + 100 ppm nanosilver	50.43 cd	50.33 cd	48.63 e	43.73 h	48.28 BC	50.03 bcd	49.20 de	45.13 jkl	41.93 n	46.57 B
wax + 100 ppm nanosilver	51.67 b	50.10 cd	47.57 f	45.57 g	48.72 AB	51.03 a	49.60 cde	47.00 hi	44.23 lm	47.97 A
Mean	51.58 A	50.32 B	47.98 C	44.53 D		50.41 A	48.73 B	45.77 C	42.67 D	
	The second season (2019)									
IMZ (Control)	50.00 de	49.10 ef	47.70 h	42.00 k	47.20 B	49.50 bc	47.87 ef	45.10 ij	40.83 m	45.82 D
Wax	51.33 bc	50.33 cd	48.23 fgh	44.90 ij	48.70 A	50.10 ab	49.33 bcd	45.90 hi	42.23 l	46.89 BC
not wax + 50 ppm nanosilver	53.00 a	50.27 cd	48.40 fgh	44.70 ij	49.09 A	50.67 a	48.63 de	46.27 gh	43.27 k	47.21 B
not wax + 100 ppm nanosilver	51.06 bcd	50.33 cd	48.83 efg	43.90 j	48.53 A	50.00 abc	49.23 cd	45.33 i	41.40 lm	46.49 C
wax + 100 ppm nanosilver	52.00 ab	50.23 cd	47.83 gh	45.63 i	48.92 A	50.00 abc	49.17 cd	47.07 fg	44.47 j	47.67 A
Mean	51.48 A	50.05 B	48.20 C	44.23 D		50.05 A	48.84 B	45.93 C	42.44 D	

IMZ = imazalil, lowercase letters in the same column indicate significant difference, while uppercase letters in the rows and columns indicate significant difference between means by LSD at a 0.05 level.

3.5. Panel Taste Index

In recent studies, 46 different mandarin varieties belonging to several natural subgroups were examined for wide genetic variability in numerous fruit-quality features, including physical, physiological (ripening period), nutritional composition, and sensory attributes [76].

Nevertheless, the effects of some postharvest treatments in the cold storage period are shown in Table 5. The data indicated that all used treatments gave an excellent taste index for Murcott mandarins, with significant differences relative to the control treatment in both seasons. The same trend was observed with the effect of used postharvest treatments throughout the shelf-life period in both study seasons (Table 5).

Table 5. Effect of some postharvest applied treatments on taste panel index (PTI) of Murcott mandarin fruits during 1, 2, 3, and 4 months of cold storage and after 6 days of shelf life during 2018 and 2019 seasons.

Treatments	Cold Storage Period (Month) (P)					6 Days Shelf Life (P)				
	The First Season (2018)									
	1	2	3	4	Mean	1	2	3	4	Mean
IMZ (Control)	4.33 abc	3.00 ef	2.33 f	1.33 g	2.75 B	3.67 cd	3.00 ef	2.00 g	1.00 h	2.42 B
Wax	4.67 ab	4.00 bcd	3.33 de	3.00 ef	3.75 A	4.00 bc	3.33 de	3.00 ef	2.67 f	3.25 A
not wax + 50 ppm nanosilver	5.00 a	4.00 bcd	4.00 bcd	3.00 ef	4.00 A	4.33 ab	3.67 cd	3.00 ef	3.00ef	3.50 A
not wax + 100 ppm nanosilver	5.00 a	4.00 bcd	4.00 bcd	3.33 de	4.08 A	4.67 a	3.00 ef	3.33 de	2.00 g	3.25 A
wax + 100 ppm nanosilver	5.00 a	4.00 bcd	3.67 cde	3.33 de	4.00 A	4.00 bc	3.67 cd	3.33 de	3.00 ef	3.50 A
Mean	4.80 A	3.80 B	3.47 B	2.80 C		4.13 A	3.33 B	2.93 C	2.33 D	
	The second season (2019)									
IMZ (Control)	4.00 bcd	3.33 def	2.33 g	1.33 h	2.75 B	4.00 abc	3.00 de	2.00 f	1.00 g	2.50 B
Wax	4.67 ab	4.00 bcd	3.67 cde	3.00 efg	3.83 A	4.33 ab	3.67 bcd	3.00 de	2.67 ef	3.42 A
not wax + 50 ppm nanosilver	5.00 a	4.33 abc	3.67 cde	2.67 fg	3.92 A	4.67 a	3.33 cde	3.00 de	2.67 ef	3.42 A
not wax + 100 ppm nanosilver	5.00 a	4.33 abc	4.00 bcd	3.33 def	4.17 A	4.67 a	3.67 bcd	3.67 bcd	2.67 ef	3.67 A
wax + 100 ppm nanosilver	5.00 a	4.33 abc	3.67 cde	3.00 efg	4.00 A	4.00 abc	3.33 cde	3.33 cde	3.00 de	3.42 A
Mean	4.73 A	4.07 B	3.47 C	2.67 C		4.33 A	3.40 B	3.00 C	2.40 D	

IMZ = imazalil, lowercase letters in the same column indicate significant difference, while uppercase letters in the rows and columns indicate significant difference between means by LSD at a 0.05 level.

Concerning the influence of the cold storage period and shelf-life period on the taste panel index, the results introduced in Table 5 showed that the "excellent" taste panel index was noted after the first month of cold storage. Subsequently, the taste panel index decreased gradually by the end of cold storage in both seasons.

3.6. Catalase Enzyme Activity

The response of catalase enzyme activity in Murcott mandarin to various postharvest treatments, regardless of the cold storage period, is displayed in Figure 1A. The results showed that the highest catalase enzyme activity was found in the control (imazalil) and wax, compared with other remaining treatments. The combined wax—100 ppm nanosilver showed the lowest activity of catalase enzymes.

Concerning the influence of the cold storage period on catalase enzyme activity, the data shown in Figure 1A indicated that catalase enzyme activity after two months of cold storage was higher than after four months of that storage.

Catalase enzyme activity is also affected by the interaction between postharvest treatments and the cold storage period, where the treatments of coating with either 100 or 50 ppm nanosilver gave high catalase enzyme activity after two months of cold storage. On the other hand, the lowest activity of the catalase enzyme was recorded with the samples coated with wax plus 100 ppm nano silver after four months of cold storage.

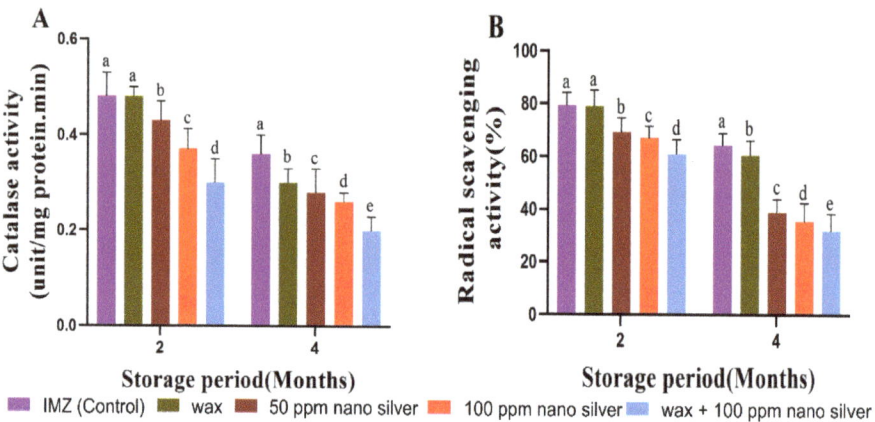

Figure 1. Studying the effect of some applied treatments after harvesting on catalase enzyme activity (**A**) and total antioxidant (**B**) of Murcott mandarin fruits during 2, and 4 months of cold storage during 2019 season during cold storage periods during the 2019 season. Data presented as mean ± SD, different lowercase letters above columns indicate significant differences.

3.7. Total Antioxidant Activities

The data in Figure 1B also indicated postharvest treatments' effect on total oxidants. The treatments coating with wax, imazalil, and the combination of wax plus 100 ppm nano silver, respectively showed that the total oxidants of Murcott mandarins after two months of cold storage were higher than after four months.

After two months of cold storage, Murcott mandarin fruits coated with wax and imazalil had higher total antioxidants than those coated with other treatments. Conversely, coating with 50 and 100 ppm nanosilver and combined wax–100 ppm nanosilver treatments gave the lowest total antioxidants. However, the combined wax—100 ppm nanosilver had the lowest total antioxidants relative to other used treatments.

3.8. Total Acidity Percentage

During the cold storage period, data were clear that the total acidity percentage (g citric acid/100 mL juice) gradually and significantly decreased with the advance in the cold storage period in the two seasons (Table 6). The lowest values were recorded four months after cold storage in the two seasons, while the highest values resulted from treatments after one month of cold storage. Coating with wax and 100 ppm nanosilver treatment retained a significantly higher acidity percentage compared with the control and other treatments in both seasons. The interaction between cold storage and the studied treatments was significant in the two seasons. The highest value in the first and second seasons (1.097 and 1.1%, respectively) came from coating with wax and 100 ppm nanosilver at one month. The lowest value in the two seasons (0.495 and 0.493% respectively) came from the control at the end of cold storage.

After the shelf-life period, the data also show that the total acidity percentage was markedly decreased with the advance in the shelf-life period (Table 6). In the two seasons, the lowest percentage was recorded during shelf life after four months of cold storage. All treatments retained significantly higher total acidity compared with the control in both seasons. Additionally, there were significant differences between all the coating with nano silver and the coating with wax-only treatments in the two seasons. The interaction between treatments and the shelf-life period was significant in both seasons. The lowest percentage during shelf life always came from treatments tested after four months of cold storage.

Table 6. Effect of Nanosilver coating on total acidity percentage (TA %) of Murcott mandarin fruits during 1, 2, 3, and 4 months of cold storage and after 6 days of shelf life during 2018 and 2019 seasons.

Treatments	Cold Storage Period (Month) (P)					6 Days Shelf Life (P)				
	The First Season (2018)									
	1	2	3	4	Mean	1	2	3	4	Mean
IMZ (Control)	0.900 c	0.650 fgh	0.600 ghi	0.495 j	0.661 D	0.750 d	0.656 fg	0.507 jk	0.467 l	0.595 E
Wax	0.947 bc	0.733 de	0.650 fgh	0.573 i	0.726 C	0.900 c	0.687 e	0.565 hi	0.503 k	0.664 D
not wax + 50 ppm nanosilver	0.980 b	0.757 d	0.670 ef	0.550 ij	0.739 C	0.947 b	0.666 efg	0.570 h	0.533 ij	0.679 C
not wax + 100 ppm nanosilver	1.100 a	0.750 d	0.650 fgh	0.593 hi	0.773 B	0.980 a	0.656 fg	0.643 g	0.580 h	0.715 B
wax + 100 ppm nanosilver	1.097 a	0.933 bc	0.750 d	0.660 fg	0.860 A	0.953 ab	0.770 d	0.686 ef	0.553 hi	0.741 A
Mean	1.005 A	0.765 B	0.664 C	0.574 D		0.906 A	0.687 B	0.594 C	0.527 D	
New LSD at 0.05%	T = 0.031		P = 0.066 T × P = 0.063			T = 0.014 P = 0.023 T × P = 0.028				
	The second season (2019)									
IMZ (Control)	0.980 c	0.633 gh	0.603 hi	0.493 j	0.677 D	0.750 c	0.603 ef	0.503 ij	0.477 j	0.583 E
Wax	1.030 b	0.710 f	0.680 fg	0.573 hi	0.766 BC	0.907 b	0.667 d	0.570 fg	0.500 ij	0.661 D
not wax + 50 ppm nanosilver	0.992 bc	0.783 e	0.687 fg	0.543 i	0.751 C	0.960 a	0.660 d	0.570 fg	0.523 hi	0.678 C
not wax + 100 ppm nanosilver	1.033 b	0.797 e	0.693 fg	0.597 hi	0.780 B	0.982 a	0.667 d	0.617 e	0.557 gh	0.705 B
wax + 100 ppm nanosilver	1.100 a	0.910 d	0.773 e	0.613 h	0.832 A	0.960 a	0.777 c	0.690 d	0.563 g	0.747 A
Mean	1.027 A	0.767 B	0.687 C	0.564 D		0.912 A	0.675 B	0.590 C	0.524 D	
New LSD at 0.05%	T = 0.022 P = 0.046 T × P = 0.044					T = 0.017 P = 0.013 T × P = 0.034				

T = Treatment, P = period, T × P = interaction between treatments and period, IMZ = imazalil, lowercase letters in the same column indicate significant difference, while uppercase letters in the rows and columns indicate significant difference between means by LSD at a 0.05 level.

3.9. Total Soluble Solids (TSSs)

During the cold storage period, it is clear that TSSs increased with the advance of the cold storage period in both seasons (Table 7). Moreover, TSSs were significantly affected by the tested treatments. The wax coated–100 ppm nanosilver treatment recorded the fewest TSSs in both seasons compared with other treatments. The interaction between the studied treatments and the cold storage period was significant in the two seasons.

After the shelf-life period, the total soluble solids increased during shelf-life as the cold storage period advanced (Table 7). The control treatment recorded the highest TSS in the two seasons compared with other treatments. The wax coated–100 ppm nanosilver treatment recorded the fewest TSSs in both seasons compared with other treatments.

Table 7. Effect of Nanosilver coating on total soluble solids (TSS) (Brix°) of Murcott mandarin fruits during 1, 2, 3, and 4 months of cold storage and after 6 days of shelf life during the 2018 and 2019 seasons.

Treatments	Cold Storage Period (Month) (P)					6 Days Shelf Life (P)				
	The First Season (2018)									
	1	2	3	4	Mean	1	2	3	4	Mean
IMZ (Control)	9.50 gh	10.07 de	10.50 ab	10.70 a	10.19 A	9.80 fg	10.33 cd	10.40 bcd	10.90 a	10.36 A
Wax	9.07 i	9.50 gh	10.30 abc	10.50 ab	9.84 BC	9.53 g	10.00 ef	10.53 bcd	10.63 abc	10.17 B
not wax + 50 ppm nanosilver	8.83 jk	9.80 e–h	10.00 def	10.20 bcd	9.71 CD	9.03 h	10.00 ef	10.47 bcd	10.87 a	10.09 B
not wax + 100 ppm nanosilver	9.40 h	9.90 d–g	10.10 cde	10.50 ab	9.97 B	9.50 g	10.00 ef	10.30 de	10.70 ab	10.12 B
wax + 100 ppm nanosilver	8.50 k	9.60 fgh	10.00 def	10.20 bcd	9.57 D	9.00 h	9.80 fg	10.27 de	10.53 bcd	9.90 C
Mean	9.06 D	9.77 C	10.18 B	10.42 A		9.37 D	10.03 C	10.39 B	10.73 A	
New LSD at 0.05%	T = 0.19 P = 0.21 T × P = 0.39					T = 0.14 P = 0.16 T × P = 0.29				
	The second season (2019)									
IMZ (Control)	9.70 fgh	10.03 cde	10.47 ab	10.77 a	10.24 A	9.87 e	10.87 a	10.30 d	10.63 b	10.42 A
Wax	9.50 hi	9.60 ghi	10.23 bcd	10.47 ab	9.95 B	9.67 f	10.00 e	10.47 bcd	10.67 ab	10.20 B
not wax + 50 ppm nanosilver	8.97 j	9.83 efg	10.07 cde	10.33 bc	9.80 B	9.03 g	10.03 e	10.53 bc	10.83 a	10.11 B
not wax + 100 ppm nanosilver	9.30 i	9.83 efg	10.00 def	10.47 ab	9.90 B	9.50 f	10.03 e	10.37 cd	10.63 b	10.13 B
wax + 100 ppm nanosilver	8.50 k	9.67 gh	10.07 cde	10.23 bcd	9.62 C	9.13 g	9.97 e	10.33 d	10.56 bc	10.00 C
Mean	9.19 D	9.79 C	10.17 B	10.45 A		9.44 D	10.18 C	10.40 B	10.67 A	
New LSD at 0.05%	T = 0.15 P = 0.17 T × P = 0.31					T = 0.09 P = 0.14 T × P = 0.19				

T = Treatment, P = period, T × P = interaction between treatments and period, IMZ = imazalil, lowercase letters in the same column indicate significant difference, while uppercase letters in the rows and columns indicate significant difference between means by LSD at a 0.05 level.

4. Discussion

One of the most important factors that negatively affect fruit quality is water loss, which reduces its commercial life after harvest [77]. For orange fruits, a 2.5% weight loss causes a contraction to begin, and a 5% loss of its original weight makes it no longer marketable [78]. Murcott mandarin fruit weight loss increased progressively in all treatments with the increasing storage period, as shown in Table 2. When separated from the tree, mature fruits undergo a number of metabolic processes such as transpiration and respiration, and there is a positive relationship between weight loss and the rate of respiration and transpiration [79]. The activity of metabolic processes in fruits leads to weight loss and fruit quality during the storage period and shelf life [12,80]. Coatings can reduce water loss and thus reduce harmful effects by trapping the moisture inside the fruits. In addition to preventing the exit of water vapor from the stomata on the peel (reducing the transpiration process) and thus maintaining the firmness of the fruit, there are many studies showing that the use of nanoparticles causes fresh weight preservation [81,82]. As a result, keeping fruit in cold and humid environments has a major impact on stomatal behavior and lowers the rate of water loss during storage. High temperatures result in peel shriveling, drying, a dull look, softening, and peel senescence in the end [83,84].

Citrus fruit firmness reveals the thickness and turgidity of the peel [85]. Fruit pulp firmness, which is a crucial factor for the quality of the fruits postharvest, was noticeably decreased during the storage period in all treatments (Table 3). In ripe fruit, reduced fruit firmness with maturing is often associated with the breakdown of the pectic components of the cell wall. Mostly, this is not the first reason for the softening of citrus fruits, in which the dissolution of pectin with ripening is very slow [86]. The reduced firmness of citrus fruits is mainly related to the loss of water from the peel, development, and senescence [87], as well as pathogens that infect the peel and secrete the enzyme that degrades the cell wall [88]. The reason the hardness of the pulp in the coated fruits was maintained is due to a decrease in the process of transpiration and respiration and a delay in rapid ripening during storage. The nano-coating material also effectively contributes to inhibiting the enzymatic and metabolic activities in the fruit and resisting the fungal infections that affect citrus [77,79,88].

Ascorbic acid (vitamin C) concentration in fruits decreases with prolonged storage, as organic acids are consumed as substrates in respiration [75,89]. Despite this, the nano-coating material plus wax (wax + 100 ppm nanosilver) was better, as it kept the level of ascorbic acid above the control level throughout storage in both seasons (Table 4). Increased water loss in fruits leads to rapid oxidation, and, therefore, a rapid loss of ascorbic acid [90]. In other studies, it was found that using high concentrations of nanomaterials in coating formulations, significantly maintained the level of ascorbic acid in coated fruits [91]. The organic acids in fruits decrease during postharvest storage as a result of their use as metabolic substrates in the respiratory system [92,93]. The combined wax–100 ppm nanosilver coating treatment retained a significantly higher acidity percentage compared with the control and other treatments in the two seasons. This may be because the coating inhibited the activity of metabolic enzymes and slowed down the rate of acidolysis in pears during storage [94].

TSSs were significantly affected by the tested treatments. The coated with wax and 100 ppm nanosilver treatment recorded the fewest TSSs in both seasons compared with other treatments. This decrease in soluble solids in the covered fruits is attributed to the slower metabolic processes, such as respiration and transpiration, compared to the untreated fruits of various postharvest treatments [95].

Previous studies showed that the taste panel of mandarin varieties coated with a low gas permeability layer has a less fresh flavor compared to those covered with a higher gas permeability layer (polyethylene and wax) [96,97]. The panel taste index found that the nanoparticle-coated fruit had more tangerine flavor than the uncoated [96]. These results were partially in agreement with those obtained by [43].

The obtained results may be due to the wax coatings contributing to the fruit shine as well as maintaining gaseous exchange and water retention. The fruit continues to respire after harvest, and although the content and composition of coatings provide high levels of wax gloss, they tend to negatively affect the permeation of gases through the peel, which might lead to the development of off-flavors [15,52,98].

The typical increased off flavor volatiles associated with anaerobic respiration in the fruit include ethanol and acetaldehyde [52,99]. Furthermore, the wax application plays an important role in prolonging fruit quality, with differing effects on some fruit quality parameters [26]. Thus, the activity of catalase enzyme and total antioxidant activities decreases. Moreover, nano silver particles, considered an antibacterial agent, promise longer durability for food [100], and nano silver particles inhibited mycelium growth of *Penicillium digitatum* and *Aspergillus niger* during storage [43], and nano silver particle formulated mucilage exhibited bactericidal activity for *Escherichia coli* and *Staphylococcus* as well as inhibited growth of *Fusarium solani* and *Aspergillus niger* [101]. Nanosilver particles significantly controlled microbial proliferation and could be considered a biocidal preservative [102]. Furthermore, as ethylene signaling inhibitors, nano silver particles effectively reduce ethylene content to increase life commercially [103].

5. Summary

The short shelf life of citrus fruits during storage has a significant impact on the determinants of fruit quality. Recently, the use of a variety of harmless and usable coatings, such as plant extracts as well as nanomaterials and others, to extend the shelf life of fruits and vegetables has been widely used. In this study, we examined several different combinations of wax and nanosilver to coat Murcott mandarin fruits during storage and shelf life, and we examined the overall effect of these coatings on quality evaluation during 1, 2, 3, and 4 months of cold storage and after 6 days of shelf life during two seasons. From the obtained data, it could be proven that the combined wax—100 ppm nanosilver and packaged in 0.005% perforated polyethylene (PPE) treatment was the most effective treatment. Therefore, these coatings could be promising alternative materials for extending mandarin fruits' postharvest life and marketing period.

Author Contributions: M.M.G. (Mohamed M. Gemail), I.E.E. and M.M.G. (Mohamed M. Gad) performed the experiments with support from I.E.E., S.M.E.-H. and M.M.G. (.Mohamed M Gemail) conceived the project. S.M.E.-H. and M.M.G. (Mohamed M. Gad) designed the experiments. I.E.E., C.C., M.M.J., B.A., W.A.A. and M.M.G. (Mohamed M. Gemail) analyzed the data and wrote the manuscript. All authors have read and agreed to the published version of the manuscript.

Funding: This research received no external funding.

Institutional Review Board Statement: Not applicable.

Informed Consent Statement: Not applicable.

Data Availability Statement: Not applicable.

Acknowledgments: All authors are grateful to the Horticulture Department, Faculty of Agriculture, Zagazig University, Egypt for providing some facilities and equipment to perform this work.

Conflicts of Interest: The authors declare no conflict of interest.

References

1. MALR. Mandarin Fruit Quality. In *Agricultural Research & Development Council*; Ministry of Agriculture and Land Reclamation, Egypt: Giza, Egypt, 2018.
2. Liu, Y.; Heying, E.; Tanumihardjo, S.A. History, global distribution, and nutritional importance of citrus fruits. *Compr. Rev. Food Sci. Food Saf.* **2012**, *11*, 530–545. [CrossRef]
3. Tomar, A.; Mall, M.; Rai, P. Pharmacological importance of citrus fruits. *Int. J. Pharm. Sci.* **2013**, *4*, 156.
4. Zou, Z.; Xi, W.; Hu, Y.; Nie, C.; Zhou, Z. Antioxidant activity of Citrus fruits. *Food Chem.* **2016**, *196*, 885–896. [CrossRef] [PubMed]
5. Diab, K.A. In Vitro Studies on Phytochemical Content, Antioxidant, Anticancer, Immunomodulatory, and Antigenotoxic Activities of Lemon, Grapefruit, and Mandarin Citrus Peels. *Asian Pac. J. Cancer Prev.* **2016**, *17*, 3559–3567. [PubMed]
6. Mannheim, C.H.; Soffer, T. Permeability of different wax coatings and their effect on citrus fruit quality. *J. Agric. Food Chem.* **1996**, *44*, 919–923. [CrossRef]
7. Ahmed, D.M.; El-Shami, S.; El-Mallah, M.H. Jojoba oil as a novel coating for exported Valencia orange fruit. Part 1. The use of trans (isomerized) jojoba oil. *Am.-Eurasian. J. Agric. Environ. Sci.* **2007**, *2*, 173–181.
8. Du Plooy, W.; Regnier, T.; Combrinck, S. Essential oil amended coatings as alternatives to synthetic fungicides in citrus postharvest management. *Postharvest Biol. Technol.* **2009**, *53*, 117–122. [CrossRef]
9. Abobatta, W.F. Citrus varieties in Egypt: An impression. *Int. J. Appl. Sci.* **2019**, *1*, 63–66.
10. Sivakumar, D.; Jiang, Y.; Yahia, E.M. Maintaining mango (*Mangifera indica* L.) fruit quality during the export chain. *Int. J. Food Res.* **2011**, *44*, 1254–1263. [CrossRef]
11. Carrillo-Lopez, A.; Ramirez-Bustamante, F.; Valdez-Torres, J.; Rojas-Villegas, R.; Yahia, E. Ripening and quality changes in mango fruit as affected by coating with an edible film. *J. Food Qual.* **2000**, *23*, 479–486. [CrossRef]
12. Hoa, T.T.; Ducamp, M.N.; Lebrun, M.; Baldwin, E.A. Effect of different coating treatments on the quality of mango fruit. *J. Food Qual.* **2002**, *25*, 471–486. [CrossRef]
13. Ridoutt, B.; Juliano, P.; Sanguansri, P.; Sellahewa, J. The water footprint of food waste: Case study of fresh mango in Australia. *J. Clean. Prod.* **2010**, *18*, 1714–1721. [CrossRef]
14. Hassan, B.; Chatha, S.A.S.; Hussain, A.I.; Zia, K.M.; Akhtar, N. Recent advances on polysaccharides, lipids and protein based edible films and coatings: A review. *Int. J. Biol. Macromol.* **2018**, *109*, 1095–1107. [CrossRef] [PubMed]
15. Hagenmaier, R.; Shaw, P. Changes in volatile components of stored tangerines and other specialty citrus fruits with different coatings. *J. Food Sci.* **2002**, *67*, 1742–1745. [CrossRef]

16. Hagenmaier, R.D. Evaluation of a polyethylene–candelilla coating for 'Valencia' oranges. *Postharvest Biol. Technol.* **2000**, *19*, 147–154. [CrossRef]
17. Nair, M.S.; Saxena, A.; Kaur, C. Effect of chitosan and alginate based coatings enriched with pomegranate peel extract to extend the postharvest quality of guava (*Psidium guajava* L.). *Food Chem.* **2018**, *240*, 245–252. [CrossRef]
18. Vyas, P.B.; Gol, N.B.; Rao, T.R. Postharvest quality maintenance of papaya fruit using polysaccharide-based edible coatings. *Int. J. Fruit Sci.* **2014**, *14*, 81–94. [CrossRef]
19. Arnon, H.; Granit, R.; Porat, R.; Poverenov, E. Development of polysaccharides-based edible coatings for citrus fruits: A layer-by-layer approach. *Food Chem.* **2015**, *166*, 465–472. [CrossRef]
20. Palou, L.; Valencia-Chamorro, S.A.; Pérez-Gago, M.B. Antifungal edible coatings for fresh citrus fruit: A review. *Coatings* **2015**, *5*, 962–986. [CrossRef]
21. Panahirad, S.; Naghshiband-Hassani, R.; Ghanbarzadeh, B.; Zaare-Nahandi, F.; Mahna, N. Shelf life quality of plum fruits (*Prunus domestica* L.) improves with carboxymethylcellulose-based edible coating. *HortScience* **2019**, *54*, 505–510. [CrossRef]
22. Naeem, A.; Abbas, T.; Ali, T.M.; Hasnain, A. Effect of guar gum coatings containing essential oils on shelf life and nutritional quality of green-unripe mangoes during low temperature storage. *Int. J. Biol. Macromol.* **2018**, *113*, 403–410. [CrossRef]
23. Saberi, B.; Golding, J.B.; Marques, J.R.; Pristijono, P.; Chockchaisawasdee, S.; Scarlett, C.J.; Stathopoulos, C.E. Application of biocomposite edible coatings based on pea starch and guar gum on quality, storability and shelf life of 'Valencia' oranges. *Postharvest Biol. Technol.* **2018**, *137*, 9–20. [CrossRef]
24. Hall, D.J. Innovations in Citrus Waxing. *Proc. Fla. State Hortic. Soc.* **1981**, *94*, 258–262.
25. Petracek, P.D.; Dou, H.; Mourer, J.; Davis, C. Measurement of gas exchange characteristics of waxes applied to citrus fruit. In Proceedings of the International Symposium on Growth and Development of Fruit Crops, East Lansing, MI, USA, 19–21 June 1997; Volume 527, pp. 73–84.
26. Njombolwana, N.S.; Erasmus, A.; Van Zyl, J.G.; Du Plooy, W.; Cronje, P.J.; Fourie, P.H. Effects of citrus wax coating and brush type on imazalil residue loading, green mould control and fruit quality retention of sweet oranges. *Postharvest Biol. Technol.* **2013**, *86*, 362–371. [CrossRef]
27. An, J.; Zhang, M.; Wang, S.; Tang, J. Physical, chemical and microbiological changes in stored green asparagus spears as affected by coating of silver nanoparticles-PVP. *LWT-Food Sci. Technol.* **2008**, *41*, 1100–1107. [CrossRef]
28. Gammariello, D.; Conte, A.; Buonocore, G.; Del Nobile, M.A. Bio-based nanocomposite coating to preserve quality of Fior di latte cheese. *J. Dairy Sci.* **2011**, *94*, 5298–5304. [CrossRef]
29. Chhipa, H. Applications of nanotechnology in agriculture. *Methods Microbiol.* **2019**, *46*, 115–142.
30. Salem, N.M.; Albanna, L.S.; Abdeen, A.O.; Ibrahim, Q.I.; Awwad, A.M. Sulfur nanoparticles improves root and shoot growth of tomato. *J. Agric. Sci.* **2016**, *8*, 179–185. [CrossRef]
31. Shankramma, K.; Yallappa, S.; Shivanna, M.; Manjanna, J. Fe_2O_3 magnetic nanoparticles to enhance *S. lycopersicum* (tomato) plant growth and their biomineralization. *Appl. Nanosci.* **2016**, *6*, 983–990. [CrossRef]
32. Lin, D.; Xing, B. Pytotoxicity of nanoparticles: Inhibition of seed germination and root growth. *Environ. Pollut.* **2007**, *150*, 245–250. [CrossRef]
33. Ghidan, A.Y.; Al-Antary, T.M.; Awwad, A.M.; Ghidan, O.Y.; Araj, S.-E.A.; Ateyyat, M.A. Comparison of different green synthesized nanomaterials on green peach aphid as aphicidal potential. *Fresenius Environ. Bull.* **2018**, *27*, 7009–7016.
34. Abumelha, H.M.; Alkhatib, F.; Alzahrani, S.; Abualnaja, M.; Alsaigh, S.; Alfaifi, M.Y.; Althagafi, I.; El-Metwaly, N. Synthesis and characterization for pharmaceutical models from Co (II), Ni (II) and Cu (II)-thiophene complexes; apoptosis, various theoretical studies and pharmacophore modeling. *J. Mol. Liq.* **2021**, *328*, 115483. [CrossRef]
35. Al Jahdaly, B.; Maghraby, Y.R.; Ibrahim, A.H.; Shouier, K.R.; Taher, M.M.; El-Shabasy, R.M. Role of Green Chemistry in Sustainable Corrosion Inhibition: A review on Recent Developments. *Mater. Today Sustain.* **2022**, *20*, 100842. [CrossRef]
36. Kim, J.S.; Kuk, E.; Yu, K.N.; Kim, J.-H.; Park, S.J.; Lee, H.J.; Kim, S.H.; Park, Y.K.; Park, Y.H.; Hwang, C.-Y. Antimicrobial effects of silver nanoparticles. *Nanomed. Nanotechnol. Biol. Med.* **2007**, *3*, 95–101. [CrossRef]
37. Lee, S.; Lee, J.; Kim, K.; Sim, S.-J.; Gu, M.B.; Yi, J.; Lee, J. Eco-toxicity of commercial silver nanopowders to bacterial and yeast strains. *Biotechnol. Bioprocess Eng.* **2009**, *14*, 490–495. [CrossRef]
38. De Azeredo, H.M. Nanocomposites for food packaging applications. *Int. Food Res. J.* **2009**, *42*, 1240–1253. [CrossRef]
39. Sarfraz, J.; Gulin-Sarfraz, T.; Nilsen-Nygaard, J.; Pettersen, M.K. Nanocomposites for food packaging applications: An overview. *J. Nanomater.* **2020**, *11*, 10. [CrossRef] [PubMed]
40. Malmiri, H.; Osman, A.; Tan, C.; Abdul Rahman, R. Developing a new antimicrobial edible coating formulation based on carboxymethylcellulose-silver nanoparticles for tropical fruits and an in vitro evaluation of its antimicrobial properties. *Acta Hortic* **2013**, *1012*, 705–710. [CrossRef]
41. Shah, S.W.A.; Jahangir, M.; Qaisar, M.; Khan, S.A.; Mahmood, T.; Saeed, M.; Farid, A.; Liaquat, M. Storage stability of kinnow fruit (Citrus reticulata) as affected by CMC and guar gum-based silver nanoparticle coatings. *Molecules* **2015**, *20*, 22645–22661. [CrossRef]
42. El-Ashry, R.M.; El-Saadony, M.T.; El-Sobki, A.E.; El-Tahan, A.M.; Al-Otaibi, S.; El-Shehawi, A.M.; Elshaer, N. Biological silicon nanoparticles maximize the efficiency of nematicides against biotic stress induced by Meloidogyne incognita in eggplant. *Saudi J. Biol. Sci.* **2022**, *29*, 920–932. [CrossRef]

43. Khaleel, R.; Al-Samarrai, G.F.; Mohammed, A. Coating of Orange Fruit with Nano-Silver Particles to Minimizing Harmful Environmental Pollution by Chemical Fungicide. *Iraqi J. Agric. Sci.* **2019**, *50*, 1668–1673.
44. Hu, Q.; Fang, Y.; Yang, Y.; Ma, N.; Zhao, L. Effect of nanocomposite-based packaging on postharvest quality of ethylene-treated kiwifruit (Actinidia deliciosa) during cold storage. *Int. Food Res. J.* **2011**, *44*, 1589–1596. [CrossRef]
45. Long, X.; Bai, L.; Xiao, W. Antibacterial and Preservation Effects of Silver-doped TiO_2 Nanoparticles on Nanfeng Citrus. *J. Agric. Biotech.* **2013**, *2*, 2164–4993.
46. Fahiminia, S.; Naseri, L. Effect of Nano-Composite Packages on Quality Properties and Shelf Life of Santa Rosa Plum (*Prunus salicina* L.). *IFSTRJ* **2015**, *11*, 88–99.
47. Hmmam, I.; Zaid, N.m.; Mamdouh, B.; Abdallatif, A.; Abd-Elfattah, M.; Ali, M. Storage Behavior of "Seddik" mango fruit coated with CMC and guar gum-based silver nanoparticles. *Horticulturee* **2021**, *7*, 44. [CrossRef]
48. Ghanem, A.M.K. Economics of bee honey and wax production in Egypt, Alexandria Journal of Agricultural Research. *J. Agric. Sci.* **1994**, *39*, 25–56.
49. Ali, M.; Ahmed, A.; Shah, S.W.A.; Mehmood, T.; Abbasi, K.S. Effect of silver nanoparticle coatings on physicochemical and nutraceutical properties of loquat during postharvest storage. *J. Food Process. Preserv.* **2020**, *44*, e14808. [CrossRef]
50. Chi, H.; Song, S.; Luo, M.; Zhang, C.; Li, W.; Li, L.; Qin, Y. Effect of PLA nanocomposite films containing bergamot essential oil, TiO_2 nanoparticles, and Ag nanoparticles on shelf life of mangoes. *Sci. Hortic.* **2019**, *249*, 192–198. [CrossRef]
51. Deng, Z.; Jung, J.; Simonsen, J.; Zhao, Y. Cellulose nanomaterials emulsion coatings for controlling physiological activity, modifying surface morphology, and enhancing storability of postharvest bananas (*Musa acuminate*). *Food Chem.* **2017**, *232*, 359–368. [CrossRef]
52. Porat, R.; Weiss, B.; Cohen, L.; Daus, A.; Biton, A. Effects of polyethylene wax content and composition on taste, quality, and emission of off-flavor volatiles in 'Mor'mandarins. *Postharvest Biol. Technol.* **2005**, *38*, 262–268. [CrossRef]
53. Rojas, C.; Pérez-Gago, M.; del Río, M. Effect of lipid incorporation to locust bean gum edible coatings on mandarin cv. Fortune. In Proceedings of the 6th International Symposium on Fruit, Nut, and Vegetable Production Engineering, Potsdam, Germany, 11–14 September 2001; pp. 11–14.
54. Cohen, E.; Shalom, Y.; Rosenberger, I. Postharvest Ethanol Buildup and Off-flavor in 'Murcott' Tangerine Fruits. *J. Am. Soc. Hortic. Sci.* **1990**, *115*, 775–778. [CrossRef]
55. Obenland, D.; Arpaia, M.L. Effect of harvest date on off-flavor development in mandarins following postharvest wax application. *Postharvest Biol. Technol.* **2019**, *149*, 1–8. [CrossRef]
56. Chien, P.-J.; Sheu, F.; Lin, H.-R. Coating citrus (Murcott tangor) fruit with low molecular weight chitosan increases postharvest quality and shelf life. *Food Chem.* **2007**, *100*, 1160–1164. [CrossRef]
57. Saad, A.M.; El-Saadony, M.T.; El-Tahan, A.M.; Sayed, S.; Moustafa, M.A.; Taha, A.E.; Taha, T.F.; Ramadan, M.M. Polyphenolic extracts from pomegranate and watermelon wastes as substrate to fabricate sustainable silver nanoparticles with larvicidal effect against Spodoptera littoralis. *Saudi J. Biol. Sci.* **2021**, *28*, 5674–5683. [CrossRef]
58. Gad, M.; Zagzog, O.; Hemeda, O. Development of nano-chitosan edible coating for peach fruits cv. Desert Red. *Int. J. Environ.* **2016**, *5*, 43–55.
59. Shah, S.; Hashim, M.S. Chitosan–aloe vera gel coating delays postharvest decay of mango fruit. *Hortic. Env. Biotechnol.* **2020**, *61*, 279–289. [CrossRef]
60. Liu, S.; Huang, H.; Huber, D.J.; Pan, Y.; Shi, X.; Zhang, Z. Delay of ripening and softening in 'Guifei' mango fruit by postharvest application of melatonin. *Postharvest Biol. Technol.* **2020**, *163*, 111136. [CrossRef]
61. Fletcher, D. An evaluation of the AOAC method of yolk color analysis. *Poult. Sci.* **1980**, *59*, 1059–1066. [CrossRef]
62. Ayesha, H.; Sinha, A.; Mishra, P. Studies on ascorbic acid (vitamin-C) content in different citrus fruits and its degradation during storage. *Sci. Cult.* **2014**, *80*, 275–278.
63. Gad, M.M.; Zagzog, O.A. Mixing xanthan gum and chitosan nano particles to form new coating for maintaining storage life and quality of elmamoura guava fruits. *IJCMAS* **2017**, *6*, 1582–1591.
64. Saad, A.M.; Mohamed, A.S.; Ramadan, M.F. Storage and heat processing affect flavors of cucumber juice enriched with plant extracts. *Int. J. Veg. Sci.* **2021**, *27*, 277–287. [CrossRef]
65. Bhakya, S.; Muthukrishnan, S.; Sukumaran, M.; Muthukumar, M. Biogenic synthesis of silver nanoparticles and their antioxidant and antibacterial activity. *Appl. Nanosci.* **2016**, *6*, 755–766. [CrossRef]
66. Fossati, P.; Prencipe, L.; Berti, G. Use of 3,5-dichloro-2-hydroxybenzenesulfonic acid/4-aminophenazone chromogenic system in direct enzymatic assay of uric acid in serum and urine. *Clin. Chem.* **1980**, *26*, 227–231. [CrossRef] [PubMed]
67. Aebi, H.; Mörikofer-Zwez, S.; von Wartburg, J.-P. Alternative molecular forms of erythrocyte catalase. In *Structure and Function of Oxidation–Reduction Enzymes*; Elsevier: Amsterdam, The Netherlands, 1972; pp. 345–351.
68. Rup, P.J.; Sohal, S.; Kaur, H. Studies on the role of six enzymes in the metabolism of kinetin in mustard aphid, *Lipaphis erysimi* (Kalt.). *J. Environ. Biol.* **2006**, *27*, 579.
69. Wallenstein, S.; Zucker, C.L.; Fleiss, J.L. Some statistical methods useful in circulation research. *Circ. Res.* **1980**, *47*, 1–9. [CrossRef]
70. Albrigo, L.; Ismail, M.; Hale, P.; Hatton, T., Jr. Shipment and storage of Florida grapefruit using unipack film barriers. In Proceedings of the International Society of Citriculture International Citrus Congress, Tokyo, Japan, 9–12 November 1981; Matsumoto, K., Ed.; International Society of Citriculture: Shimizu, Japan, 1982.

71. Ben-Yehoshua, S. Gas Exchange, Transpiration and the Commercial Deterioration of Orange Fruit in Storage. *J. Am. Soc. Hortic. Sci.* **1969**, *49*, 524–528. [CrossRef]
72. Kaufmann, M.R. Water potential components in growing citrus fruits. *Plant Physiol.* **1970**, *46*, 145–149. [CrossRef]
73. Hagenmaier, R. The flavor of mandarin hybrids with different coatings. *Postharvest Biol. Technol.* **2002**, *24*, 79–87. [CrossRef]
74. Perez-Gago, M.; Rojas, C.; Del Rio, M. Effect of hydroxypropyl methylcellulose-lipid edible composite coatings on plum (cv. *Autumn giant*) quality during storage. *J. Food Sci.* **2003**, *68*, 879–883. [CrossRef]
75. Del Caro, A.; Piga, A.; Vacca, V.; Agabbio, M. Changes of flavonoids, vitamin C and antioxidant capacity in minimally processed citrus segments and juices during storage. *Food Chem.* **2004**, *84*, 99–105. [CrossRef]
76. Goldenberg, L.; Yaniv, Y.; Kaplunov, T.; Doron-Faigenboim, A.; Porat, R.; Carmi, N. Genetic diversity among mandarins in fruit-quality traits. *J. Agric. Food Chem.* **2014**, *62*, 4938–4946. [CrossRef] [PubMed]
77. Kawada, K.; Albrigo, L. Effects of film packaging, in-carton air filters, and storage temperatures on the keeping quality of Florida grapefruit. *Proc. Fla. State Hortic. Soc.* **1979**, *92*, 209–212.
78. Grierson, W.; Wardowski, W. Relative humidity effects on the postharvest life of fruits and vegetables. *HortScience* **1978**, *13*, 570–574. [CrossRef]
79. Silva, G.M.C.; Silva, W.B.; Medeiros, D.B.; Salvador, A.R.; Cordeiro, M.H.M.; da Silva, N.M.; Santana, D.B.; Mizobutsi, G.P. The chitosan affects severely the carbon metabolism in mango (*Mangifera indica* L. cv. Palmer) fruit during storage. *Food Chem.* **2017**, *237*, 372–378. [CrossRef] [PubMed]
80. Wills, R.; Lee, T.; Graham, D.; McGlasson, W.; Hall, E. Postharvest. In *An Introduction to the Physiology and Handling of Fruit and Vegetables*; CABI: Granada, Spain, 1981.
81. Gardesh, A.S.K.; Badii, F.; Hashemi, M.; Ardakani, A.Y.; Maftoonazad, N.; Gorji, A.M. Effect of nanochitosan based coating on climacteric behavior and postharvest shelf-life extension of apple cv. Golab Kohanz. *LWT-Food Sci. Technol.* **2016**, *70*, 33–40. [CrossRef]
82. Nawab, A.; Alam, F.; Hasnain, A. Mango kernel starch as a novel edible coating for enhancing shelf-life of tomato (*Solanum lycopersicum*) fruit. *Int. J. Biol. Macromol.* **2017**, *103*, 581–586. [CrossRef]
83. Levy, Y. Effect of evaporative demand on water relations of Citrus limon. *Ann. Bot.* **1980**, *46*, 695–700. [CrossRef]
84. El-Otmani, M. Growth regulator improvement of postharvest quality. In *Fresh Citrus Fruits*; Wardowski, W.E., Miller, W.M., Hall, D.J., Grierson, W., Eds.; Florida Science Source, Inc.: Longboak Key, FL, USA, 2006; pp. 67–104.
85. Baldwin, E. Citrus fruit. In *Biochemistry of Fruit Ripening*; Springer: Berlin/Heidelberg, Germany, 1993; pp. 107–149.
86. Eaks, I.L.; Sinclair, W.B. Cellulose-hemicellulose fractions in the alcohol-insoluble solids of Valencia orange peel. *J. Food Sci.* **1980**, *45*, 985–988. [CrossRef]
87. Spiegel-Roy, P.; Goldschmidt, E.E. *The Biology of Citrus*; Cambridge University Press: Cambridge, UK, 1996.
88. Eckert, J.W. Postharvest disorders and diseases of citrus fruit. In *The Citrus Industry*; University of California Press: Berkeley, CA, USA, 1989; pp. 179–260.
89. Abu-Rayyan, A.; Al Jahdaly, B.A.; AlSalem, H.S.; Alhadhrami, N.A.; Hajri, A.K.; Bukhari, A.A.H.; Waly, M.M.; Salem, A.M. A Study of the Synthesis and Characterization of New Acrylamide Derivatives for Use as Corrosion Inhibitors in Nitric Acid Solutions of Copper. *J. Nanomater.* **2022**, *12*, 3685. [CrossRef]
90. Maftoonazad, N.; Ramaswamy, H. Postharvest shelf-life extension of avocados using methyl cellulose-based coating. *LWT-Food Sci. Technol.* **2005**, *38*, 617–624. [CrossRef]
91. Emamifar, A.; Mohammadizadeh, M. Preparation and application of LDPE/ZnO nanocomposites for extending shelf life of fresh strawberries. *Food Technol. Biotechnol.* **2015**, *53*, 488–495. [CrossRef] [PubMed]
92. Petriccione, M.; Mastrobuoni, F.; Pasquariello, M.S.; Zampella, L.; Nobis, E.; Capriolo, G.; Scortichini, M. Effect of chitosan coating on the postharvest quality and antioxidant enzyme system response of strawberry fruit during cold storage. *Foods* **2015**, *4*, 501–523. [CrossRef] [PubMed]
93. Sheng, L.; Shen, D.; Luo, Y.; Sun, X.; Wang, J.; Luo, T.; Zeng, Y.; Xu, J.; Deng, X.; Cheng, Y. Exogenous γ-aminobutyric acid treatment affects citrate and amino acid accumulation to improve fruit quality and storage performance of postharvest citrus fruit. *Food Chem.* **2017**, *216*, 138–145. [CrossRef] [PubMed]
94. Aşık, E.; Candoğan, K. Effects of chitosan coatings incorporated with garlic oil on quality characteristics of shrimp. *J. Food Qual.* **2014**, *37*, 237–246. [CrossRef]
95. Khaliq, G.; Mohamed, M.T.M.; Ali, A.; Ding, P.; Ghazali, H.M. Effect of gum arabic coating combined with calcium chloride on physico-chemical and qualitative properties of mango (*Mangifera indica* L.) fruit during low temperature storage. *Sci. Hortic.* **2015**, *190*, 187–194. [CrossRef]
96. Miranda, M.; Sun, X.; Ference, C.; Plotto, A.; Bai, J.; Wood, D.; Assis, O.B.G.; Ferreira, M.D.; Baldwin, E. Nano-and micro-carnauba wax emulsions versus shellac protective coatings on postharvest citrus quality. *J. Am. Soc. Hortic. Sci.* **2021**, *146*, 40–49. [CrossRef]
97. Saad, A.M.; Elmassry, R.A.; Wahdan, K.M.; Ramadan, F.M. Chickpea (*Cicer arietinum*) steep liquor as a leavening agent: Effect on dough rheology and sensory properties of bread. *Acta Perio. Technol.* **2015**, *46*, 91–102. [CrossRef]
98. El-Saadony, M.T.; Alkhatib, F.M.; Alzahrani, S.O.; Shafi, M.E.; Abdel-Hamid, S.E.; Taha, T.F.; Aboelenin, S.M.; Soliman, M.M.; Ahmed, N.H. Impact of mycogenic zinc nanoparticles on performance, behavior, immune response, and microbial load in Oreochromis niloticus. *Saudi J. Biol. Sci.* **2021**, *28*, 4592–4604. [CrossRef]

99. Chen, S.; Nussinovitch, A. Permeability and roughness determinations of wax-hydrocolloid coatings, and their limitations in determining citrus fruit overall quality. *Food Hydrocoll.* **2001**, *15*, 127–137. [CrossRef]
100. von Goetz, N.; Fabricius, L.; Glaus, R.; Weitbrecht, V.; Günther, D.; Hungerbühler, K. Migration of silver from commercial plastic food containers and implications for consumer exposure assessment. *Food Addit. Contam.* **2013**, *30*, 612–620. [CrossRef]
101. Shahid, M.; Anjum, F.; Iqbal, Y.; Khan, S.G.; Pirzada, T. Modification of date palm mucilage and evaluation of their nutraceutical potential. *Pak. J. Agric. Sci.* **2020**, *57*, 401–411.
102. Jowkar, M.; Hassanzadeh, N.; Kafi, M.; Khalighi, A. Comprehensive microbial study on biocide application as vase solution preservatives for cut 'Cherry Brandy'rose flower. *Int. J. Hortic. Sci. Technol.* **2017**, *4*, 89–103.
103. Kamiab, F.; Shahmoradzadeh Fahreji, S.; Zamani Bahramabadi, E. Antimicrobial and physiological effects of silver and silicon nanoparticles on vase life of lisianthus (*Eustoma grandiflora* cv. Echo) flowers. *Int. J. Hortic. Sci. Technol.* **2017**, *4*, 135–144.

Disclaimer/Publisher's Note: The statements, opinions and data contained in all publications are solely those of the individual author(s) and contributor(s) and not of MDPI and/or the editor(s). MDPI and/or the editor(s) disclaim responsibility for any injury to people or property resulting from any ideas, methods, instructions or products referred to in the content.

Article

Screening of Essential Oils and Effect of a Chitosan-Based Edible Coating Containing Cinnamon Oil on the Quality and Microbial Safety of Fresh-Cut Potatoes

Sarengaowa [1], Liying Wang [1], Yumeng Liu [1], Chunmiao Yang [1], Ke Feng [2,3,*] and Wenzhong Hu [1,*]

[1] School of Pharmacy and Food Science, Zhuhai College of Science and Technology, Zhuhai 519041, China
[2] LiveRNA Therapeutics Inc., Zhuhai 519041, China
[3] College of Life Science and Technology, Huazhong Agricultural University, Wuhan 430070, China
* Correspondence: fengkesky@163.com (K.F.); wenzhongh@sina.com (W.H.)

Abstract: Fresh-cut potatoes (*Solanum tuberosum* L.) are a popular food owing to their freshness, convenience, and health benefits. However, they might present a potentially high health risk to consumers during transportation, processing, and marketing. In the current study, 18 essential oils (EOs) were screened to test their antimicrobial activity against *Listeria monocytogenes* (LM), *Salmonella typhimurium* (ST), *Staphylococcus aureus* (SA), and *Escherichia coli* O157:H7 (EC O157:H7). The antibacterial effectiveness of a chitosan edible coating (EC) containing cinnamon oil was evaluated against microorganisms on fresh-cut potatoes. Fresh-cut potatoes were treated with chitosan EC and chitosan EC containing different concentrations (0, 0.2, 0.4, and 0.6%, v/v) of cinnamon oil, and uncoated samples served as the control. The viability of naturally occurring microorganisms and artificially inoculated LM on fresh-cut potatoes was evaluated, as were the colour, weight loss, and firmness of potatoes, every 4 days for a total of 16 days at 4 °C. The results demonstrate that the inhibition zones of cinnamon, oregano, and pomelo oils were 16.33–30.47 mm, 22.01–31.31 mm, and 21.75–35.18 mm, respectively. The cinnamon oil exhibits the lowest MIC (0.313 µL/mL) for four foodborne pathogens compared with oregano and pomelo oils. The chitosan EC containing 0.2% cinnamon oil effectively maintains the quality of fresh-cut potatoes including inhibiting the browning, preventing the weight loss, and maintaining the firmness. The decline of total plate counts, yeast and mould counts, total coliform counts, lactic acid bacteria count, and *Listeria monocytogenes* in EC containing 0.2% cinnamon oil were 2.14, 1.92, 0.98, 0.73, and 1.94 log cfu/g, respectively. Therefore, the use of chitosan EC containing cinnamon oil might be a promising approach for the preservation of fresh-cut potatoes.

Keywords: fresh-cut potatoes; chitosan edible coating; cinnamon oil; quality; *Listeria monocytogenes*

1. Introduction

The potato (*Solanum tuberosum* L.) is a type of grain and vegetable crop with high starch, high protein, high vitamin, and low calorie contents that plays an important role in our daily life [1,2]. Potato has been the vital economic crop in many countries. China, India, the Russian Federation, Ukraine, and the United States of America were the top five highest average potato production countries during 2000 to 2019 [3]. Potato processing has shown a tremendous growth in the recent past. Post-harvest processing of sweet potato involves grading and sorting, cleaning, peeling, drying or secondary processing, and storage [4]. Fresh-cut fruits and vegetables, having been cleaned, peeled, and cut, with the maintenance of the characteristics of freshness and nutrition, are therefore healthy and ready to eat and use [5]. Hence, they are increasingly becoming the focus of consumer attention. In this context, an increase in fresh-cut potatoes as a type of fresh-cut fruit and vegetable appears an inevitable future trend in the potato industry. The potato is the largest vegetable crop, grown in 79% of the countries of the world [6]. In 2016, approximately 21 kg

(46 lbs) of potatoes was consumed per capita in the United States, with almost half of that consumption represented by fresh potatoes [7]. However, cutting potatoes causes the cells and tissues to rupture due to mechanical damage [8]. In terms of appearance, they show a softening of texture; there is a loss of flavour, there is microbial infection, and there are other problems, resulting in a shortened shelf life and reduced quality [9]. Therefore, it is urgent to develop effective and environmentally friendly methods to maintain the quality of fresh-cut potatoes during their shelf life.

An edible coating (EC) is an oxygen and moisture barrier that can improve the shelf life and quality of fresh-cut fruits and vegetables [10]. The gas exchange, migration of moisture and solutes, respiration, and oxidative reactions of the surface of fruits and vegetables can be reduced through coating [11]. Different types of ECs, such as chitosan, pectin, starch, alginates, gums, and carrageenan, have been widely studied for the preservation of fresh-cut fruits and vegetables [12–15]. Among these, chitosan EC has received a great deal of attention from the food industries. This derivative of chitin is a copolymer of N-acetylglucosamine and glucosamine residues linked by β-1,4-glycosidic bonds that is insoluble in dilute acids. The main properties of chitosan include film forming, antimicrobial activity, its nontoxic nature, biodegradability, and biocompatibility [16]. As a preservative coating for fruits and vegetables, chitosan has been proven to be edible and biologically safe [17]. In the form of a semipermeable film, chitosan EC can alter the internal atmosphere, reduce transpiration loss, and delay the ripening of fruits and vegetables. These properties make chitosan a superior edible coating [18]. Several recent studies have also indicated that chitosan has beneficial effects in food preservation [19].

According to the Food and Drug Administration, essential oils (EOs) are safe for human consumption [20]. It is widely recognised that EOs contain a broad range of antiparasitic, antibacterial, antifungal, and antiviral constituents [21]. Cinnamon oil is mainly extracted from the bark, leaves, and twigs of cassia that is predominantly grown in China, Indonesia, Vietnam, and Sri Lanka [22,23]. The main components of cinnamon oil are cinnamaldehyde, cinnamyl ester, salicylaldehyde, eugenol, vanillin, and other components [24]. It exhibits antimicrobial activity against several bacteria, yeasts, and fungi. Some studies have indicated that cinnamon oil can damage the cell wall and cell membrane of bacteria, leading to the increased permeability of cell membranes, leakage of cell contents, and reduced rates of bacterial survival [25]. In addition, the volatilisation properties of cinnamon oils can easily enhance the antibacterial activity against pathogenic and spoilage organisms on fresh-cut fruits and vegetables for preservation. However, the strong smell of cinnamon oil can affect the flavour of fresh-cut fruits and vegetables and their acceptability in terms of sensory evaluation. It is, therefore, necessary to combine cinnamon oil with other technologies to reduce its adverse effects on fresh-cut fruits and vegetables. The preservation effects of lemongrass, rosemary, cinnamon, tea tree, mint, palmarosa, oregano, and vanilla oils incorporated into EC have been evaluated on some fresh-cut fruits such as Fuji apples, pineapple, melon, and citrus [26–30].

In this study, the antibacterial activity of eighteen essential oils was screened against *Salmonella typhimurium, Staphylococcus aureus, Listeria monocytogenes*, and *Escherichia coli* O157:H7, and cinnamon oil showed the strongest antibacterial activity. In the preservation of fruits and vegetables, the chitosan edible coating has great prospects in maintaining the quality of fruits and vegetables among various types of edible coatings. However, no published studies have reported on the effects of the incorporation of cinnamon oil into chitosan edible coating for fresh-cut potatoes. Therefore, the current study aimed at evaluating the effects of different concentrations of cinnamon oil incorporated into chitosan edible coating on the quality, naturally occurring microorganisms, and artificially inoculated *Listeria monocytogenes* (Figure 1). The quality change of fresh-cut potatoes was assessed by measuring the colour, weight loss, and firmness. The safety of fresh-cut potatoes was evaluated through analysing the number of total plate counts, yeast and mould counts, total coliform counts, and lactic acid bacteria counts; and *Listeria monocytogenes*. The antibacterial and preservation agents of chitosan EC containing cinnamon oil were developed to provide a safe, convenient, and low-cost preservative for fresh-cut potatoes.

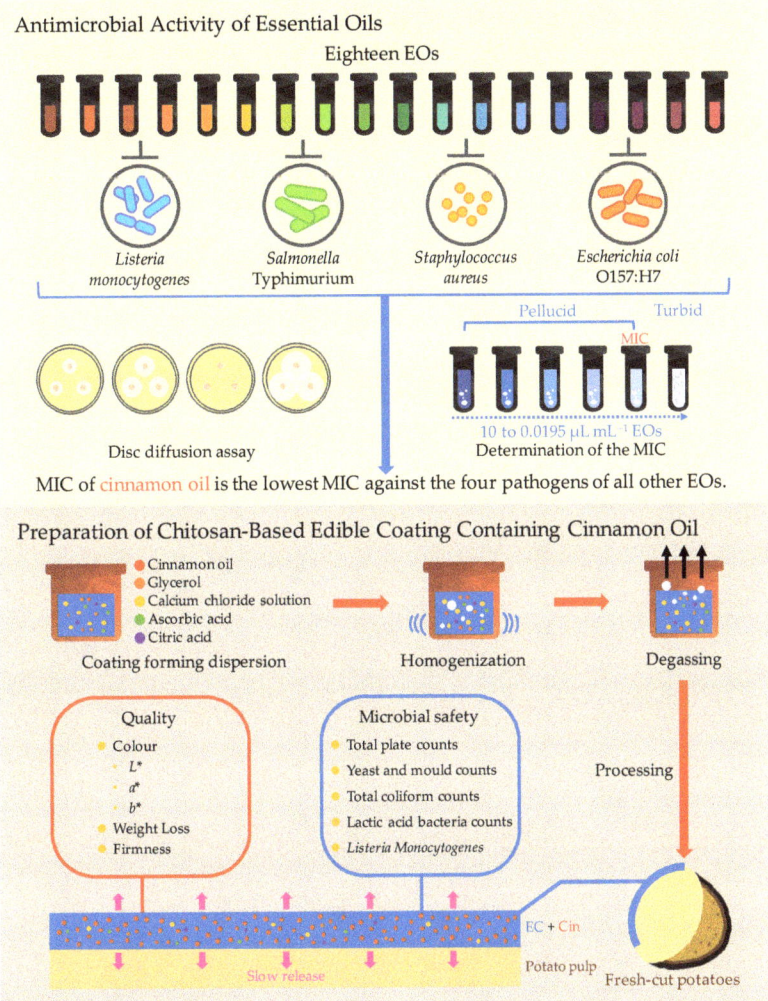

Figure 1. Diagram of experimental method.

2. Materials and Methods

2.1. Bacterial Strains and Preparation of Bacterial Inoculum

Salmonella typhimurium (ST, CICC 21484), *Staphylococcus aureus* (SA, CICC 21600), *Listeria monocytogenes* (LM, CICC 21633), and *Escherichia coli* O157:H7 (EC O157:H7, CICC 21530) were obtained from China Center of Industrial Culture Collection (CICC, Beijing, China). LM was cultured in tryptic soy broth with yeast extract (TSB-YE) for 12 h at 37 °C. ST, EC O157:H7, and SA were cultured in tryptic soy broth (TSB) for 12 h at 37 °C. The suspension of LM was centrifuged at 2795× g for 5 min at 4 °C and washed with 0.1% (w/v) peptone water (Aobox Biotechnology, Beijing, China) three times. The suspension was diluted 1:10 with 0.1% (w/v) peptone water to obtain the proper inoculum. The bacterial counts are represented as log cfu/mL [31].

2.2. Vegetables

Potatoes of uniform size and colour, and free of defects, were purchased from a New-Mart in Dalian City (China). All samples were kept at low temperature (approximately 4 °C) before processing [32].

2.3. Antimicrobial Activity of Essential Oils

2.3.1. Essential Oils

Eighteen EOs were used for this study. Cinnamon oil (*Cinnamomum cassia*), oregano oil (*Origanum vulgare*), clove oil (*Eugenia caryophyllus*), and tea tree oil (*Melaleuca alternifolia*), pomelo oil (*Citrus maxima* (Burm.) Merr.), jasmine oil (*Jasminum sambac* (L.) Ait.), eucalyptus oil (*Eucalyptus globulus*), sweet orange oil (*Citrus sinensis* (Linn.) Osbeck), sea buckthorn pulp oil (*Hippophae rhamnoides* L.), sweet osmanthus oil (*Osmanthus fragrans* (Thunb.) Lour.), lavender oil (*Lavandula angustifolia*), petitgrain oil (*Citurs sinensis* L. Osbeck), grapefruit oil (*Citrus paradisi* Macf.), rose oil (*Rosa rugosa* Thunb.), and citrus oil (*Citrus reticulata* Blanco) were obtained from Ji'an (Ji'an Zhongxiang Natural Plant, Ji'an, China). Some EOs obtained from Tongren (Miaoyao Biotech Co., Ltd., GuiZhou, China) were blumea oil (*Blumea balsamifera*), rosemary oil (*Rosmarinus officinalis*), and valeriana oil (*Valeriana officinalis*). Prior to their use in experiments, each EO was filtered using a 0.22 mm filter membrane (Millex-GP Filter Unit; Merck Millipore, Darmstadt, Germany) and stored at room temperature.

2.3.2. Disc Diffusion Assay

Paper disc agar plate assays were used to determine the antimicrobial activities of eighteen EOs against ST, SA, LM, and EC O157:H7 [33]. The bacterial suspension of 8 log cfu/mL was uniformly spread on a TSA plate. Sterile filter paper discs (6 mm) impregnated with EOs of 5 µL were placed on the surface of the TSA plate. After 24 h of incubation at 37 °C, the inhibition zone diameter (mm) on the TSA plate was measured. Inhibition zones were classified according to size as low (<12 mm), moderate (12–20 mm), and strong ($\geq$20 mm) [34]. Three measurements were taken to determine the average results.

2.3.3. Determination of the Minimal Inhibitory Concentration (MIC)

The MIC was determined using a modified method [35]. Briefly, EOs were mixed in dimethylsulfoxide (DMSO, 0.8% v/v) (Kemiou, Tianjin, China) and added into the tube containing TSB using a two-fold dilution method (10 to 0.0195 µL/mL) [36]. Then, 100 µL bacterial suspensions (5 log cfu/mL) were respectively added to each tube. The MIC was determined after enrichment by measuring the turbidity of the culture media.

2.4. Preparation of Chitosan Edible Coating

The chitosan EC was prepared by mixing 2% (w/v) chitosan (food-grade, 50 KD, Henan Qiang Li Chemical Products Co., Ltd., Zhengzhou, China), 1.5% (w/v) glycerol, and 2% (w/v) calcium chloride solution (food-grade) containing 1% (w/v) ascorbic acid (food-grade) and 1% (w/v) citric acid (food-grade) in ultrapure water and stirred at 70 °C until the solution became transparent [28,37]. Citric acid and ascorbic acid were mixed into chitosan EC as both antioxidants and colour fixatives. Cinnamon oil was incorporated into chitosan EC at different concentrations (0.2%, 0.4%, and 0.6% v/v). Homogenisation of the final solutions was achieved at 12,500 rpm for three minutes using an Ultra Turrax T25 mixer (IKA® WERKE, Staufen, Germany).

2.5. Processing and Packaging of Fresh-Cut Potatoes

The surface of fresh potatoes was sterilised with 75% (v/v) alcohol after washing. The sample was air-dried inside a biosafety cabinet for 10 min at 25 °C. Potato cubes (1 cm^3) were prepared using a sterile knife. Samples were soaked in chitosan solution for two minutes. Samples coated with chitosan EC or chitosan EC containing cinnamon oil of 0.2%, 0.4%, and 0.6% were subsequently assessed. Uncoated potatoes were evaluated as a control.

For 16 days at 4 °C, the fresh-cut potatoes were packaged on polystyrene trays (255 mL) wrapped in PVC films.

2.6. Determination of Potato Colour, Weight Loss, and Firmness

2.6.1. Colour

The colour parameters of potato cubes, including L^* (lightness), a^* ($+a^*$ = redness, $-a^*$ = greenness), and b^* ($+b^*$ = yellowness, $-b^*$ = blueness), were detected using a CR400/CR410 colorimeter (Minolta, Tokyo, Japan) [38]. The experiments were conducted in triplicate.

2.6.2. Weight Loss

Potato cubes were placed on polystyrene trays and weighed using a digital balance (PL-2002, METTLER TOLEDO, Greifensee, Switzerland) during storage [39]. The weight loss rate equation is given by Equation (1):

$$\text{Weight loss rate (\%)} = [(m_1 - m_2)/m_1] \times 100 \quad (1)$$

where m_1 is the initial weight (g), and m_2 is the weight at the specified time point (g).

2.6.3. Firmness

The firmness of potato cubes was measured using a TA.XT texture analyser (Stable Micro Systems Ltd., Godalming, UK). We measured the firmness of the cube based on the force (N) exerted on the slices in triplicates using the compression probe P5 (5 mm diameter), at a speed of 1.0 mm s^{-1}, and a penetration distance of 8 mm [40]. Each sample was duplicated three times.

2.7. Naturally Occurring Microorganisms

Each sample treated with chitosan EC containing cinnamon oil underwent microbiological analysis after 0, 4, 8, 12, and 16 days of storage. For the analysis, potato cubes were disrupted in a sterile blender containing 90 mL of 0.1% peptone water.

Suspensions of 0.1 mL from the potato cubes were cultured and counted on plate count agar (PCA) for total plate counts at 37 °C for 48 h, on potato dextrose agar (PDA) at 28 °C for 48–96 h for yeast and mould counts, on violet-red bile dextrose agar (VRBDA) at 37 °C for 24 h for total coliform counts, and on Lactobacilli MRS agar at 37 °C for 48 h for *Lactobacillus* counts [41,42]. All culturing media were purchased from Qingdao Hopebio Giotechnology Co., Ltd. (Qingdao, China).

2.8. Inoculation and Analysis of Listeria monocytogenes

The potatoes were cut into cubes (approximately 10 g per cube) and uniformly inoculated in petri dishes. The entire top surface of the potato cubes was inoculated with suspensions of LM (8.84 log cfu/mL, 500 µL each cube) for a challenge study. In a biosafety cabinet, the samples were air-dried at 25 °C for 1 h. Then, the samples were coated with EC or EC containing cinnamon oil as previously described. The control group consisted of fresh-cut potatoes without a coating. Each potato cube was placed in a blender bag and stored at 4 °C for 16 days. A triplicate of each experiment was conducted, with samples being analysed every 4 days for 16 days. The population of LM inoculated into fresh-cut potatoes was counted on an Oxford agar base (Qingdao Hopebio Giotechnology Ltd. Company, Qingdao, China). All the plates were incubated at 37 °C for 24 h. The number of microorganisms is expressed as log cfu/g [43].

2.9. Statistical Analysis

All experiments were performed in triplicate, and the data are presented as mean ± standard deviation. SPSS software was used to analyse the data (Version 14.0; SPSS, Chicago, IL, USA). The significance of the differences between variables was tested using one-way

ANOVA (between groups) and repeated-measures ANOVA (within group). The means were compared using Duncan's multiple range test. The statistical significance was determined at $p < 0.05$.

3. Results and Discussion

3.1. Antimicrobial Assay of Essential Oils

The method of agar disc diffusion was used to evaluate the antibacterial activity of eighteen EOs against four pathogens. There were differences among these EOs in terms of their antibacterial properties (Table 1). The size of the inhibition zones of cinnamon, oregano, and pomelo oils ranged from 16.33 to 35.18 mm, representing the strongest antibacterial activity among the eighteen EOs. Cinnamon oil had the strongest inhibition effect on SA with an inhibition zone diameter of 30.47 mm. Oregano oil strongly inhibited SA and LM with an inhibition zone diameter of 31.31 mm for SA and 30.01 mm for LM. Pomelo oil demonstrated the highest antibacterial activity against ST, with an inhibition zone diameter of 35.18 mm. Some essential oils including clove, eucalyptus, sweet orange, and blumea oils exhibited moderate antibacterial activity against foodborne pathogens (inhibition zone is 12–20 mm). There was no antibacterial activity against foodborne pathogens from jasmine, sea buckthorn pulp, sweet osmanthus, and citrus oils. Other essential oils showed a low to strong antibacterial activity against four foodborne pathogens. Some reports have demonstrated that essential oils exhibited different antibacterial effects, and different foodborne pathogens have different resistance against essential oils [44]. The antibacterial effect of essential oils depends on the type and content of antibacterial components contained in essential oils [45]. Therefore, the individual components and concentrations of EOs play an important role in the aspect of antimicrobial activity. Some studies have demonstrated that the principal constituents of cinnamon, oregano, and pomelo oils are cinnamaldehyde, thymol, and limonene, respectively [46–48]. Cinnamaldehyde and thymol usually have a broad spectrum of antibacterial activity, and they have a significant antibacterial effect on pathogenic bacteria such as *Escherichia coli, Staphylococcus aureus*, and *Listeria monocytogenes* [49,50]. The antimicrobial mechanism of the EOs including cinnamon, oregano, and pomelo oils is related to the disturbance of membrane permeability, which results in the release of cellular contents in the form of some inhibition enzymes such as ATPase, histidine decarboxylase, and amylase [51].

Table 1. Diameter (mm) of inhibition zones of essential oils against the four pathogenic strains.

Essential Oil	Latin Name	Strains Tested (mm)			
		SA	ST	LM	EC O157:H7
Cinnamon	*Cinnamomum cassia*	30.47 ± 0.14 [b]	26.07 ± 0.63 [bc]	26.8 ± 0.18 [b]	16.33 ± 0.19 [f]
Oregano	*Origanum vulgare*	31.31 ± 0.23 [a]	26.87 ± 0.53 [b]	30.01 ± 0.25 [a]	22.01 ± 0.91 [b]
Clove	*Eugenia caryophyllus*	14.76 ± 0.01 [f]	15.98 ± 0.51 [h]	13.99 ± 0.62 [e]	12.43 ± 0.94 [g]
Tea tree	*Melaleuca alternifolia*	10.75 ± 0.47 [h]	18.65 ± 0.93 [f]	12.74 ± 0.04 [f]	20.89 ± 0.26 [c]
Pomelo	*Citrus maxima* (Burm.) Merr.	21.75 ± 0.30 [c]	35.18 ± 0.01 [a]	30.05 ± 0.85 [a]	27.01 ± 0.99 [a]
Jasmine	*Jasminum sambac* (L.) Ait.	-	-	-	-
Eucalyptus	*Eucalyptus globulus*	17.84 ± 0.95 [e]	19.86 ± 0.10 [e]	12.45 ± 0.37 [g]	16.33 ± 0.55 [f]
Rosemary	*Rosmarinus officinalis*	9.49 ± 0.99 [i]	10.75 ± 0.51 [j]	7.75 ± 0.76 [h]	12.11 ± 0.75 [g]
Sweet orange	*Citrus sinensis* (Linn.) Osbeck	11.99 ± 0.29 [g]	17.67 ± 0.51 [g]	17.65 ± 0.93 [d]	17.49 ± 0.48 [d]
Sea buckthorn pulp	*Hippophae rhamnoides* L.	-	-	-	-
Sweet osmanthus	*Osmanthus fragrans* (Thunb.) Lour.	-	-	-	-
Lavender	*Lavandula angustifolia*	8.65 ± 0.54 [j]	25.06 ± 0.83 [d]	-	-
Petitgrain	*Citurs sinensis* L. Osbeck	7.78 ± 0.52 [k]	7.83 ± 0.59 [l]	-	-
Grapefruit	*Citrus paradisi* Macf.	8.91 ± 0.45 [j]	14.19 ± 0.91 [i]	8.13 ± 0.98 [g]	8.29 ± 0.01 [hi]
Rose	*Rosa rugosa* Thunb.	17.35 ± 0.89 [e]	17.7 ± 0.002 [g]	12.09 ± 0.87 [fg]	8.85 ± 0.05 [h]
Citrus	*Citrus reticulata* Blanco	-	-	-	-
Blumea	*Blumea balsamifera*	18.11 ± 0.21 [d]	17.97 ± 0.04 [g]	20.36 ± 0.17 [c]	16.66 ± 0.59 [e]
Valerian	*Valeriana officinalis*	10.64 ± 0.84 [h]	9.96 ± 0.46 [k]	12.41 ± 0.04 [f]	7.88 ± 0.82 [i]

Data are means of diameters of inhibition zones ± standard deviation. Values in the same column not followed by the same lowercase letter are significantly different ($p < 0.05$). LM, *Listeria monocytogenes*; ST, *Salmonella typhimurium*; SA, *Staphylococcus aureus*; EC O157:H7, *Escherichia coli* O157:H7. -, no inhibition zones.

3.2. Determination of Minimal Inhibitory Concentration

Cinnamon, oregano, and pomelo oils were evaluated for their MIC against four pathogens (Table 2). The results show that the MIC of cinnamon oil was 0.313 µL/mL for LM, ST, SA, and EC O157:H7, which is the lowest MIC against the four pathogens of all other tested EOs. The MIC of oregano oil was 0.625 µL/mL against ST and 1.25 µL/mL against LM, SA, and EC O157:H7. The MIC of pomelo oil was 1.25 µL/mL against ST and 2.5 µL/mL against LM, SA, and EC O157:H7. In this study, cinnamon oil has a lower MIC for four foodborne pathogens compared with oregano and pomelo oils. As reported in other studies, cinnamon oil also exhibits strong antimicrobial properties [52]. Cinnamon oil among six essential oils (rosemary, cinnamon, ginger, pepper mint, sweet orange, and tahiti lemon oils) showed the lowest MIC values of 6.25%, 3.12%, and 3.12% (v/v) for *Staphylococcus aureus*, *Escherichia coli*, and *Salmonella enterica*, respectively [33]. Another research reported that cinnamon oil showed also the lowest MIC values for four fungal species, four yeasts species, and two bacteria species, thereby confirming its higher inhibitory activity compared with clove oils [53]. The mechanism of cinnamon oil inhibiting the microorganism is mainly through denaturing proteins in cell membranes, interfering with the activity of enzymes in cell walls [54]. In many studies, cinnamon oil as a bioactive component positively inhibited microbial growth in food matrices. It also indicated that cinnamon oil could be applied for inhibiting microorganism growth on food and ensuring safety [55]. However, high concentrations of essential oils produce more intense flavour due to their volatility, which might affect the acceptability of food (such as fruits and vegetables) for the consumer. Therefore, a cinnamon essential oil with the lowest MIC was chosen for evaluating further the preservation of fresh-cut potatoes in this study.

Table 2. Minimal inhibitory concentrations (MIC) of cinnamon, oregano, and pomelo peel oils.

Essential Oil	Strains Tested	Concentrations of Essential Oils (µL/mL)										MIC
		10	5	2.5	1.25	0.625	0.313	0.156	0.078	0.039	0.020	
Cinnamon oil	LM	-	-	-	-	-	-	+	+	+	++	0.313
	ST	-	-	-	-	-	-	+	+	+	++	0.313
	SA	-	-	-	-	-	-	+	+	+	++	0.313
	EC O157:H7	-	-	-	-	-	+	+	+	+	++	0.313
Oregano oil	LM	-	-	-	-	+	++	++	++	++	++	1.25
	ST	-	-	-	-	-	+	+	++	++	++	0.625
	SA	-	-	-	-	+	+	++	++	++	++	1.25
	EC O157:H7	-	-	-	-	+	++	++	++	++	++	1.25
Pomelo peel oil	LM	-	-	-	+	+	+	+	++	++	++	2.5
	ST	-	-	-	-	+	+	+	++	++	++	1.25
	SA	-	-	+	+	+	+	++	++	++	++	2.5
	EC O157:H7	-	-	-	+	+	+	+	++	++	++	2.5

LM, *Listeria monocytogenes*; ST, *Salmonella typhimurium*; SA, *Staphylococcus aureus*; EC O157:H7, *Escherichia coli* O157:H7. -: no growth; +: minor growth; ++: major growth.

3.3. Effects of Cinnamon Oil on the Quality of Fresh-Cut Potatoes

3.3.1. Colour

Colour is a key determinant of consumer acceptability in fruit products. The L^* in colour represents the brightness of fresh-cut potatoes. L^* is one of the indicators of surface darkening caused by enzymatic browning or pigment gathered during storage [56]. The lower the L^* value, the greater the browning. L^* and b^* showed a significant decrease, and a^* showed a significant increase with the extension of storage time (Figure 2A–C) ($p < 0.05$). The L^* of fresh-cut potatoes treated with chitosan-based EC was higher than that of control. The L^* was higher for fresh-cut potatoes treated with chitosan-based EC containing 0.2% cinnamon oil than for the other groups, and the decline (7.70) in L^* was the slowest during storage time. a^* is the lower (0.99) and b^* is the higher (16.25) on fresh-cut potatoes treated with chitosan-based EC containing 0.2% cinnamon oil compared with that in other groups.

The L^* of fresh-cut potatoes treated with the chitosan-based EC containing 0.4% and 0.6% cinnamon oil was reduced 23.5 and 25.4, respectively. The a^* of fresh-cut potatoes in EC containing 0.4% and 0.6% cinnamon oil was increased 7.71 and 9.08, respectively. The b^* of fresh-cut potatoes in EC containing 0.4% and 0.6% cinnamon oil was reduced 7.65 and 8.10, respectively. In addition, the appearance changes of fresh-cut potatoes were observed on the 16th day in this study (Figure 2D). The appearance of fresh-cut potatoes treated with chitosan-based EC, or chitosan-based EC containing 0.2% cinnamon oil, was better than other treatments. The obvious dark browning of fresh-cut potatoes was observed in potatoes treated with chitosan-based EC containing 0.4% and 0.6% cinnamon oil.

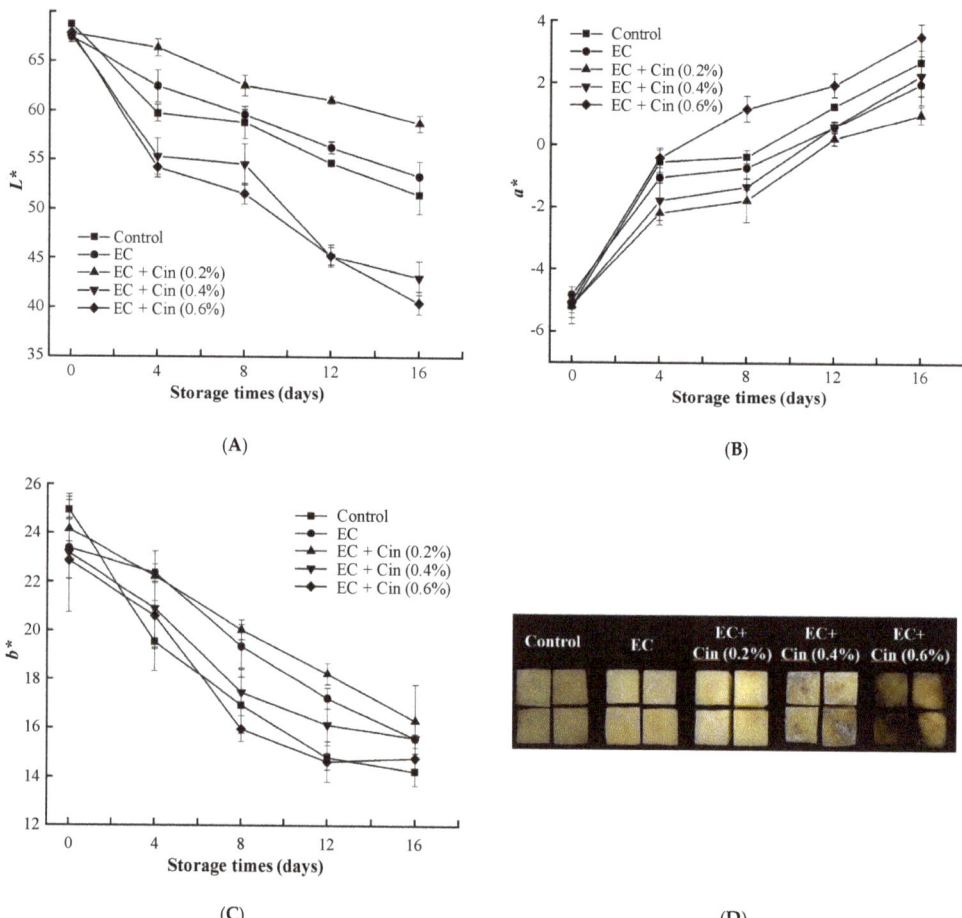

Figure 2. Changes in colour of fresh-cut potatoes coated with chitosan-based EC containing cinnamon oil. (**A**) L^*; (**B**) a^*; (**C**) b^*; (**D**) appearance on the 16th day. Control: uncoated; EC: edible coating; Cin: cinnamon oil. Bars represent means ± SD (n = 3, p < 0.05).

When potatoes are cut, the tissue cells are broken, and enzymes such as polyphenol oxidases (PPOs) are liberated and brought into contact with their substrates, causing browning [57]. The browning depends on the characteristics of the samples, amount of endogenous phenolic compound, oxygen condition, and activity of relevant enzymes [58]. The browning has a slight change from L^* and appearance of fresh-cut potatoes coated with chitosan-based EC compared with the control. These results were in agreement with other studies in which chitosan coatings delayed browning in fresh-cut rose apple and litchi in comparison with noncoating [59]. The reason may be that chitosan-based EC prevents oxygen from reaching the surface of the potato and reduces browning [60]. However, the browning has caused a strong change from L^* and appearance of fresh-cut potatoes treated with chitosan-based EC containing high concentration of cinnamon oil (0.4% and 0.6%). One study demonstrated that a high concentration of cinnamon oil damages the tissue structure of fresh-cut potatoes and causes serious browning [61]. The reason was may be that the high concentration of cinnamon oil accelerated the browning of fresh-cut potatoes. PPO oxidises phenolics in the presence of oxygen on the cut surface of potatoes, producing quinones, which autopolymerise to form brown coloured pigments [62]. The other reason was that a high concentration of cinnamon oil might produce phytotoxic effect for fresh-cut potatoes. Several studies reported similar results; the phytotoxic effects of EOs might affect fresh-cut lettuce and fresh-cut apples [36,63]. Therefore, the chitosan-based EC containing lower doses of cinnamon oil would be recommended for maintaining the colour of fresh-cut potatoes.

3.3.2. Weight Loss

Weight loss is an important indicator for evaluating the quality of fresh-cut fruits and vegetables during storage times. The weight loss was evaluated in fresh-cut potatoes treated with chitosan-based EC and chitosan-based EC containing different concentration of cinnamon oil during storage time (Figure 3). The weight loss of fresh-cut potatoes significantly increased during storage times ($p < 0.05$). There is no significant difference in weight loss of fresh-cut potatoes among chitosan-based EC, chitosan-based EC containing 0.2% and 0.4% cinnamon oil, and control ($p > 0.05$). However, the weight loss of the fresh-cut potatoes treated with the chitosan-based EC containing 0.6% cinnamon oil was significantly higher than that in other groups ($p < 0.05$). Results also indicate that the concentration of cinnamon oil affected the weight loss. Fresh-cut potatoes treated with a low concentration of cinnamon oil incorporated into chitosan-based EC lost less water than samples treated with that EC with 0.6% cinnamon oil. The weight loss of the fresh-cut potatoes treated with chitosan-based EC containing 0.6% cinnamon rapidly increased after 4 days of storage, possibly due to the fact that the high concentration of EOs can cause potential toxicity of fresh-cut potatoes and accelerate the decay of samples [64].

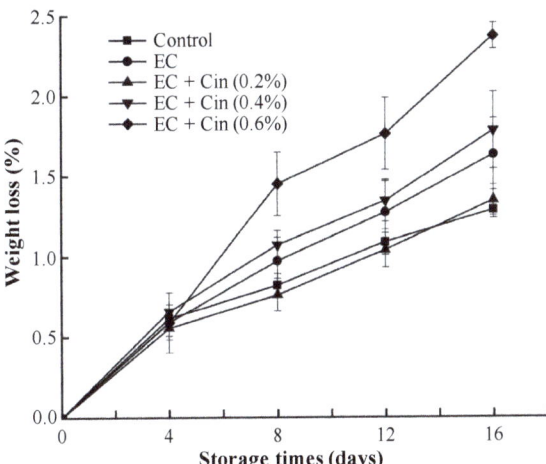

Figure 3. Changes in weight loss of fresh-cut potatoes coated with chitosan-based EC containing cinnamon oil. Control: uncoated; EC: edible coating; Cin: cinnamon oil. Bars represent means ± SD ($n = 3$, $p < 0.05$).

3.3.3. Firmness

Fruit firmness is closely related to the cell composition and cell-wall structure. Fruit softening is a consequence of the disassembly of the middle lamella and primary cell-wall structures [65]. Fruit softening is a process of starch hydrolysis to sugar and pectin degradation. It is an important factor in the quality of fresh-cut fruits and vegetables and their acceptability to consumers. The firmness of fresh-cut potatoes showed a significant decrease in different treatment groups during storage times ($p < 0.05$) (Figure 4). The decline of firmness is 3.80 N in chitosan coating and 8.20 N in the control during storage time. The result demonstrates that chitosan-coating treatments mitigated the firmness decrease to a greater degree than the control. There is no significant difference in the firmness of fresh-cut potatoes treated with chitosan-based EC and chitosan-based EC containing 0.2% and 0.4% cinnamon oil at 16 days ($p > 0.05$). However, the firmness of fresh-cut potatoes in the group treated with the chitosan-based EC containing 0.6% cinnamon oil was reduced 10.20 N during storage time. In this study, the result agreed with other studies that reported that the application of chitosan-based coatings inhibited the fruit and vegetable softening process [66–68]. Cutting operation might cause the increase in pectinase activity in potatoes' tissue. Under the action of pectinase, pectin in the cell wall is decomposed and tissue is softened. The coating treatments may allow firmness to be maintained by inhibiting water loss due to the activities of pectin-degrading enzymes and by reducing the rate of metabolic processes during senescence [69]. On the other hand, high concentrations of cinnamon oils damage the tissue of fresh-cut potatoes, causing them to be more susceptible to spoilage and fruit softening [64].

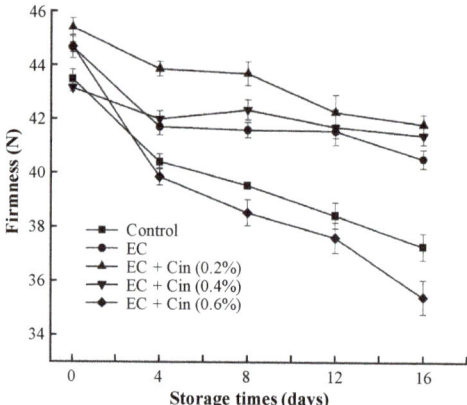

Figure 4. Changes in firmness of fresh-cut potatoes coated with chitosan-based EC containing cinnamon oil. Control: uncoated; EC: edible coating; Cin: cinnamon oil. Bars represent means ± SD ($n = 3$, $p < 0.05$).

3.4. Microbiological Analysis

The population of naturally occurring microorganisms on fresh-cut potatoes treated with chitosan EC with or without cinnamon oil was evaluated (Figure 5). The population of total plate counts, yeast and mould counts, total coliform counts, and lactic acid bacteria counts on the fresh-cut potatoes considerably increased with the prolongation of storage time ($p < 0.05$); the increment is 3.57, 3.37, 2.14, and 1.07 log cfu/g, respectively. This may be because nutrients released from the fresh-cut potatoes after cutting provide suitable growth conditions for microorganisms. Some reports have also shown that pathogens and spoilage microorganisms can grow on fresh, frozen, dried, ready-to-serve, and minimally processed potato products [70,71]. The total plate counts, yeast and mould counts, total coliform counts, and lactic acid bacteria counts during storage time were significantly lower for fresh-cut potatoes treated with chitosan EC than those for fresh-cut potatoes in the control group (*p* < 0.05), the decrease is 1.44, 1.72, 0.57, and 0.56 log cfu/g, respectively. Some studies have also demonstrated that chitosan has antibacterial activity, and involving chitosan coatings reduced microbial growth on mangoes, papaya, and strawberry [72–74]. The mechanism of chitosan EC is mainly the leakage of electrolytes and intracellular protein constituents caused by interactions between chitosan with positive charge and the surface of bacterial cells with negative charge [75,76]. According to the total plate counts, yeast and mould counts, total coliform counts, and lactic acid bacteria counts, the populations were significantly lower in chitosan-based EC containing 0.2% cinnamon oil among the different treatment groups at 16 days (*p* < 0.05), the decrease is 2.14, 1.92, 0.98, and 0.73 log cfu/g, respectively. The population of the decrement of naturally occurring microorganisms on chitosan-based EC containing 0.2% cinnamon oil is more than that on chitosan-based EC. It demonstrated that the combination of chitosan EC and cinnamon oil exhibited a synergetic antibacterial effect against naturally occurring microorganisms. Other studies reported that chitosan coating incorporating several common essential oils can enhance antimicrobial activity. It also showed that the compatibility of cinnamon oil with chitosan in film formation was better than that of other essential oils with chitosan [37]. However, the populations showed a gradual increase with the increase in cinnamon oil concentration (0.4% and 0.6%). It might be that cinnamon oil at a higher concentration of 0.4% and 0.6% damages the cell structure of fresh-cut potatoes. The pulp of fruits and vegetables provides rich nutrient content for microorganism growth [77]. Therefore, microorganisms can easily grow on fresh-cut potatoes treated with chitosan-based EC containing 0.4% and 0.6% cinnamon oil during storage. Interestingly, no coliform nor lactic acid bacteria were observed on the fresh-cut potatoes treated with chitosan-based EC containing cinnamon oil for 4 or 8 days. However, the populations of coliform and lactic acid bacteria on the fresh-cut potatoes in the control group significantly increased after 4 and 8 days. This demonstrates that chitosan-based EC and chitosan-based EC containing cinnamon oil had antibacterial activity against coliform and lactic acid bacteria and inhibited their growth on fresh-cut potatoes. Moreover, according to the standard of the Institute of Food Science and Technology (IFST), 6 log cfu/g of natural microorganisms is considered the limit of acceptance for the shelf life of a fruit product [78]. This is the reason why toxic substances may be produced when microbiological counts exceed 6.0 log cfu/g [79]. In this study, the population of total plate counts on fresh-cut potatoes is less than 6.0 log cfu/g in chitosan EC containing cinnamon oil during 16 days. Therefore, the acceptance of fresh-cut potatoes treated with chitosan EC containing 0.2% cinnamon oil was extended to 16 days.

3.5. Listeria monocytogenes Analysis

The growth of *L. monocytogenes* on the fresh-cut potatoes treated with chitosan EC and chitosan EC containing cinnamon oil was evaluated (Figure 6). The number of *L. monocytogenes* on the fresh-cut potatoes treated with chitosan-based EC and containing cinnamon oil was reduced approximately 1 log cfu/g compared with that on the control at the initial day. That reason is that *L. monocytogenes* inoculated on the surface of fresh-cut potatoes may have been removed after soaking with the chitosan-based EC treatment [80]. *L. monocytogenes* was reduced 2.17 log cfu/g on fresh-cut potatoes treated with chitosan-based EC at 16 days (*p* < 0.5). It indicated that chitosan EC exhibits antimicrobial activity against *L. monocytogenes*. This result is in agreement with those of other research studies in which the chitosan-based film reduced 2 log cfu and 1.3 log cfu against *E. coli* and *L. monocytogenes*, respectively [81]. *L. monocytogenes* was reduced at 1.94, 2.44, and 2.92 log cfu/g in chitosan-based EC containing 0.2, 0.4, and 0.6% cinnamon oil, respectively, compared with that in the control group during storage time (*p* < 0.05). In this study, higher antibacterial activity was shown in EC-added cinnamon oil (the decrease is 2.92 log cfu/g) than only EC (the decrease is 2.17 log cfu/g) against *L. monocytogenes*. Therefore, the combination of EC and essential oils shows a synergistic effect of antibacterial activity against *L. monocytogenes*. The volatile component of cinnamon oil may have strong antibacterial activity. The cinnamon oil destroyed membrane phospholipids and permeability of cell membranes, eventually causing the cytoplasm to leak and the cell to die [54]. The cinnamaldehyde in cinnamon oil is an aromatic aldehyde that can reduce the survival of foodborne pathogens through the inhibition of amino acid decarboxylase activity [82]. Some studies reported that cinnamon incorporated into polymer-based films (fish gelatin films, chitosan gelatin blend films, polymer film) has enhanced the antimicrobial effect of the film [83–85].

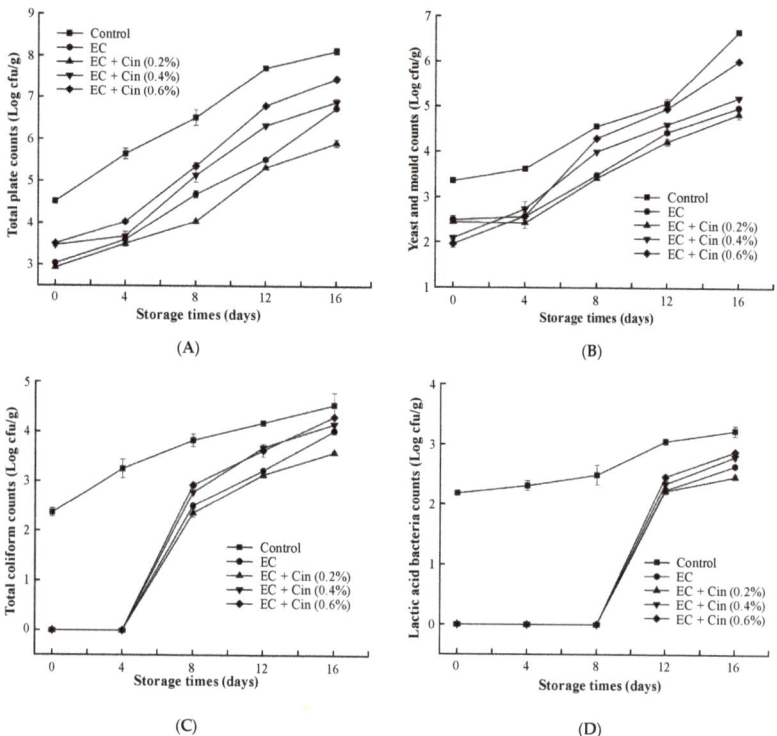

Figure 5. Population of naturally occurring microorganisms on fresh-cut potatoes coated with chitosan-based EC containing cinnamon oil. (**A**) Total plate counts; (**B**) yeast and mould counts; (**C**) total coliform counts; (**D**) lactic acid bacteria counts. Control: uncoated; EC: edible coating; Cin: cinnamon oil. Bars represent means ± SD (n = 3, $p < 0.05$).

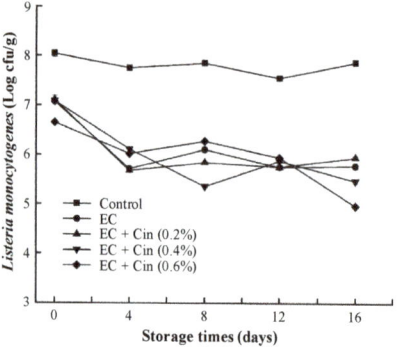

Figure 6. Reduction in *Listeria monocytogenes* on fresh-cut potatoes coated with chitosan-based EC containing cinnamon oil. Control: uncoated; EC: edible coating; Cin: cinnamon oil. Bars represent means ± SD (n = 3, $p < 0.05$).

4. Conclusions

We demonstrated that the use of chitosan EC containing cinnamon oil maintained the quality, reduced the deterioration, and thus extended the shelf life of fresh-cut potatoes. Chitosan EC containing 0.2% cinnamon oil reduced the degree of browning and delayed the weight loss and softening of the fresh-cut potatoes. Moreover, the addition of cinnamon oil increased the antibacterial activity of chitosan EC against naturally occurring microorganisms and LM.

Accordingly, the effective antibacterial activity of chitosan EC incorporated with cinnamon oil indicates its potential and extended application in fresh-cut fruits and vegetables preservation. It can further be applied to other types of fresh-cut fruits and vegetables, owing to the characteristics of cinnamon oil as a GRAS compound and being easily obtainable. Higher concentrations of cinnamon oil can inhibit a wide range of microorganisms. However, the EO components may impact on the quality such as colour, firmness, taste, and odour of the coated fruit. Therefore, chitosan EC incorporated with lower concentrations of cinnamon oil may be the optimum formula for maintaining the quality of fresh-cut potatoes. It also provides a strategy for designing new preservation agents and achieving the ultimate goal of commercialization of food products.

Author Contributions: Conceptualization, K.F. and W.H.; methodology, S. and K.F.; software, S. and K.F.; validation, W.H.; investigation, Y.L.; resources, W.H.; data curation, L.W. and C.Y.; writing—original draft preparation, S.; writing—review and editing, W.H. and K.F.; supervision, W.H.; project administration, K.F. and W.H.; funding acquisition, S. All authors have read and agreed to the published version of the manuscript.

Funding: This study was supported by Zhuhai College of Science and Technology Innovation Capability Cultivation Project (Grant No. 2020XJCQ018), Doctor Promotion Program of Zhuhai College of Science and Technology, and Young Innovative Talents Project of "Innovation and Improving School Project" of Education Department of Guangdong Province (Grant No. 2019KQNCX197).

Institutional Review Board Statement: Not applicable.

Informed Consent Statement: Not applicable.

Data Availability Statement: The data used to support the findings of this study are available from the corresponding author upon request.

Conflicts of Interest: The authors declare no conflict of interest.

References

1. Li, Y.; Zhao, D. Transcriptome analysis of scions grafted to potato rootstock for improving late blight resistance. *BMC Plant Biol.* **2021**, *21*, 272. [CrossRef]
2. Dogbe, W.; Revoredo-Giha, C. Nutritional implications of trade-offs between fresh and processed potato products in the United Kingdom (UK). *Front. Nutr.* **2021**, *7*, 614176. [CrossRef] [PubMed]
3. Zheng, Z.Q.; Zhao, H.B.; Liu, Z.D.; He, J.; Liu, W.Z. Research progress and development of mechanized potato planters: A review. *Agriculture* **2021**, *11*, 521. [CrossRef]
4. Vithu, P.; Sanjaya, K.D.; Kalpana, R. Post-harvest processing and utilization of sweet potato: A review. *Food Rev. Int.* **2019**, *35*, 726–762.
5. Van Haute, S.; Sampers, I.; Holvoet, K.; Uyttendaele, M. Physicochemical quality and chemical safety of chlorine as a reconditioning agent and wash water disinfectant for fresh-cut lettuce washing. *Appl. Environ. Microbiol.* **2013**, *79*, 2850–2861. [CrossRef] [PubMed]
6. Batool, T.; Ali, S.; Seleiman, M.F.; Naveed, N.H.; Ali, A.; Ahmed, K.; Abid, M.; Rizwan, M.; Shahid, M.R.; Alotaibi, M.; et al. Plant growth promoting rhizobacteria alleviates drought stress in potato in response to suppressive oxidative stress and antioxidant enzymes activities. *Sci. Rep.* **2020**, *10*, 16975. [CrossRef] [PubMed]
7. Ciccone, M.; Chambers, D.; Iv, E.C.; Talavera, M. Determining which cooking method provides the best sensory differentiation of potatoes. *Foods* **2020**, *9*, 451. [CrossRef] [PubMed]
8. Kang, W.; Robitaille, M.C.; Merrill, M.; Teferra, K.; Kim, C.; Raphael, M.P. Mechanisms of cell damage due to mechanical impact: An in vitro investigation. *Sci. Rep.* **2020**, *10*, 12009. [CrossRef] [PubMed]

9. Agriopoulou, S.; Stamatelopoulou, E.; Sachadyn-Król, M.; Varzakas, T. Lactic acid bacteria as antibacterial agents to extend the shelf life of fresh and minimally processed fruits and vegetables: Quality and safety aspects. *Microorganisms* **2020**, *8*, 952. [CrossRef] [PubMed]
10. Blancas-Benitez, F.J.; Montaño-Leyva, B.; Aguirre-Güitrón, L.; Moreno-Hernández, C.L.; Fonseca-Cantabrana, Á.; Romero-Islas, L.; González-Estrada, R. Impact of edible coatings on quality of fruits: A review. *Food Control* **2022**, *139*, 109063. [CrossRef]
11. Valencia-Chamorro, S.A.; Palou, L.; Del Río, M.A.; Pérez-Gago, M.B. Antimicrobial edible films and coatings for fresh and minimally processed fruits and vegetables: A review. *Crit. Rev. Food Sci.* **2011**, *51*, 872–900. [CrossRef] [PubMed]
12. Chen, G.; Zhang, B.; Zhao, J. Dispersion process and effect of oleic acid on properties of cellulose sulfate-oleic acid composite film. *Materials* **2015**, *8*, 2346–2360. [CrossRef]
13. Petriccione, M.; Mastrobuoni, F.; Pasquariello, M.S.; Zampella, L.; Nobis, E.; Capriolo, G.; Scortichini, M. Effect of chitosan coating on the postharvest quality and antioxidant enzyme system response of strawberry fruit during cold storage. *Foods* **2015**, *4*, 501–523. [CrossRef] [PubMed]
14. Duan, C.; Meng, X.; Meng, J.; Khan, M.I.H.; Dai, L.; Khan, A.; An, X.; Zhang, J.; Huq, T.; Ni, Y. Chitosan as a preservative for fruits and vegetables: A review on chemistry and antimicrobial properties. *J. Bioresour. Bioprod.* **2019**, *4*, 11–21. [CrossRef]
15. Oyom, W.; Zhang, Z.; Bi, Y.; Tahergorabi, R. Application of starch-based coatings incorporated with antimicrobial agents for preservation of fruits and vegetables: A review. *Prog. Org. Coat.* **2022**, *166*, 106800. [CrossRef]
16. Kumar, P.; Sethi, S.; Sharma, R.R.; Singh, S.; Varghese, E. Improving the shelf life of fresh-cut 'Royal Delicious' apple with edible coatings and anti-browning agents. *J. Food Sci. Technol.* **2018**, *55*, 3767–3778. [CrossRef] [PubMed]
17. Tharanathan, R.N.; Kittur, F.S. Chitin—The undisputed biomolecule of great potential. *Crit. Rev. Food Sci. Nutr.* **2003**, *43*, 61–87. [CrossRef]
18. Tokatlı, K.; Demirdöven, A. Effects of chitosan edible film coatings on the physicochemical and microbiological qualities of sweet cherry (*Prunus avium* L.). *Sci. Hortic.* **2020**, *259*, 108656. [CrossRef]
19. Jeon, Y.I.; Kamil, J.Y.V.A.; Shahidi, F. Chitosan as an edible invisible film for quality preservation of herring and Atlantic cod. *J. Agric. Food Chem.* **2002**, *20*, 5167–5178. [CrossRef] [PubMed]
20. Daglia, M. Polyphenols as antimicrobial agents. *Curr. Opin. Biotechnol.* **2012**, *23*, 174–181. [CrossRef] [PubMed]
21. Zhang, Y.; Ma, Q.; Critzer, F.; Davidson, P.M.; Zhong, Q. Organic thyme oil emulsion as an alternative washing solution to enhance the microbial safety of organic cantaloupes. *Food Control* **2016**, *67*, 31–38. [CrossRef]
22. Liu, Y.; Liang, X.; Zhang, R.; Lan, W.; Qin, W. Fabrication of electrospun polylactic acid/cinnamaldehyde/β-cyclodextrin fibers as an antimicrobial wound dressing. *Polymers* **2017**, *9*, 464. [CrossRef]
23. Kawatra, P.; Rajagopalan, R. Cinnamon: Mystic powers of a minute ingredient. *Pharmacogn. Res.* **2015**, *7*, S1–S6. [CrossRef] [PubMed]
24. Kosari, F.; Taheri, M.; Moradi, A.; Alni, R.H.; Alikhani, M.Y. Evaluation of cinnamon extract effects on *clbB* gene expression and biofilm formation in *Escherichia coli* strains isolated from colon cancer patients. *BMC Cancer* **2020**, *20*, 267. [CrossRef]
25. Vasconcelos, N.G.; Croda, J.; Simionatto, S. Antibacterial mechanisms of cinnamon and its constituents: A review. *Microb. Pathog.* **2018**, *120*, 198–203. [CrossRef] [PubMed]
26. Rojas-Graü, M.A.; Raybaudi-Massilia, R.M.; Soliva-Fortuny, R.C.; AvenaBustillos, R.J.; McHugh, T.H.; Martní-Belloso, O. Apple puree-alginate edible coating as carrier of antimicrobial agents to prolong shelf-life of fresh-cut apples. *Postharvest Biol. Technol.* **2007**, *45*, 254–264. [CrossRef]
27. Raybaudi-Massilia, R.M.; Mosqueda-Melgar, J.; Martín-Belloso, O. Edible alginate-based coating as carrier of antimicrobials to improve shelf-life and safety of fresh-cut melon. *Int. J. Food Microbiol.* **2008**, *121*, 313–327. [CrossRef]
28. Azarakhsha, N.; Osmana, A.; Ghazalia, H.M.; Tan, C.P.; Adzahan, N.M. Lemongrass essential oil incorporated into alginate-based edible coating for shelf-life extension and quality retention of fresh-cut pineapple. *Postharvest Biol. Technol.* **2014**, *88*, 1–7. [CrossRef]
29. Salvia-Trujillo, L.; Rojas-Graü, M.A.; Soliva-Fortuny, R.; Martín-Belloso, O. Use of antimicrobial nanoemulsions as edible coatings: Impact on safety and quality attributes of fresh-cut Fuji apples. *Postharvest Biol. Technol.* **2015**, *105*, 8–16. [CrossRef]
30. Zhang, W.; Lin, M.; Feng, X.; Yao, Z.; Wang, T.; Xu, C. Effect of lemon essential oil-enriched coating on the postharvest storage quality of citrus fruits. *Food Sci. Technol.* **2022**, *42*, e125421. [CrossRef]
31. Joshua, P.V.; Di, L.; Linda, J.H.; Donald, W.S.; Michelle, D.D. Fate of *Escherichia coli* O157:H7, *Listeria monocytogenes*, and *Salmonella* on fresh-cut celery. *Food Microbiol.* **2013**, *34*, 151–157.
32. Shen, X.; Zhang, M.; Devahastin, S.; Guo, Z. Effects of pressurized argon and nitrogen treatments in combination with modified atmosphere on quality characteristics of fresh-cut potatoes. *Postharvest Biol. Technol.* **2019**, *149*, 159–165. [CrossRef]
33. Ferreira, L.R.; Rosário, D.K.A.; Silva, P.I.; Carneiro, J.C.S.; Pimentel Filho, N.J.; Bernardes, P.C. Cinnamon essential oil reduces adhesion of food pathogens to polystyrene. *Int. Food Res.* **2019**, *26*, 1103–1110.
34. Rota, M.C.; Herrera, A.; Martínez, R.M.; Sotomayor, J.A.; Jordán, M.J. Antimicrobial activity and chemical composition of *Thymus vulgaris*, *Thymus zygis* and *Thymus hyemalis* essential oils. *Food Control* **2008**, *19*, 681–687. [CrossRef]
35. Somrani, M.; Inglés, M.C.; Debbabi, H.; Abidi, F.; Palop, A. Garlic, Onion, and Cinnamon Essential Oil Anti-biofilms' Effect against *Listeria monocytogenes*. *Foods* **2020**, *9*, 567. [CrossRef] [PubMed]
36. Sarengaowa; Hu, W.Z.; Jiang, A.L.; Xiu, Z.L.; Feng, K. Effect of thyme oil–alginate-based coating on quality and microbial safety of fresh-cut apples. *J. Sci. Food Agric.* **2018**, *98*, 2302–2311. [CrossRef] [PubMed]

37. Wang, L.; Liu, F.; Jiang, Y.; Chai, Z.; Li, P.; Cheng, Y.; Jing, H.; Leng, X. Synergistic antimicrobial activities of natural essential oils with chitosan film. *J. Agric. Food Chem.* **2011**, *59*, 12411–12419. [CrossRef] [PubMed]
38. Liu, P.; Xu, N.; Liu, R.; Liu, J.; Peng, Y.; Wang, Q. Exogenous proline treatment inhibiting enzymatic browning of fresh-cut potatoes during cold storage. *Postharvest Biol. Technol.* **2022**, *184*, 111754. [CrossRef]
39. Zhao, S.; Han, X.; Liu, B.; Wang, S.; Guan, W.; Wu, Z.; Theodorakis, P.E. Shelf-life prediction model of fresh-cut potato at different storage temperatures. *J. Food Eng.* **2022**, *317*, 110867. [CrossRef]
40. Ji, Y.; Hu, W.; Liao, J.; Jiang, A.; Xiu, Z.; Saren, G.; Guan, Y.; Yang, X.; Feng, K.; Liu, C. Effect of atmospheric cold plasma treatment on antioxidant activities and reactive oxygen species production in postharvest blueberries during storage. *J. Sci. Food Agric.* **2020**, *100*, 5586–5595. [CrossRef] [PubMed]
41. Gómez, P.L.; Salvatori, D.M. Pulsed light treatment of cut apple: Dose effect on color, structure, and microbiological stability. *Food Bioprocess Technol.* **2012**, *5*, 2311–2322. [CrossRef]
42. Siroli, L.; Patrignani, F.; Serrazanetti, D.I.; Tabanelli, G.; Montanari, C.; Gardini, F.; Lanciotti, R. Lactic acid bacteria and natural antimicrobials to improve the safety and shelf-life of minimally processed sliced apples and lamb's lettuce. *Food Microbiol.* **2015**, *47*, 74–84. [CrossRef]
43. Sarengaowa; Hu, W.; Feng, K.; Xiu, Z.; Jiang, A.; Lao, Y. Efficacy of thyme oil-alginate-based coating in reducing foodborne pathogens on fresh-cut apples. *Int. J. Food Sci. Technol.* **2019**, *54*, 3128–3137. [CrossRef]
44. Antunes, M.D.C.; Cavaco, A.M. The use of essential oils for postharvest decay control: A review. *Flavour Frag. J.* **2010**, *25*, 351–366. [CrossRef]
45. Yuan, Y.; Huang, M.; Pang, Y.X.; Yu, F.L.; Chen, C.; Liu, L.W.; Chen, Z.X.; Zhang, Y.B.; Chen, X.L.; Hu, X. Variations in essential oil yield, composition, and antioxidant activity of different plant organs from *Blumea balsamifera* (L.) DC. at different growth times. *Molecules* **2016**, *21*, 1024. [CrossRef] [PubMed]
46. He, J.; Wu, D.; Zhang, Q.; Chen, H.; Li, H.; Han, Q.; Lai, X.; Wang, H.; Wu, Y.; Yuan, J.; et al. Efficacy and mechanism of cinnamon essential oil on inhibition of colletotrichum acutatum isolated from 'hongyang' kiwifruit. *Front. Microbiol.* **2018**, *9*, 1288. [CrossRef] [PubMed]
47. Simirgiotis, M.J.; Burton, D.; Parra, F.; López, J.; Muñoz, P.; Escobar, H.; Parra, C. Antioxidant and antibacterial capacities of *origanum vulgare* L. essential oil from the arid andean region of Chile and its chemical characterization by GC-MS. *Metabolites* **2020**, *10*, 414. [CrossRef] [PubMed]
48. He, W.; Li, X.; Peng, Y.; He, X.; Pan, S. Anti-oxidant and anti-melanogenic properties of essential oil from peel of Pomelo cv. Guan Xi. *Molecules* **2019**, *24*, 242. [CrossRef]
49. Osaili, T.M.; Hasan, F.; Dhanasekaran, D.K.; Obaid, R.S.; Al-Nabulsi, A.A.; Ayyash, M.; Karam, L.; Savvaidis, I.N.; Holley, R. Effect of active essential oils added to chicken tawook on the behaviour of *Listeria monocytogenes*, *Salmonella* spp. and *Escherichia coli* O157:H7 during storage. *Int. J. Food Microbiol.* **2021**, *337*, 108947. [CrossRef] [PubMed]
50. Kosakowska, O.; Węglarz, Z.; Pióro-Jabrucka, E.; Przybył, J.L.; Kraśniewski, K.; Gniewosz, M.; Bączek, K. Antioxidant and antibacterial activity of essential oils and hydroethanolic extracts of Greek oregano (*O. vulgare* L. subsp. *hirtum (link) ietswaart*) and Common oregano (*O. vulgare* L. subsp. *vulgare*). *Molecules* **2021**, *26*, 988. [PubMed]
51. Scollard, J.; McManamon, O.; Schmalenberger, A. Inhibition of *Listeria monocytogenes* growth on fresh-cut produce with thyme essential oil and essential oil compound verbenone. *Postharvest Biol. Technol.* **2016**, *120*, 61–68. [CrossRef]
52. Zhang, Y.; Liu, X.; Wang, Y.; Jiang, P.; Quek, S.Y. Antibacterial activity and mechanism of cinnamon essential oil against *Escherichia coli* and *Staphylococcus aureus*. *Food Control* **2016**, *59*, 282–289. [CrossRef]
53. Matan, N.; Rimkeeree, H.; Mawson, A.J.; Chompreeda, P.; Haruthaithanasan, V.; Parker, M. Antimicrobial activity of cinnamon and clove oils under modified atmosphere conditions. *Int. J. Food Microbiol.* **2006**, *107*, 180–185. [CrossRef] [PubMed]
54. Burt, S. Essential oils: Their antibacterial properties and potential applications in foods—A review. *Int. J. Food Microbiol.* **2004**, *94*, 223–253. [CrossRef] [PubMed]
55. Nabavi, S.F.; Di Lorenzo, A.; Izadi, M.; Sobarzo-Sánchez, E.; Daglia, M.; Nabavi, S.M. Antibacterial effects of cinnamon: From farm to food, cosmetic and pharmaceutical industries. *Nutrients* **2015**, *7*, 7729–7748. [CrossRef] [PubMed]
56. Dovene, A.K.; Wang, L.; Bokhary, S.U.F.; Madebo, M.P.; Yonghua, Z.; Jin, P. Effect of cutting styles on quality and antioxidant activity of stored fresh-cut sweet potato (*Ipomoea batatas* L.) cultivars. *Foods* **2019**, *8*, 674. [CrossRef] [PubMed]
57. Olivas, G.I.; Mattinson, D.S.; Barbosa-Cánovas, G.V. Alginate coatings for preservation of minimally processed 'Gala' apples. *Postharvest Biol. Technol.* **2007**, *45*, 89–96. [CrossRef]
58. Ru, Z.H.; Lai, Y.Y.; Xu, C.J.; Li, L. Polyphenol oxidase (PPO) in early stage of browning of phalaenopsis leaf explants. *J. Agric. Sci.* **2013**, *5*, 57–64. [CrossRef]
59. Zhang, D.; Quantick, P.C. Effects of chitosan coating on enzymatic browning and decay during postharvest storage of litchi (*Litchi chinensis* Sonn.) fruit. *Postharvest Biol. Technol.* **1997**, *12*, 195–202. [CrossRef]
60. Sahraee, S.; Milani, J.M.; Regenstein, J.M.; Kafil, H.S. Protection of foods against oxidative deterioration using edible films and coatings: A review. *Food Biosci.* **2019**, *32*, 100451. [CrossRef]
61. Nea, F.; Kambiré, D.A.; Genva, M.; Tanoh, E.A.; Wognin, E.L.; Martin, H.; Brostaux, Y.; Tomi, F.; Lognay, G.C.; Tonzibo, Z.F.; et al. Composition, seasonal variation, and biological activities of *Lantana camara* essential oils from Côte d'Ivoire. *Molecules* **2020**, *25*, 2400. [CrossRef] [PubMed]

62. Massolo, J.F.; Concellón, A.; Chaves, A.R.; Vicente, A.R. 1-methylcyclopropene (1-MCP) delays senescence, maintains quality and reduces browning of non-climacteric eggplant (*Solanum melongena* L.) fruit. *Postharvest Biol. Technol.* **2011**, *59*, 10–15. [CrossRef]
63. Scollard, J.; Francis, G.A.; O'Beirne, D. Some conventional and latent antilisterial effects of essential oils, herbs: Carrot and cabbage in fresh-cut vegetable systems. *Postharvest Biol. Technol.* **2013**, *77*, 87–93. [CrossRef]
64. Sánchez-González, L.; Vargas, M.; González-Martínez, C.; Chiralt, A.; Cháfer, M. Use of essential oils in bioactive edible coatings: A review. *Food Eng. Rev.* **2011**, *3*, 1–16. [CrossRef]
65. Jackmen, R.L.; Stanley, D.W. Perspectives in the textural evaluation of plant foods. *Trends Food Sci. Technol.* **1995**, *6*, 187–194. [CrossRef]
66. Zhang, L.; Chen, F.; Lai, S.; Wang, H.; Yang, H. Impact of soybean protein isolate-chitosan edible coating on the softening of apricot fruit during storage. *LWT-Food Sci. Technol.* **2018**, *96*, 604–611. [CrossRef]
67. Eshghi, S.; Hashemi, M.; Mohammadi, A.; Badii, F.; Mohammadhoseini, Z.; Ahmadi, K. Effect of nanochitosan-based coating with and without copper loaded on physicochemical and bioactive components of fresh strawberry fruit (*Fragaria x ananassa* Duchesne) during storage. *Food Bioprocess Technol.* **2014**, *7*, 2397–2409. [CrossRef]
68. Gardesh, A.S.K.; Badii, F.; Hashemi, M.; Ardakani, A.Y.; Maftoonazad, N.; Gorji, A.M. Effect of nanochitosan based coating on climacteric behavior and postharvest shelf-life extension of apple cv *Golab Kohanz*. *LWT-Food Sci. Technol.* **2016**, *70*, 33–40. [CrossRef]
69. Zhou, R.; Mo, Y.; Li, Y.; Zhao, Y.; Zhang, G.; Hu, Y. Quality and internal characteristics of Huanghua peers (*Pyrus pyrifolia* Nakai, cv. *Huanghua*) treated with different kinds of coatings during storage. *Postharvest Biol. Technol.* **2008**, *49*, 171–179.
70. Tamminga, S.K.; Beumer, R.R.; Keijbets, M.J.H.; Kampelmacher, E.M. Microbial spoilage and development of food poisoning bacteria in peeled, completely or partly cooked vacuum- packed potatoes. *Arch. Lemensmittelhyg* **1978**, *29*, 215–219.
71. Doan, C.H.; Davidson, P.M. Microbiology of potatoes and potato products: A review. *J. Food Prot.* **2000**, *63*, 668–683. [CrossRef]
72. Chien, P.; Sheu, F.; Yang, F. Effects of edible chitosan coating on quality and shelf life of sliced mango fruit. *J. Food Eng.* **2007**, *78*, 225–229. [CrossRef]
73. Gonzalez-Aguilar, G.A.; Valenzuela-Soto, E.; Lizardi-Mendoza, J.; Goycoole, F.; Martinez-Tellez, M.A.; Villegas-Ochoa, M.A.; Monroy-Garcia, I.N.; Ayala-Zavala, J.F. Effect of chitosan coating in preventing deterioration and preserving the quality of fresh-cut papaya 'Maradol'. *J. Sci. Food Agric.* **2009**, *89*, 15–23. [CrossRef]
74. Hernandez-Munoz, P.; Almenar, E.; Ocio, M.J.; Gavara, R. Effect of calcium dips and chitosan coatings on postharvest life of strawberries (*Fragaria x ananassa*). *Postharvest Biol. Technol.* **2006**, *39*, 247–253. [CrossRef]
75. Devlieghere, F.; Vermeulen, A.; Debevere, J. Chitosan: Antimicrobial activity, interactions with food components and applicability as a coating on fruit and vegetables. *Food Microbiol.* **2004**, *21*, 703–714. [CrossRef]
76. Jung, E.J.; Youn, D.K.; Lee, S.H.; No, H.K.; Ha, J.G.; Prinyawiwatkul, W. Antibacterial activity of chitosans with different degrees of deacetylation and viscosities. *Int. J. Food Sci. Technol.* **2010**, *45*, 676–682. [CrossRef]
77. Feng, K.; Hu, W.Z.; Jiang, A.L.; Sarengaowa; Xu, Y.P.; Ji, Y.R.; Shao, W.J. Growth of *Salmonella* spp. and *Escherichia coli* O157:H7 on fresh-cut fruits stored at different temperatures. *Foodborne Pathog. Dis.* **2017**, *14*, 510–517. [CrossRef] [PubMed]
78. IFST. *Development and Use of Microbiological Criteria for Foods*; Institute of Food Science and Technology: London, UK, 1999; 76p.
79. Wu, Z.S.; Zhang, M.; Wang, S. Effects of high pressure argon treatments on the quality of fresh-cut apples at cold storage. *Food Control* **2012**, *23*, 120–127. [CrossRef]
80. Zhang, C.L.; Cao, W.; Hung, Y.C.; Li, B.M. Application of electrolyzed oxidizing water in production of radish sprouts to reduce natural microbiota. *Food Control* **2016**, *67*, 177–182. [CrossRef]
81. Roy, S.; Rhim, J. Fabrication of bioactive binary composite film based on gelatin/chitosan incorporated with cinnamon essential oil and rutin. *Colloids Surf. B* **2021**, *204*, 111830. [CrossRef]
82. Wendakoon, C.N.; Sakaguchi, M. Inhibition of amino acid decarboxylase activity of Enterobacter aerogenes by active components of spices. *J. Food Prot.* **1995**, *58*, 280–283. [CrossRef]
83. Wu, J.; Sun, X.; Guo, X.; Ge, S.; Zhang, Q. Physicochemical properties, antimicrobial activity and oil release of fish gelatin films incorporated with cinnamon essential oil. *Aquac. Fish.* **2017**, *2*, 185–192. [CrossRef]
84. Haghighi, H.; Biard, S.; Bigi, F.; De Leo, R.; Bedin, E.; Pfeifer, F.; Siesler, H.W.; Licciardello, F.; Pulvirenti, A. Comprehensive characterization of active chitosan gelatin blend films enriched with different essential oils. *Food Hydrocoll.* **2019**, *95*, 33–42. [CrossRef]
85. Sharma, S.; Barkauskaite, S.; Jaiswal, S.; Duffy, B.; Jaiswal, A.K. Development of essential oil incorporated active film based on biodegradable blends of poly (lactide)/poly (butylene adipate-co-terephthalate) for food packaging application. *J. Packag. Technol. Res.* **2020**, *4*, 235–245. [CrossRef]

Article

The Application of *Aloe vera* Gel as Coating Agent to Maintain the Quality of Tomatoes during Storage

Ignasius Radix A. P. Jati *, Erni Setijawaty, Adrianus Rulianto Utomo and Laurensia Maria Y. D. Darmoatmodjo

Department of Food Technology, Widya Mandala Surabaya Catholic University, Jl. Dinoyo 42-44, Surabaya 60265, Indonesia
* Correspondence: radix@ukwms.ac.id

Abstract: *Aloe vera* is widely used to manufacture medicinal products, cosmetics, and hair treatments. The polysaccharide components in *A. vera* gel can be used as ingredients for edible films or coatings. The edible film can also be applied to fresh fruits and vegetables using the coating principle. Tomatoes are one of the fruit commodities that can be maintained in terms of quality during storage using an edible coating. This study aims to determine the effect of an edible coating made from *A. vera* on tomatoes' physical, chemical, and organoleptic properties during storage. The *A. vera* gel was prepared and used for coating the tomatoes, and the tomatoes were then stored for twelve days. The analysis was conducted every three days, and a comparison with non-coated tomatoes was performed for tomatoes' physicochemical and organoleptic properties. The results show that the application of A. vera as a coating agent could prolong the shelf life of tomatoes, as described in the ability to decrease moisture content and weight loss. The coated tomatoes had lower titratable acidity value, pH, and total soluble solid contents than the non-coated tomatoes. From the organoleptic test, the non-coated tomatoes were preferred by the panelists for color, but the glossiness, skin appearance, and texture of the coated tomatoes were preferred. The coating process could maintain the hardness of tomatoes and prevent the production of phenolic compounds, flavonoids, and lycopene; thus, the antioxidant activity could be conserved.

Keywords: tomato; *Aloe vera*; edible coating; storage; postharvest

1. Introduction

Aloe vera is a Liliaceae family plant extensively distributed in the Middle East and Africa. This plant is widely grown in tropical and subtropical areas, including Indonesia. Its resistance to dry conditions is because of its ability to absorb and store water for a longer time. Therefore, *A. vera* can live in drought and extreme dry conditions [1]. *A. vera* is widely used to manufacture medicinal products, cosmetics, and hair treatments [2]. Meanwhile, on a small scale, it is also processed for food products such as nata de *A. vera*, drinks, and snack mixes. However, the utilization of *A. vera* is limited to food products because it naturally tastes bitter when consumed [3].

The most significant component of *A. vera* gel is water (99.20%). The remaining solids consist of carbohydrates, monosaccharides comprising mainly glucomannan and small amounts of arabinan and galactan, and polysaccharides such as D-glucose, D-mannose, arabinose, galactose, and xylose [4]. According to Gupta et al. [5], the active chemical components contained in *A. vera* are vitamins, minerals, lignin, saponins, salicylic acid, and amino acids, which could act as antimicrobials and antioxidants.

The presence of polysaccharide components in *A. vera* gel can be used as an ingredient for edible films or coatings. Polysaccharide components can provide hardness, density, quality, viscosity, adhesiveness, and gelling ability [6]. An edible film or coating is a thin layer made of hydrocolloids (proteins, polysaccharides, and alginates), lipids (fatty acids, glycerol, and wax), and emulsifiers that function as coatings of or packaging for food products and at the same time can be directly consumed [7]. The main goal of developing edible films or coatings is to create an environmentally friendly packaging or protector for food and food products to replace plastic or other harmful substances to extend the product's shelf life. In addition, the advanced research of edible film and coating allows them to become carriers of beneficial compounds such as vitamins, minerals, antioxidants, and antimicrobials. As a result, the film or coating are able to actively protect the food and food products from damage [8]. Moreover, the edible film and coating can also carry preservative agents, flavoring agents, and colorants to extend the shelf life, enhance the flavor, and improve the appearance of food and food products [9]. Some food products that often found using edible packaging are candy, chocolate, sausage, dried fruit, and bakery products [10].

The edible film can also be applied to fresh fruits and vegetables using the coating principle. An enormous percentage of postharvest losses, especially for fruits and vegetables, is a major challenge in developing countries to ensuring food security status [11]. In contrast to edible films that are in a solid layer form when used to wrap food products, edible coatings are applied in a liquid state to coat fruits or vegetables by dipping or spraying. The coating agent will then dry and form a thin layer that protects the product. As a result, the edible coating can extend the shelf life of fresh fruits and vegetables because it decreases the contact with oxygen, as well as the respiration rate, and generally affects the metabolism of fruits and vegetables, thereby preventing the spoilage of fruits [12]. In addition, the presence of an edible coating also inhibits the transpiration of water vapor from the commodity to the environment, reducing the risk of wilting and weight loss and minimizing the vulnerability to insects or other animals, known as postharvest losses [13]. Due to their functionality and environmentally friendly nature, research on edible coatings has been increasing rapidly, especially characterization based on different materials and formulation, for example the use of starch, soy protein isolate, carboxymethyl cellulose, alginate, chitosan, agar, chlorine, ascorbic acid as an antioxidant, pectin, and essential oil coatings, and their application on food and food products, such as strawberries, blueberries, apples, and several types of cut fruit [14].

Tomatoes (*Solanum lycopersicum* Mill.) are one of the fruit commodities that can be maintained in terms of quality during storage using the edible coating. Tomato, as a climacteric fruit, is susceptible to postharvest damage [15]. The skin and flesh of the fruit are soft, increasing the risk of physical damage due to friction and impact. Wounds on the surface of the fruit skin will trigger damage due to the increase in respiration rate and the growth of microbes, thus accelerating spoilage [16]. Proper storage for tomatoes at 10 °C could extend the shelf life by 14 days. Meanwhile, tomatoes which are stored at room temperature (25 °C) undergo a rapid quality decrease on the fifth day of storage [17]. Research on the application of edible coatings on tomatoes has been reported [18–20], generally using various starch and hydrocolloids. However, limited research is available on the edible coatings made from *A. vera* to maintain the physical, chemical, and organoleptic qualities of tomato during storage. Therefore, this study aims to determine the effect of an edible coating made from *A. vera* on tomatoes' physical, chemical, and organoleptic properties during storage.

2. Materials and Methods

A. vera was grown in Madiun District, East Java, and purchased through a national *A. vera* supplier in Sidoarjo District, East Java Province, Indonesia. Meanwhile, the tomatoes were obtained from local farmers in Malang District, East Java Province. The tomatoes (cv. Ratna) were harvested 90 days after sowing in July 2021. A total of 150 tomatoes were selected, 5 tomatoes for each coating and non-coating treatment and for 3 replications. The tomatoes were chosen within the turning level of maturity, which means that more than 10% but not more than 30% of the surface in the aggregate shows a definite change in color from green to tannish-yellow, pink, red, or a combination thereof. The average diameter of the tomatoes was 2.5 ± 0.25 cm, weight 20 ± 2 g for each tomato, and they had a slightly acidic taste with the absence of injury. Meanwhile, the *A. vera* was harvested at six months (July 2021), possessed a clean green skin color, was approximately 45 ± 4.5 cm long, weighed around 350 ± 35 g for each rind, and had the absence of injury on the surface of the rind. Moreover, the chemicals used for analysis (NaOH, phenolphthalein indicator, H_2SO_4, $FeCl_3$, Folin Ciocalteau, Na_2CO_3, gallic acid, $NaNO_2$, $AlCl_3$, hexane, acetone, ethanol, DPPH, BHT, $FeSO_4 \cdot 7H_2O$) were purchased from Merck, Darmstadt, Germany, and Sigma Aldrich, Singapore, unless otherwise stated.

2.1. Preparation of A. vera Coating Gel and Coating Process

The *A. vera* rind was washed to remove the impurities. Then, it was trimmed, and the thick outer skin was peeled. Next, the gel fraction was washed with warm water to remove the yellow sap. The gel was then crushed using a blender and filtered through 80 mesh sieves to separate the gel from the solid fraction. The gel was then heated in an iron cast pot using a stove at 80 °C for 5 min. After heating, the *A. vera* gel was allowed to cool to room temperature. Meanwhile, the tomato was washed to remove the impurities, soaked in the *A. vera* gel for 5 min, and placed in an open tray at room temperature to let the *A. vera* gel dry. The coated tomato was then kept in the open space at room temperature for 12 days. The observation was conducted at the interval of 3 days.

2.2. Moisture Content

The thermogravimetric method was used to determine the tomato's moisture content. Briefly, the sample was cut, and 1 g of the sample was put in a weighing bottle. The sample was then placed in the drying oven at 105 °C for 2 h. After that, the sample was cooled in a desiccator for 10 min before weighing. This step was repeated until the constant weight of the sample was achieved. Finally, the sample's moisture content was expressed as the moisture percentage within the sample.

2.3. Weight Loss

The weight loss of the sample was monitored during the storage period. The weight of the tomatoes was measured at the beginning of the experiment (day 0) after the air drying. Then, the sample was weighed every 3 days of observation for 12 days. The weight loss was expressed as a percentage of loss to the initial weight.

2.4. Titratable Acidity

The titratable acidity of tomatoes was measured according to [21]. Briefly, the sample was crushed. Then, 10 g of sample was placed in a 100 mL volumetric flask, filled with distilled water, and mixed thoroughly. After that, the sample solution was filtered using Whatman no. 42 filter paper. Then, 10 mL of sample was placed in an Erlenmeyer flask, and three drops of 1% phenolphthalein indicator were added. Finally, the titration was performed using 0.1 N NaOH until the pale pink color was observed. The result was expressed as a percentage of titratable acidity.

2.5. The pH

The pH was examined using a pH meter. First, 10 mL of tomato filtrate was placed in a glass beaker. Next, the electrode was simmered in the sample until the stable pH value was observed.

2.6. Total Soluble Solid

The total soluble solid of tomato was determined using a refractometer. In brief, three drops of the tomato filtrate were placed in the refractometer prism, which was cleaned beforehand using distilled water and lens paper, and the measurement was performed. The result was expressed as °Brix.

2.7. Color

The color profiles of tomatoes were determined using the color reader Konica Minolta CR-10 (Konica Minolta, Osaka, Japan). The results were expressed as lightness (L*), redness (a*), yellowness (b*), hue (°h), and chroma (C).

2.8. Hardness

The hardness of the tomato was measured using texture profile analyzer equipment (TA-XT Plus, Stable Micro Systems Ltd, Surrey, United Kingdom) [22]. The probe used was a cylindrical probe with a diameter of 36 mm. The hardness of the sample was determined as the highest peak identified from the curve produced by the equipment. The result was expressed as Force (N).

2.9. Organoleptic Test

The organoleptic test was performed to determine sensory properties of tomato preferred by the panelists. The quality parameters tested were color, glossy, skin appearance, texture, and aroma. The scoring methods (1–5 score) were used for all parameters. In this test, the coated and non-coated tomato stored after 9 days was chosen because it reflects the optimum condition of tomatoes after storage. A total of 120 semi-trained panelists participated in the organoleptic test. The Hedonic Scale Scoring method (preference test) with a scale ranging from 1 (strongly disliked) to 7 (strongly liked) was used for the organoleptic test.

2.10. Extraction of Tomatoes

A 50 g piece of tomato was sliced and blended for 30 s. Then, 50 g of distilled water was added as a solvent for extraction. The extraction process was conducted using a beaker with a magnetic stirrer for 3 h. Then, the tomato slurry was filtered using a smooth fabric cloth. Finally, the filtrate was collected and freeze-dried for 72 h. A 0.25 g freeze-dried sample was diluted in 25 mL of distilled water for analysis.

2.11. Qualitative Analysis

Qualitative analysis was performed for phytochemicals, such as alkaloids, saponin, tannin, and cardiac glycoside. In addition, reducing sugar was also examined qualitatively. The result is expressed as a numbering scale. The highest number represents the highest content of phytochemicals and reducing sugar in the sample, as indicated by the strong color intensity formed by the chemical reaction.

a. *Alkaloids*

In brief, 1 mL of extract was placed in a test tube. Then, 1 mL of chloroform containing one drop of ammonia and five drops of 5 M H_2SO_4 was added. The tube was then vortexed, and the mixture was pipetted into two spot plates with three drops for each spot. Finally, the Mayer and Wagner reagents were added to spot plates I and II. For spot plate I, the result is positive if the white color is formed. Meanwhile, the brown color indicates a positive test result for spot plate II [23].

b. *Saponin and Tannin*

Two test tubes were prepared with 3 mL of extract added for each tube. For the saponin test, the test tube was vertically sonicated for 10 s and let rest for 10 min. The existence of saponins in the extract can be observed from the presence of a stable foam. Meanwhile, the test tube was heated for 10 min for the tannin test, and 5 mL of $FeCl_3$ solution was added. If the sample contains tannin, the solution will turn to dark blue color [23].

c. *Cardiac glycoside and reducing sugar*

Briefly, 1 mL of extract was placed in a test tube, and 1 mL each of Fehling A and Fehling B were added. The tube was then vortexed and heated for 10 min in a water bath.

The resulted color was observed visually [23]. Meanwhile, for reducing sugar, a similar sample volume was added to 2 mL of Benedict reagent, and then the mixture was boiled for 5 min in the water bath. The brick-red cuprous oxide precipitate will be observed [24].

2.12. Total Phenolic Content

The phenolic compound was measured according to [25]. In brief, 0.5 mL of extract was placed in a test tube, and 1 mL of Folin Ciocalteau reagent was added. The mixture was vortexed and stored for 5 min. After that, 2 mL of 2.5% Na_2CO_3 and 4 mL of distilled water were added to the mixture, immediately vortexed, and stored in a dark place for 30 min. The absorbance of the mixture was measured at 760 nm. The result of absorbance was plotted in a gallic acid standard curve. The result was expressed as mg gallic acid equivalent/100 g sample.

2.13. Total Flavonoid Content

The flavonoid content was examined based on a previous report by [26]. An amount of 0.5 mL of extract was mixed with 0.3, 0.3, and 2 mL of 5% $NaNO_2$, 10% $AlCl_3$, and 1 M NaOH, respectively, in a 10 mL volumetric flask. After that, the distilled water was added to the volume. The mixture was then homogenized. The absorbance of the mixture was measured at 510 nm. The catechin and distilled water were used as standard and blank, respectively, and the result was expressed as mg catechin equivalent/g sample.

2.14. Lycopene Content

The lycopene content of the sample was measured spectrophotometrically [27]. In brief, the fresh tomatoes were blended, and 5 g of tomato puree was placed in a beaker glass covered with aluminum foil. Then, 50 mL of hexane: acetone: ethanol (2:1:1) solvent was added. The mixture was homogenized using a magnetic stirrer. After that, the mixture was placed into a separating funnel, and 10 mL of distilled water was added. The mixture was shaken vigorously for 15 min. The upper layer of the mixture was collected, placed in a 50 mL volumetric flask, and filled up with a similar solvent. The mixture was then homogenized, and absorbance was measured at 513 nm. The lycopene content was express as mg/kg sample.

2.15. Antioxidant Activity

a. DPPH Method

The capacity of extract in the scavenge DPPH radical was determined according to [28]. Briefly, the mixture of 1 mL of extract, 2 mL of 0.2 M DPPH, and 2 mL of methanol was homogenized and stored for one h in a dark room. After that, the absorbance was determined using a spectrophotometer at 517 nm. BHT was used as a control. The result of the scavenging capacity of the extract was expressed as follows: % radical scavenging capacity = ((Absorbance of control − Absorbance of the sample)/absorbance of control) × 100%.

b. Ferric Reducing Antioxidant Power FRAP

The FRAP method was performed according to [25]. Briefly, 60 μL extract, 180 μL distilled water, and 1.8 mL FRAP reagent was mixed in a centrifuge tube and homogenized. The mixture was then incubated at 37 °C for 30 min. The absorbance of the mixture was measured spectrophotometrically at 593 nm. Meanwhile, Fe [II] ($FeSO_4.7H_2O$, with the range of 100–2000 mM) was used to create a standard curve. The result of FRAP was expressed as mmol Fe[II]/g.

2.16. Statistical Analysis

The experiments were carried out using a completely randomized design with three replications. Data were expressed as means ± SD. The Student's t test was performed to determine the significant differences in parameters between the coated and non-coated tomatoes. The analysis was performed using SPSS v23, IBM, New York, United States with statistical significance set at $p < 0.05$.

3. Results and Discussion

Respiration produces energy that the tomato can use to carry out metabolic processes in the ripening stage to reach the fully matured stage and leads to the senescence stage [29]. Providing an edible coating as the outer layer of tomatoes could potentially prolong the shelf life of tomatoes.

Based on the determination, the moisture content of both coated and non-coated tomatoes decreased during storage. Nevertheless, there was a difference in the amount of moisture content decrease between coated and non-coated tomatoes (Figure 1A). Non-coated tomatoes had an initial moisture content of 94.44 ± 0.08%, and after being stored for 12 days, the moisture content reached 92.97 ± 0.34%. Meanwhile, tomatoes with edible coating did not lose as much moisture content as non-coated tomatoes. Tomato fruit coated with *A. vera* gel had an initial moisture content of 95.11 ± 0.04%, and after being stored for 12 days, the moisture content of the tomato fruit became 94.24 ± 0.29%. The result shows that the decrease in moisture content of non-coated tomatoes (1.47%) is higher than that of coated tomatoes (0.87%). The statistical analysis performed observed a significant difference in the loss of moisture between the coated and non-coated tomatoes. Therefore, the *A. vera* gel was shown as an effective coating agent in maintaining the moisture content of tomatoes during storage.

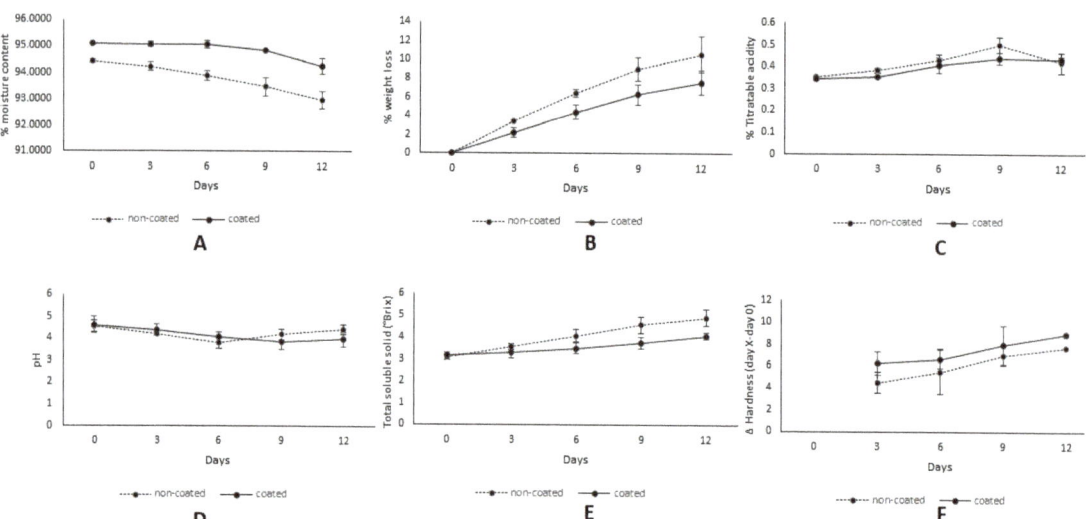

Figure 1. The effect of *A. vera* edible coating on (**A**) moisture content, (**B**) weight loss, (**C**) titratable acidity, (**D**) pH, (**E**) total soluble solid, and (**F**) hardness of tomatoes.

The decrease in moisture content in tomatoes was caused by the respiration and transpiration processes during storage. The water content of fruit will reduce during storage caused by the transpiration process, which evaporates water in the fruit tissue [30]. A thin coating layer of *A. vera* gel on the surface of tomatoes can inhibit the exposure of fruit to oxygen, thus delaying the respiration process. In addition, the *A. vera* gel coating layer could act as a barrier and reduce the water evaporating from the fruit due to transpiration, thus maintaining the water content of the fruit [31]. This result is in line with a previous report that the edible coating can modify the surrounding atmosphere of the fruit by forming a semipermeable layer, protecting the fruit from excessive water losses and exposure to oxygen [32]. Meanwhile, Allegra et al. [33], who applied *A. vera* gel as an edible coating on fig fruit, which is also a climacteric fruit, suggested a significant decrease in moisture content during storage. Therefore, the presence of an edible coating could lower the reduction rate of moisture content. Moreover, Mendy et al. [34] worked on papaya fruit stored at room temperature. A smaller decrease was observed on papaya coated with *A. vera* gel.

The percentage of weight loss is the decrease in the weight of the tomato during storage compared to the initial weight. Weight loss is a crucial parameter for the quality of tomatoes. The weight loss of tomatoes caused by the decrease in moisture content could negatively influence the sensory properties of tomatoes, especially their fresh appearance [35]. The more significant moisture loss gave a negative appearance to the wrinkled skin of the tomato, which could decrease consumer acceptance. The results showed that non-coated tomatoes had a higher weight loss percentage (10.59%) than coated tomatoes (7.62%) (Figure 1B). Furthermore, a significant difference was observed between non-coated and coated tomatoes on the weight loss percentage during storage. *A. vera* gel as an edible coating can prevent excessive weight loss by inhibiting the transpiration process and limiting the oxygen contact with the fruit so that the respiration rate of tomatoes can be inhibited [36]. Meanwhile, a positive correlation between the percentage of weight loss and the moisture content indicates that the evaporation of water mainly contributes to the weight loss of tomatoes during storage.

Figure 1C illustrates the change in total titratable acidity of coated and non-coated tomatoes during storage. An increased trend in titratable acidity was observed until the ninth day of storage, which was 0.34 to 0.43% for the coated group and 0.35–0.49% for the non-coated group. After nine days, the titratable acidity was decreased to 0.43 and 0.41% for the coated and non-coated tomatoes, respectively. Even though on the 12th day, the non-coated tomatoes experienced a higher decrease than the coated tomatoes, there was no significant difference observed. The change in total acid can describe the respiration pattern of tomatoes. If the respiration rate of tomatoes increases, the total acidity of tomatoes can increase, and vice versa. As a climacteric fruit, during storage, the respiration rate of the tomato is increasing, which influences the titratable acidity [37]. After a certain number of days, the respiration rate decreased, and the organic acids declined. A decrease in the respiration rate caused a decrease in the percentage of total acid and the use of organic acids for metabolic processes. Therefore, the titratable acidity was decreased. The application of *A. vera* gel can reduce the fruit's respiration rate because it minimizes tomatoes' exposure to O_2. *A. vera* gel can create a wax-like layer on the surface of the fruit so that it can reduce the penetration of gases such as O_2 and CO_2, thus reducing the respiration rate, ethylene production, and ripening stage and inhibiting senescence [38].

The pattern of pH change in coated and non-coated tomatoes is shown in Figure 1D. The pH of non-coated tomatoes decreased from 4.56 to 3.39 from day 0 to day 6, respectively. Meanwhile, a slight increase was observed on day 9 and day 12. A similar pattern was observed for coated tomatoes. Nevertheless, until day 6, the decrease in pH value was lower compared to non-coated tomatoes. Further storage on days 9 and 12 showed a lower pH value (3.85 and 3.89, respectively). According to Mohammadi et al. [39], the increase in pH could be due to the decline of the organic acid available and the low rate of formation. From the result, it can be suggested that non-coated tomatoes have a faster respiration rate, thus entering the post-climacteric stage earlier. Furthermore, Adiletta et al. [40] reported that the pH of non-coated figs is higher compared to coated figs because organic acids are used as substrates for enzymatic reactions in the respiration process. Therefore, the non-coated fruit has a faster respiration rate, indicated by the higher increase in pH [41].

The total soluble solids (TSS) determination can reflect the fruit's maturity level. Soluble solids widely found in fruits are glucose, fructose, and maltose. The results (Figure 1E) showed that during storage, an increase in total soluble solids was observed for both treatments and with the coated tomatoes and was found to be lower. Coated tomatoes' TSS increased from 3.17 on day 0 to 4.08 on day 12. Meanwhile, for non-coated tomatoes, the pH increased from 3.08 to 4.92 from day 0 to day 12, respectively. The result indicates that the ripening process of coated tomatoes is slower than non-coated tomatoes. During ripening, the polysaccharides are hydrolyzed into their simple form, such as reducing sugar and other water-soluble compounds and used as the respiration substrate [42]. Therefore, the higher the maturity level of the tomatoes, the higher the TSS value, which means that the tomatoes become sweeter. On the other hand, the *A. vera* gel coating caused the minor incline of the TSS of tomatoes, which could be due to the inhibition of respiration, which reduces the energy uptake that consequently decreases the hydrolysis of polysaccharides into a soluble solid [43].

Meanwhile, the result of the hardness of the tomatoes is presented in Figure 1F. Both treatments show a decrease in hardness during storage. The data present the difference between hardness in days of storage with initial hardness (day 0). For coated tomatoes, the differences on day 3 and day 12 were 6.27 and 8.89, respectively. Meanwhile, for non-coated tomatoes, the difference between day 3 and day 0 was 4.53, and day 12 and day 0 was 7.76. The longer storage time resulted in the continuous decrease in hardness due to the ripening process. The hardness decrease needs to be carefully monitored because the further decline of hardness is associated with the low quality of tomatoes. The reduction in tomato fruit hardness is caused by respiration and transpiration processes. These processes break down carbohydrates into simpler compounds and cause a tissue rupture, thus leading to a softer texture [44]. Moreover, the metabolism of tomatoes can degrade the pectin, a substance responsible for wall integrity of fruit, into more minor water-soluble compounds with the help of the enzymes polygalacturonases and pectinmethylesterases, resulting in the texture softening of the fruit wall [45]. The non-coated treatment had a higher hardness decrease due to the tomatoes' metabolism. The *A. vera* coating agent inhibits the metabolism process, significantly reducing the work of enzyme-converting protopectin into water-soluble pectin [46]. Esmaeili et al. [47] reported that coating strawberries with *A. vera* gel could prevent the softening of the fruit tissue.

The changes in the color of the fruit are affected by metabolic activity. In this research, the lightness, redness, yellowness, hue, and chroma were determined, and the results are presented in Table 1. The lightness result shows a decrease in the coated and non-coated tomatoes due to the increase in the ripeness. The data are presented as the difference in lightness between certain days of storage with the initial (day 0) value. For coated tomatoes, values on day 3 were 1.24, increased gradually, and reached 6.13 on day 12. Meanwhile, for non-coated tomatoes, the value increased from 2.2 on day 3 to 16.5 on day 12. This result is supported by a previous finding, which reported a decrease in the lightness value of mango during storage, with the uncoated one having a lower lightness than the coated one [48]. Meanwhile, the redness result (a*) shows an increase in the tomatoes redness value during storage, with the uncoated tomatoes having a higher redness value than the coated tomatoes. It can be concluded that the changes in color in uncoated tomatoes are faster. The presence of an edible coating can inhibit the formation of redness in tomatoes. Fruit coatings can reduce the ethylene formation rate, thus delaying the maturity, chlorophyll degradation, anthocyanin accumulation, and carotenoid synthesis [36]. The color changes in tomatoes were in line with the duration of storage as the ripening stage occurred. During ripening, the chlorophyll present in the thylakoids is degraded, and lycopene accumulates in the chromoplasts [49]. Previous research observed that *A. vera* gel as a coating agent of mango could inhibit the chlorophyll degradation, thus delaying the red color formation [50]. In contrast with the redness, the yellowness of tomatoes (b*) declined in both treatments. The non-coated tomatoes show a higher yellowness decrease than the coated group. For example, on day 0, the yellowness value was 1.23; on day 12, the difference in the yellowness value was larger, at 6.68. Meanwhile, for non-coated tomatoes, the difference in the yellowness value was larger, with 6.51 for day 3 and 15.94 for day 12. The non-coated tomatoes show a higher yellowness decrease than the coated group. The edible coating could inhibit the yellowness formation of tomato. The metabolic process of tomatoes during storage leads to the red color formation given by lycopene. The dominance of lycopene outdoes the contribution of carotenoids and xanthophyll in providing the yellow color of a tomato. The °Hue in coated tomatoes was decreased for both treatments. The edible coating significantly inhibited the respiration and transpiration rate of tomatoes, thus minimizing color changes. A similar trend was observed for chroma value. Aghdam et al. [51] observed a decrease in chroma during storage.

Table 1. Color changes in tomato during storage.

Parameters	Treatment	Δ Color (Day X-Day 0)			
		3	6	9	12
Lightness	Coated	1.24 ± 0.29	1.57 ± 0.48	3.72 ± 1.11	6.13 ± 1.11
	Non-Coated	2.24 ± 0.73	5.38 ± 0.48	14.82 ± 1.10	16.5 ± 1.10
Redness	Coated	1.23 ± 0.61	2.57 ± 0.67	3.69 ± 0.79	4.23 ± 0.46
	Non-Coated	3.11 ± 0.73	5.17 ± 1.02	6.35 ± 1.20	6.71 ± 0.53
Yellowness	Coated	2.46 ± 0.91	4.42 ± 1.23	5.31 ± 0.80	6.68 ± 0.76
	Non-Coated	6.57 ± 0.872	9.80 ± 1.25	14.08 ± 1.82	15.95 ± 1.32
°Hue	Coated	2.07 ± 0.40	4.23 ± 0.37	5.83 ± 0.69	7.43 ± 0.80
	Non-Coated	4.94 ± 1.01	8.47 ± 1.40	11.70 ± 1.91	13.18 ± 0.63
Chroma	Coated	2.02 ± 1.03	3.46 ± 1.33	3.92 ± 0.96	4.85 ± 1.02
	Non-Coated	5.80 ± 0.71	8.46 ± 1.14	12.04 ± 1.61	13.79 ± 1.36

In this research, the organoleptic test was also performed. The results in Table 2 show that on day 9, the non-coated tomatoes were preferred by the panelists for the color because they had a more intense red color than the coated tomatoes. The presence of an edible coating could inhibit the maturity stage, thus preventing the red color formation of tomatoes. Meanwhile, for appearance, gloss, and texture, the coated tomatoes were chosen by the panelists because the coating could delay the shrinkage of the fruit wall and thus create a pleasant overall appearance of the tomatoes. At the same time, applying an edible coating could create a glossy surface for fruit [52]. Furthermore, the inhibition of tomato metabolism by the edible coating could retain the rigid texture of the tomatoes preferred by the panelists.

Table 2. Organoleptic properties of tomato stored for 9 days.

Parameters	Treatment	Score
Color	Coated	3.64 ± 0.24
	Non-Coated	4.44 ± 0.31
Skin appearance	Coated	2.71 ± 0.18
	Non-Coated	1.54 ± 0.11
Glossy	Coated	2.88 ± 0.27
	Non-Coated	2.19 ± 0.14
Texture	Coated	3.05 ± 0.33
	Non-Coated	1.98 ± 0.17

Tomato is well known as a healthy food commodity because it possesses various bioactive compounds that could act as antioxidants. Phytochemical components can act as antioxidants because they can inhibit the free radical reaction of oxidation, which is responsible for the cell damage that leads to various diseases [53]. In this research, the bioactive compound of coated and non-coated tomatoes, which were stored for twelve days, was quantified and examined for its antioxidant capacity. Identification of phytochemical compounds was performed qualitatively before the quantitative analysis. Several studies have stated that phytochemical compounds contained in tomatoes include saponins, alkaloids, flavonoids, phenols, and carotenoids [54]. The results of phytochemical identification can be seen in Table 3. The tomato sample possesses alkaloid, phenolic, flavonoid, and saponin contents. Meanwhile, triterpenoids, sterol, and tannin were absent. The longer storage time increased such compounds, and the non-coated tomatoes indicate higher phytochemical contents. In addition, reducing sugar was also observed to increase with the storage time.

The rise in reducing sugar content was due to the breakdown of polysaccharides into simple sugars used for metabolism [55].

Table 3. The qualitative identification of phytochemical compounds in tomato *.

Compounds	Day 0		Day 3		Day 6		Day 9		Day 12	
	C	NC	C	NC	C	NC	C	NC	C	NC
Alkaloids	1	1	2	2	2	2	2	2	2	2
Phenolic	1	1	2	3	2	2	2	2	2	2
Flavonoid	1	1	2	2	2	2	2	2	2	2
Triterpenoids	-	-	-	-	-	-	-	-	-	-
Sterol	-	-	-	-	-	-	-	-	-	-
Saponin	1	1	2	3	3	4	4	5	5	6
Tannin	-	-	-	-	-	-	-	-	-	-
Reducing Sugar	1	1	2	3	3	4	4	5	5	6

C: coated tomato; NC: non-coated. * The highest number represents the highest content of phytochemicals and reducing sugar in the sample.

The increase in phenolic content was observed on the third day (5.88 mg GAE/g and 5.60 mg GAE/g, for non-coated and coated tomatoes, respectively) and started to reduce on the sixth day of storage (5.43 mg GAE/g and 5.51 mg GAE/g for non-coated and coated tomatoes, respectively (Figure 2A). Even though the phenolic compound of coated tomatoes was lower compared to the non-coated, = there was no significant difference found. The decline in phenolic content in non-coated tomatoes was higher compared to the coated group. The phenolic content in climacteric fruit was lessened during the ripening process [56]. Meanwhile, the rise in phenolic contents could be due to the breakdown of cell wall components. Therefore, the phenolic compounds initially located in the vacuole in the form of bound phenolics become accessible as free phenolics [57]. As a result, the total phenol of the coated tomatoes was slightly lower than the non-coated group. This result is in line with a previous report by Riaz et al. [58], where the phenolic content of non-coated fruit was higher compared to the coated group. The edible coating acts as a barrier from the surrounding environment, which could inhibit the catabolism reaction used for energy for the ripening stage. A previous report suggested that the decrease in phenolic compounds can also be due to the autoxidation reaction of phenol compounds by oxygen and light [59].

The individual flavonoid compounds of tomato include naringenin, the flavanone group, rutin, kaempferol, and quercetin [60]. A similar pattern with phenolic content was observed in the flavonoid content of tomatoes (Figure 2B). On day 3 and day 6, the coated tomatoes had a total flavonoid of 0.8066 mg CE/g and 0.8116 mg CE/g, respectively. Meanwhile, for non-coated tomatoes, the flavonoid content on days 3 and 6 was 0.8648 mg CE/g and 0.7812 mg CE/g, respectively. The analysis confirmed that there was no significant difference observed between coated and non-coated tomatoes on flavonoid content. A similar result could be explained by flavonoids being the most prominent components of the phenol group. Therefore, the edible coating could decelerate the tomato metabolism, thus reducing the flavonoid content. Meanwhile, the edible coating could inhibit the rapid decrease in flavonoid content during storage. Such functions are related to the capability of the coating as the barrier between the air and moisture from the environment [61].

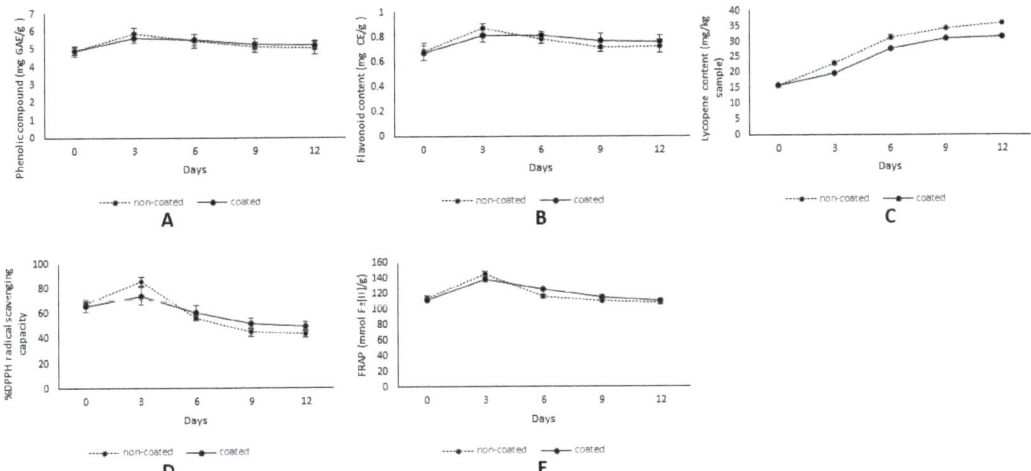

Figure 2. The effect of *A. vera* coating on (**A**) phenolic content, (**B**) flavonoid content, (**C**) lycopene content, (**D**) DPPH radical scavenging capacity, and (**E**) Ferric Reducing Antioxidant Power of tomatoes.

Results in Figure 2C showed an increase in lycopene content during storage. For coated tomatoes, the lycopene content increased from 15.77 mg/kg on day 0 to 31.48 mg/kg on day 12 of storage. Meanwhile, for non-coated tomatoes, the lycopene content raised from 15.74 mg/kg on day 0 to 35.74 mg/kg on day 12. There was a significant difference observed between coated and non-coated tomatoes in flavonoid content. During the ripening stage, lycopene content was increased due to degradation of chlorophyll and accumulation of lycopene in fruit [62]. Previous reports observed the increase in lycopene in stored tomatoes. During storage, the non-coated tomatoes exhibited a higher increase in lycopene content than the coated group and the delay of color change in the *A. vera*-coated fruit. The application of *A. vera* as a coating agent prevents the degradation of chlorophyll and the accumulation of lycopene in the ripening stage. In addition, the *A. vera* coating act as a barrier to air and moisture, thus decreasing the respiration rate of fruit [63,64].

Furthermore, the antioxidant activity of tomatoes was examined using DPPH and FRAP methods. The result shows that the tomato extract can scavenge DPPH radicals (Figure 2D). The coated tomatoes had a 65.6% radical scavenging activity on day 0 and slightly increased on day 3 to 74.12%. Further storage resulted in decreased antioxidant activity. On day 12, the antioxidant activity of tomatoes reached 49.57%. A similar pattern was observed for non-coated tomatoes. The highest antioxidant activity was possessed by tomatoes on day 3, with 85.57%. A positive correlation (R = 0.3281) was observed between the extract's phenolic content and antioxidant activity. The phenolic compound was reported to have high antioxidant activity, mainly due to its ability as a hydrogen donor to stabilize free radicals [65]. However, after the third day of storage, the antioxidant activity of the tomatoes declined. The result is also in line with the decrease in phenolic content. In addition to the lower phenolic compound content, the decrease in DPPH radical scavenging activity during storage could be due to the bioactive compound in fruit being susceptible to degradation when stored in an open environment. Such storage exposes the fruit to oxidation, which is also accelerated by the presence of light and high-temperature storage. Meanwhile, a similar trend was observed for the FRAP method (Figure 2E). The tomato extract could reduce the ferric to ferrous ion. The coated tomatoes on day 0 had 111.02 mmol Fe[II]/g and increased to 138.21 mmol Fe[II]/g on day 3. Further storage decreased the antioxidant activity to 110.21 mmol Fe[II]/g on day 12. A similar pattern was found for non-coated tomatoes, with tomatoes stored for 3 days having the highest antioxidant activity (145.43 mmol Fe[II]/g) and the tomatoes stored for 12 days having the lowest antioxidant activity (107.64 mmol Fe[II]/g). The phenolic content plays a vital role in the antioxidant capacity of tomato extract by acting as a chelating agent. Even though the lycopene content was increased, it does not contribute significantly to the antioxidant capacity due to its nature as a lipophilic substance. The hydrophilic substance is dominant in acting as an antioxidant compared to the lipophilic [66].

4. Conclusions

The application of *A. vera* gel edible coating could prolong the shelf life of tomatoes, as observed from the color measurement and organoleptic test. In addition, the *A. vera* edible coating could decrease the loss of moisture content and weight of tomatoes, which further affects the freshness of tomatoes. Furthermore, the edible coating can inhibit the maturity stage, as shown in the titratable acidity, pH, and total soluble solids. Meanwhile, the coating process could retain the hardness of the tomato. From the organoleptic test, the non-coated tomatoes were preferred by the panelists for the color, but for the glossiness, skin appearance, and texture, the coated tomatoes were preferred. Moreover, the presence of *A. vera* gel could minimize the degradation of phenolic and flavonoid compounds while inhibiting lycopene production, thus protecting the ability of tomatoes to act as an antioxidant and affecting the color of tomatoes that may influence the consumer acceptance. Based on these properties, *A. vera* could potentially be used for coating other fruit commodities. It could also be mixed with hydrocolloids to construct a film suitable for food packaging applications.

Author Contributions: Conceptualization, A.R.U., E.S., I.R.A.P.J.; methodology, I.R.A.P.J., A.R.U., E.S.; software, L.M.Y.D.D.; formal analysis, I.R.A.P.J., A.R.U., E.S.; resources, I.R.A.P.J., L.M.Y.D.D.; writing—original draft preparation, I.R.A.P.J., A.R.U., E.S.; writing—review and editing, A.R.U., E.S., I.R.A.P.J.; visualization, L.M.Y.D.D.; supervision, A.R.U.; project administration, E.S.; funding acquisition, I.R.A.P.J. All authors have read and agreed to the published version of the manuscript.

Funding: This research was funded by the Directorate of Research and Community Services, Deputy of Research Empowerment and Development, The Ministry of Education, Culture, Research and Technology, Republic of Indonesia, grant number 260K/WM01.5/N/2022. The APC was funded by the Directorate of Research and Community Services, Deputy of Research Empowerment and Development, The Ministry of Education, Culture, Research and Technology, Republic of Indonesia.

Data Availability Statement: Data are available upon request.

Conflicts of Interest: The authors declare no conflict of interest.

References

1. Sánchez, M.; González-Burgos, E.; Iglesias, I.; Gómez-Serranillos, M.P. Pharmacological Update Properties of Aloe Vera and its Major Active Constituents. *Molecules* **2020**, *25*, 1324. [CrossRef] [PubMed]
2. Kumar, R.; Singh, A.K.; Gupta, A.; Bishayee, A.; Pandey, A.K. Therapeutic potential of Aloe vera—A miracle gift of nature. *Phytomedicine* **2019**, *60*, 152996. [CrossRef] [PubMed]
3. Shakib, Z.; Shahraki, N.; Razavi, B.M.; Hosseinzadeh, H. *Aloe vera* as an herbal medicine in the treatment of metabolic syndrome: A review. *Phytother. Res.* **2019**, *33*, 2649–2660. [CrossRef] [PubMed]
4. Govindarajan, S.; Babu, S.N.; Vijayalakshmi, M.A.; Manohar, P.; Noor, A. *Aloe vera* carbohydrates regulate glucose metabolism through improved glycogen synthesis and downregulation of hepatic gluconeogenesis in diabetic rats. *J. Ethnopharmacol.* **2021**, *281*, 114556. [CrossRef]
5. Gupta, V.K.; Yarla, N.S.; de Lourdes Pereira, M.; Siddiqui, N.J.; Sharma, B. Recent Advances in Ethnopharmacological and Toxicological Properties of Bioactive Compounds from *Aloe barbadensis* (Miller), *Aloe vera*. *CBC* **2021**, *17*, e010621184955. [CrossRef]
6. Sarker, A.; Grift, T.E. Bioactive properties and potential applications of Aloe vera gel edible coating on fresh and minimally processed fruits and vegetables: A review. *Food Measure* **2021**, *15*, 2119–2134. [CrossRef]
7. Salehi, F. Edible Coating of Fruits and Vegetables Using Natural Gums: A Review. *Int. J. Fruit Sci.* **2020**, *20*, S570–S589. [CrossRef]
8. Ganiari, S.; Choulitoudi, E.; Oreopoulou, V. Edible and active films and coatings as carriers of natural antioxidants for lipid food. *Trends Food Sci. Technol.* **2017**, *68*, 70–82. [CrossRef]
9. Chen, W.; Ma, S.; Wang, Q.; McClements, D.J.; Liu, X.; Ngai, T.; Liu, F. Fortification of edible films with bioactive agents: A review of their formation, properties, and application in food preservation. *Crit. Rev. Food Sci. Nutr.* **2022**, *62*, 5029–5055. [CrossRef]

10. Kumar, L.; Ramakanth, D.; Akhila, K.; Gaikwad, K.K. Edible films and coatings for food packaging applications: A review. *Environ. Chem. Lett.* **2022**, *20*, 875–900. [CrossRef]
11. Porat, R.; Lichter, A.; Terry, L.A.; Harker, R.; Buzby, J. Postharvest losses of fruit and vegetables during retail and in consumers' homes: Quantifications, causes, and means of prevention. *Postharvest Biol. Technol.* **2018**, *139*, 135–149. [CrossRef]
12. Nair, M.S.; Tomar, M.; Punia, S.; Kukula-Koch, W.; Kumar, M. Enhancing the functionality of chitosan- and alginate-based active edible coatings/films for the preservation of fruits and vegetables: A review. *Int. J. Biol. Macromol.* **2020**, *164*, 304–320. [CrossRef] [PubMed]
13. Maringgal, B.; Hashim, N.; Mohamed Amin Tawakkal, I.S.; Muda Mohamed, M.T. Recent advance in edible coating and its effect on fresh/fresh-cut fruits quality. *Trends Food Sci. Technol.* **2020**, *96*, 253–267. [CrossRef]
14. Valencia, G.A.; Luciano, C.G.; Monteiro Fritz, A.R. Smart and Active Edible Coatings Based on Biopolymers. In *Polymers for Agri-Food Applications*; Gutiérrez, T.J., Ed.; Springer International Publishing: Cham, Switzerland, 2019; pp. 391–416. ISBN 978-3-030-19415-4.
15. Al-Dairi, M.; Pathare, P.B.; Al-Yahyai, R. Effect of Postharvest Transport and Storage on Color and Firmness Quality of Tomato. *Horticulturae* **2021**, *7*, 163. [CrossRef]
16. Abera, G.; Ibrahim, A.M.; Forsido, S.F.; Kuyu, C.G. Assessment on post-harvest losses of tomato (*Lycopersicon esculentem* Mill.) in selected districts of East Shewa Zone of Ethiopia using a commodity system analysis methodology. *Heliyon* **2020**, *6*, e03749. [CrossRef]
17. Jung, J.-M.; Shim, J.-Y.; Chung, S.-O.; Hwang, Y.-S.; Lee, W.-H.; Lee, H. Changes in quality parameters of tomatoes during storage: A review. *Koraen J. Agric. Sci.* **2019**, *46*, 239–256. [CrossRef]
18. Yadav, A.; Kumar, N.; Upadhyay, A.; Sethi, S.; Singh, A. Edible coating as postharvest management strategy for shelf-life extension of fresh tomato (*Solanum lycopersicum* L.): An overview. *J. Food Sci.* **2022**, *87*, 2256–2290. [CrossRef]
19. Chrysargyris, A.; Nikou, A.; Tzortzakis, N. Effectiveness of *Aloe vera* gel coating for maintaining tomato fruit quality. *N. Z. J. Crop Horticultural Sci.* **2016**, *44*, 203–217. [CrossRef]
20. Athmaselvi, K.A.; Sumitha, P.; Revathy, B. Development of Aloe vera based edible coating for tomato. *Int. Agrophys.* **2013**, *27*, 369–375. [CrossRef]
21. Tyl, C.; Sadler, G.D. pH and Titratable Acidity. In *Food Analysis*; Nielsen, S.S., Ed.; Food Science Text Series; Springer International Publishing: Cham, Switzerland, 2017; pp. 389–406. ISBN 978-3-319-45774-1.
22. Lázaro, A.; Ruiz-Aceituno, L. Instrumental Texture Profile of Traditional Varieties of Tomato (*Solanum lycopersicum* L.) and its Relationship to Consumer Textural Preferences. *Plant Foods Hum. Nutr.* **2021**, *76*, 248–253. [CrossRef]
23. Sorescu, A.-A.; Nuta, A.; Ion, R.-M.; Iancu, L. Qualitative Analysis of Phytochemicals from Sea Buckthorn and Gooseberry. In *Phytochemicals—Source of Antioxidants and Role in Disease Prevention*; Asao, T., Asaduzzaman, M., Eds.; IntechOpen: London, UK, 2018; ISBN 978-1-78984-377-4.
24. Hernández-López, A.; Sánchez Félix, D.A.; Zuñiga Sierra, Z.; García Bravo, I.; Dinkova, T.D.; Avila-Alejandre, A.X. Quantification of Reducing Sugars Based on the Qualitative Technique of Benedict. *ACS Omega* **2020**, *5*, 32403–32410. [CrossRef] [PubMed]
25. Jati, I.R.A.P.; Nohr, D.; Konrad Biesalski, H. Nutrients and antioxidant properties of Indonesian underutilized colored rice. *Nutr. Food Sci.* **2014**, *44*, 193–203. [CrossRef]
26. Huang, R.; Wu, W.; Shen, S.; Fan, J.; Chang, Y.; Chen, S.; Ye, X. Evaluation of colorimetric methods for quantification of citrus flavonoids to avoid misuse. *Anal. Methods* **2018**, *10*, 2575–2587. [CrossRef]
27. Anthon, G.; Barrett, D.M. Standardization of A Rapid Spectrophotometric Method for Lycopene Analysis. *Acta Hortic.* **2007**, *758*, 111–128. [CrossRef]
28. Astadi, I.R.; Astuti, M.; Santoso, U.; Nugraheni, P.S. In vitro antioxidant activity of anthocyanins of black soybean seed coat in human low density lipoprotein (LDL). *Food Chem.* **2009**, *112*, 659–663. [CrossRef]
29. Zhong, T.-Y.; Yao, G.-F.; Wang, S.-S.; Li, T.-T.; Sun, K.-K.; Tang, J.; Huang, Z.-Q.; Yang, F.; Li, Y.-H.; Chen, X.-Y.; et al. Hydrogen Sulfide Maintains Good Nutrition and Delays Postharvest Senescence in Postharvest Tomato Fruits by Regulating Antioxidative Metabolism. *J. Plant Growth Regul.* **2021**, *40*, 2548–2559. [CrossRef]
30. Díaz-Pérez, J.C. Transpiration. In *Postharvest Physiology and Biochemistry of Fruits and Vegetables*; Elsevier: Amsterdam, The Netherlands, 2019; pp. 157–173. ISBN 978-0-12-813278-4.
31. Salama, H.E.; Abdel Aziz, M.S. Development of active edible coating of alginate and aloe vera enriched with frankincense oil for retarding the senescence of green capsicums. *LWT* **2021**, *145*, 111341. [CrossRef]
32. Miteluţ, A.C.; Popa, E.E.; Drăghici, M.C.; Popescu, P.A.; Popa, V.I.; Bujor, O.-C.; Ion, V.A.; Popa, M.E. Latest Developments in Edible Coatings on Minimally Processed Fruits and Vegetables: A Review. *Foods* **2021**, *10*, 2821. [CrossRef]
33. Allegra, A.; Farina, V.; Inglese, P.; Gallotta, A.; Sortino, G. Qualitative traits and shelf life of fig fruit ('Melanzana') treated with Aloe vera gel coating. *Acta Hortic.* **2021**, *1310*, 87–92. [CrossRef]
34. Mendy, T.K.; Misran, A.; Mahmud, T.M.M.; Ismail, S.I. Application of *Aloe vera* coating delays ripening and extend the shelf life of papaya fruit. *Sci. Horticult.* **2019**, *246*, 769–776. [CrossRef]
35. Kaewklin, P.; Siripatrawan, U.; Suwanagul, A.; Lee, Y.S. Active packaging from chitosan-titanium dioxide nanocomposite film for prolonging storage life of tomato fruit. *Int. J. Biol. Macromol.* **2018**, *112*, 523–529. [CrossRef] [PubMed]
36. Shah, S.; Hashmi, M.S. Chitosan–aloe vera gel coating delays postharvest decay of mango fruit. *Hortic. Environ. Biotechnol.* **2020**, *61*, 279–289. [CrossRef]

37. Yan, J.; Luo, Z.; Ban, Z.; Lu, H.; Li, D.; Yang, D.; Aghdam, M.S.; Li, L. The effect of the layer-by-layer (LBL) edible coating on strawberry quality and metabolites during storage. *Postharvest Biol. Technol.* **2019**, *147*, 29–38. [CrossRef]
38. Maan, A.A.; Reiad Ahmed, Z.F.; Iqbal Khan, M.K.; Riaz, A.; Nazir, A. *Aloe vera* gel, an excellent base material for edible films and coatings. *Trends Food Sci. Technol.* **2021**, *116*, 329–341. [CrossRef]
39. Mohammadi, L.; Ramezanian, A.; Tanaka, F.; Tanaka, F. Impact of *Aloe vera* gel coating enriched with basil (*Ocimum basilicum* L.) essential oil on postharvest quality of strawberry fruit. *Food Measure* **2021**, *15*, 353–362. [CrossRef]
40. Adiletta, G.; Zampella, L.; Coletta, C.; Petriccione, M. Chitosan Coating to Preserve the Qualitative Traits and Improve Antioxidant System in Fresh Figs (*Ficus carica* L.). *Agriculture* **2019**, *9*, 84. [CrossRef]
41. Maringgal, B.; Hashim, N.; Mohamed Amin Tawakkal, I.S.; Mohamed, M.T.M.; Hamzah, M.H.; Mohd Ali, M. Effect of *Kelulut* Honey Nanoparticles Coating on the Changes of Respiration Rate, Ascorbic Acid, and Total Phenolic Content of Papaya (*Carica papaya* L.) during Cold Storage. *Foods* **2021**, *10*, 432. [CrossRef]
42. John, A.; Yang, J.; Liu, J.; Jiang, Y.; Yang, B. The structure changes of water-soluble polysaccharides in papaya during ripening. *Int. J. Biol. Macromol.* **2018**, *115*, 152–156. [CrossRef]
43. Nourozi, F.; Sayyari, M. Enrichment of *Aloe vera* gel with basil seed mucilage preserve bioactive compounds and postharvest quality of apricot fruits. *Sci. Horticult.* **2020**, *262*, 109041. [CrossRef]
44. Pan, Y.-W.; Cheng, J.-H.; Sun, D.-W. Inhibition of fruit softening by cold plasma treatments: Affecting factors and applications. *Crit. Rev. Food Sci. Nutr.* **2021**, *61*, 1935–1946. [CrossRef]
45. Huang, X.; Pan, S.; Sun, Z.; Ye, W.; Aheto, J.H. Evaluating quality of tomato during storage using fusion information of computer vision and electronic nose. *J. Food Process Eng.* **2018**, *41*, e12832. [CrossRef]
46. Shakir, M.S.; Ejaz, S.; Hussain, S.; Ali, S.; Sardar, H.; Azam, M.; Ullah, S.; Khaliq, G.; Saleem, M.S.; Nawaz, A.; et al. Synergistic effect of gum Arabic and carboxymethyl cellulose as biocomposite coating delays senescence in stored tomatoes by regulating antioxidants and cell wall degradation. *Int. J. Biol. Macromol.* **2022**, *201*, 641–652. [CrossRef] [PubMed]
47. Esmaeili, Y.; Zamindar, N.; Paidari, S.; Ibrahim, S.A.; Mohammadi Nafchi, A. The synergistic effects of aloe vera gel and modified atmosphere packaging on the quality of strawberry fruit. *J. Food Process. Preserv.* **2021**, *45*, e16003. [CrossRef]
48. Rastegar, S.; Hassanzadeh Khankahdani, H.; Rahimzadeh, M. Effectiveness of alginate coating on antioxidant enzymes and biochemical changes during storage of mango fruit. *J. Food Biochem.* **2019**, *43*. [CrossRef] [PubMed]
49. Li, Y.; Liu, C.; Shi, Q.; Yang, F.; Wei, M. Mixed red and blue light promotes ripening and improves quality of tomato fruit by influencing melatonin content. *Environ. Exp. Bot.* **2021**, *185*, 104407. [CrossRef]
50. Hajebi Seyed, R.; Rastegar, S.; Faramarzi, S. Impact of edible coating derived from a combination of *Aloe vera* gel, chitosan and calcium chloride on maintain the quality of mango fruit at ambient temperature. *Food Measure* **2021**, *15*, 2932–2942. [CrossRef]
51. Aghdam, M.S.; Flores, F.B.; Sedaghati, B. Exogenous phytosulfokine α (PSKα) application delays senescence and relieves decay in strawberry fruit during cold storage by triggering extracellular ATP signaling and improving ROS scavenging system activity. *Sci. Horticult.* **2021**, *279*, 109906. [CrossRef]
52. Saxena, A.; Sharma, L.; Maity, T. Enrichment of edible coatings and films with plant extracts or essential oils for the preservation of fruits and vegetables. In *Biopolymer-Based Formulations*; Elsevier: Amsterdam, The Netherlands, 2020; pp. 859–880. ISBN 978-0-12-816897-4.
53. Yu, M.; Gouvinhas, I.; Rocha, J.; Barros, A.I.R.N.A. Phytochemical and antioxidant analysis of medicinal and food plants towards bioactive food and pharmaceutical resources. *Sci. Rep.* **2021**, *11*, 10041. [CrossRef]
54. Rouphael, Y.; Corrado, G.; Colla, G.; De Pascale, S.; Dell'Aversana, E.; D'Amelia, L.I.; Fusco, G.M.; Carillo, P. Biostimulation as a Means for Optimizing Fruit Phytochemical Content and Functional Quality of Tomato Landraces of the San Marzano Area. *Foods* **2021**, *10*, 926. [CrossRef]
55. Williams, R.S.; Benkeblia, N. Biochemical and physiological changes of star apple fruit (*Chrysophyllum cainito*) during different "on plant" maturation and ripening stages. *Sci. Horticult.* **2018**, *236*, 36–42. [CrossRef]
56. Guofang, X.; Xiaoyan, X.; Xiaoli, Z.; Yongling, L.; Zhibing, Z. Changes in phenolic profiles and antioxidant activity in rabbiteye blueberries during ripening. *Int. J. Food Prop.* **2019**, *22*, 320–329. [CrossRef]
57. Allegro, G.; Pastore, C.; Valentini, G.; Filippetti, I. The Evolution of Phenolic Compounds in *Vitis vinifera* L. Red Berries during Ripening: Analysis and Role on Wine Sensory—A Review. *Agronomy* **2021**, *11*, 999. [CrossRef]
58. Riaz, A.; Aadil, R.M.; Amoussa, A.M.O.; Bashari, M.; Abid, M.; Hashim, M.M. Application of chitosan-based apple peel polyphenols edible coating on the preservation of strawberry (*Fragaria ananassa* cv Hongyan) fruit. *J Food Process. Preserv.* **2021**, *45*, e15018. [CrossRef]
59. Zhou, X.; Iqbal, A.; Li, J.; Liu, C.; Murtaza, A.; Xu, X.; Pan, S.; Hu, W. Changes in Browning Degree and Reducibility of Polyphenols during Autoxidation and Enzymatic Oxidation. *Antioxidants* **2021**, *10*, 1809. [CrossRef]
60. Liu, C.; Zheng, H.; Sheng, K.; Liu, W.; Zheng, L. Effects of postharvest UV-C irradiation on phenolic acids, flavonoids, and key phenylpropanoid pathway genes in tomato fruit. *Sci. Horticult.* **2018**, *241*, 107–114. [CrossRef]
61. Panahirad, S.; Naghshiband-Hassani, R.; Bergin, S.; Katam, R.; Mahna, N. Improvement of Postharvest Quality of Plum (*Prunus domestica* L.) Using Polysaccharide-Based Edible Coatings. *Plants* **2020**, *9*, 1148. [CrossRef] [PubMed]
62. Kapoor, L.; Simkin, A.J.; George Priya Doss, C.; Siva, R. Fruit ripening: Dynamics and integrated analysis of carotenoids and anthocyanins. *BMC Plant Biol.* **2022**, *22*, 27. [CrossRef]

63. Georgiadou, E.C.; Antoniou, C.; Majak, I.; Goulas, V.; Filippou, P.; Smolińska, B.; Leszczyńska, J.; Fotopoulos, V. Tissue-specific elucidation of lycopene metabolism in commercial tomato fruit cultivars during ripening. *Sci. Horticult.* **2021**, *284*, 110144. [CrossRef]
64. Nguyen, H.T.; Boonyaritthongchai, P.; Buanong, M.; Supapvanich, S.; Wongs-Aree, C. Chitosan- and κ-carrageenan-based composite coating on dragon fruit (*Hylocereus undatus*) pretreated with plant growth regulators maintains bract chlorophyll and fruit edibility. *Sci. Horticult.* **2021**, *281*, 109916. [CrossRef]
65. Zeb, A. Concept, mechanism, and applications of phenolic antioxidants in foods. *J. Food Biochem.* **2020**, *44*, e13394. [CrossRef]
66. Zacarías-García, J.; Rey, F.; Gil, J.-V.; Rodrigo, M.J.; Zacarías, L. Antioxidant capacity in fruit of Citrus cultivars with marked differences in pulp coloration: Contribution of carotenoids and vitamin C. *Food Sci. Technol. Int.* **2021**, *27*, 210–222. [CrossRef] [PubMed]

Article

Application of Chitosan-Lignosulfonate Composite Coating Film in Grape Preservation and Study on the Difference in Metabolites in Fruit Wine

Boran Hu [1,*], Lan Lin [1], Yujie Fang [1], Min Zhou [1] and Xiaoyan Zhou [1,2,*]

[1] School of Food Science and Engineering, Yangzhou University, Yangzhou 225127, China; linlan_yang2023@163.com (L.L.); fang.yujie431@mail.kyutech.jp (Y.F.); lovelylovelymia@163.com (M.Z.)
[2] Key Laboratory of Chinese Cuisine Intangible Cultural Heritage Technology Inheritance, Ministry of Culture and Tourism, Yangzhou 225127, China
* Correspondence: huboran@yzu.edu.cn (B.H.); yzuxyz@163.com (X.Z.)

Abstract: In order to solve the global problem of fruit rotting due to microbial infection and water loss after harvest, which leads to a large amount of food waste, this experiment uses degradable biological composite coating to prolong the preservation period of grapes. Chitosan (CH) and Lignosulfonate (LS) were used as Bio-based film materials, CH films, 1% CH/LS films and 2% CH/LS biomass composite films were synthesized by the classical casting method and applied to grape preservation packaging. Its preservation effect was tested by grape spoilage rate, water loss rate, hardness, soluble solids, titratable acid, and compared with plastic packaging material PE film. At the same time, ^{1}H NMR technology combined with pattern recognition analysis (PCA) and partial least squares discriminant analysis (PLS-DA) was used to determine the nuclear magnetic resonance (NMR) of Cabernet Sauvignon, Chardonnay and Italian Riesling wines from the eastern foothills of Helan Mountain to explore the differences in metabolites of wine. The results of preservation showed that the grapes quality of CH films and 2% CH/LS coating package is better than the control group, the decay rates decreased from 37.71% to 21.63% and 18.36%, respectively, the hardness increased from 6.83 to 10.4 and 12.78 and the soluble solids increased from 2.1 in the control group to 3.0 and 3.2. In terms of wine metabolites, there are similar types of metabolites between cabernet Sauvignon dry red wine and Chardonnay and Italian Riesling dry white wine, but there are significant differences in content. The study found that 2% CH/LS coating package could not only reduce the spoilage rate of grapes, inhibit the consumption of soluble solids and titratable acids, but also effectively extend the shelf life of grapes by 6 days.

Keywords: grapes; food preservation; chitosan; biomass composite coating; wine; metabolites; proton nuclear magnetic resonance technology

Citation: Hu, B.; Lin, L.; Fang, Y.; Zhou, M.; Zhou, X. Application of Chitosan-Lignosulfonate Composite Coating Film in Grape Preservation and Study on the Difference in Metabolites in Fruit Wine. *Coatings* 2022, *12*, 494. https://doi.org/10.3390/coatings12040494

Academic Editor: Elena Poverenov

Received: 28 February 2022
Accepted: 30 March 2022
Published: 7 April 2022

Publisher's Note: MDPI stays neutral with regard to jurisdictional claims in published maps and institutional affiliations.

Copyright: © 2022 by the authors. Licensee MDPI, Basel, Switzerland. This article is an open access article distributed under the terms and conditions of the Creative Commons Attribution (CC BY) license (https://creativecommons.org/licenses/by/4.0/).

1. Introduction

Grapes are fresh and juicy, bright in color, sweet and sour and rich in nutritional value [1]. However, because they are harvested in the high-temperature season, the fruit stem is a typical respiratory climacteric type [2], the physiological metabolism is vigorous after harvest, and the fruit tissue is crisp and tender. During storage, it is very prone to water loss, stem withering, browning and easily infected by bacteria and rot, resulting in a short storage cycle, which seriously affects the appearance quality and commodity value of fresh grapes, causing huge economic losses and food waste. At present, the preservation measures of grapes at home and abroad mainly use sulfur dioxide (SO_2) preservatives, coating preservation, low-temperature refrigeration and other technologies to prolong the storage period of grapes [3]. Although SO_2 has a good effect on maintaining the quality of grapes, it is easy to cause certain bleaching damage to the fruit, affect the original flavor, and the SO_2 remaining in the fruit will be harmful to human health [4]. Packaged food preservation is one of the effective solutions. However, the packaging material commonly used in the preservation process is plastic; for example, PE, PS, or PET [5]. Chitosan is a kind of natural edible macromolecular polysaccharide that is non-toxic, has no smell, with good film forming characteristics [6,7]. After coating treatment, a colorless and transparent biological film is formed on the grape surface, which can prevent water loss and reduce the rate of weight loss in the fruit. It can also effectively prevent microbial invasion, reduce the respiratory intensity of the fruit [8–10] to a certain extent, and reduce the decay of the fruit, thus prolonging its storage time [11]. However, the coating effect of single chitosan is not stable [12], the drying time is long, the water permeability is high, and the coating toughness is poor. Sodium lignosulfonate is an anionic polymer surfactant extracted from papermaking waste liquor [13]. Because of its similar structure to lignin, it is often used as a bio-based polymer material [14]. LS has a good function but cannot form a film by itself. We tried adding LS to CH to make films suitable for fruit preservation, because the use of bio-based materials for packaging can not only be used as cheap packaging for fresh fruits and vegetables, but also reduce the pollution of plastic packaging to the environment. It achieves the requirements of highly efficient anticorrosion, low carbon and environmental protection.

Wine is fresh grapes or grape juice as raw materials, through all or part of the fermentation brewing, containing a certain degree of alcohol fermented wine [15]. It is not only a nutritious beverage, but moderate drinking can prevent various chronic diseases and enhance human health [16–18]. Fresh grapes are the key raw materials for wine brewing, and the style of wine is closely related to the variety of wine grapes, the climate of the place of origin, soil conditions and the distinctive brewing processes, making the wines produced in different producing areas have different flavors [19]. The eastern foot of Helan Mountain in Ningxia is located in the "golden" area of grape planting at 30 to 40 latitudes from north to south. The superior geographical location, unique landform features, suitable soil and climate conditions mean that wine grapes in the eastern foot of Helan Mountain in Ningxia fully possess excellent brewing characteristics.

With the rapid development of the world economy and the improvement of people's quality of life, the wine industry has developed rapidly and has become an increasingly popular product for consumers [20], but the consumer's ability to distinguish is limited. Usually, physical and chemical indexes and sensory evaluation are used to identify the quality of wine grapes and wine, but it is difficult to reflect various metabolites in wine grapes and wine that affect their quality and are beneficial to human health through these indicators [21]. In studies on wine metabolites, a "metabolic fingerprint" is generally provided based on ^{1}H-NMR technology [22,23]. Gregory et al. [24] used NMR to analyze and study metabolites in French red wine in the region of Bordeaux, so as to better distinguish red wine of Bordeaux from other red wines produced in French wine regions. In chemometrics, pattern recognition is the main method used to solve the attribution problem and marker search in complex systems, among which PLS-DA is the most important pattern recognition method applied in metabonomics. It is widely used in plant, drug metabolomics and food source determination and classification research [25]. Godelmann R. et al. [26] used nuclear magnetic resonance technology combined with multivariate statistics and principal component analysis to analyze and study the target compounds and non-target compounds of wine metabolites, and the results showed that the accuracy of variety identification reached 95%, the accuracy of age identification reached 97% and 96%, respectively, and the accuracy of origin identification reached 89%. All these results showed that ^{1}H-NMR combined with multivariate analysis was an extraordinary effective method to identify different wine varieties and region.

This paper investigated the preparation method of separate 2% chitosan film (CH), chitosan—1% sodium lignosulfonate film (1% CH/LS) and chitosan—2% sodium lignosulfonate film (2% CH/LS), and the film forming properties of the three films were studied. The preservation effect of 2% CH/LS film on grape berry was then studied further and compared with PE film (Control) and Chitosan film (CH). The differences in metabolites in Cabernet Sauvignon, Chardonnay and Italian Riesling wines at the eastern foot of Helan

Mountain in Ningxia were also studied to determine the biomarkers that contribute to the differences, so as to provide the cornerstone for wine quality control, variety identification and protection, Additionally, we also provide a scientific theoretical basis for consumers to choose high quality wine.

2. Materials and Methods

2.1. Reagents and Equipment

Chitosan($C_6H_{11}NO_4$, Average molecular weight (MW): 150 kDa; Degree of deacetylation $\geq$ 90%) were purchased from Shanghai Lanji Biological Technology Co., Ltd. (Shanghai, China); Sodium-Lignosulfonate (Content $\geq$ 98%) were purchased from Hefei BOSF Biotechnology Co., Ltd. (Hefei, China); Ascorbic acid (Vc) was obtained from Sinopharm Group Chemical Reagent Co., Ltd. (Shanghai, China); Oxalic acid, Sodium oxalate were produced by Su Yi Chemical Reagent Co., Ltd., Shanghai, China; DSS was produced by Qingdao Tenglong Microwave Technology Co., Ltd. (Qingdao, China); Heavy water (D_2O deuterium degree > 99.9%) is from Tenglong Weibo Technology Co., Ltd., Qingdao, China.

SNL315SV-230 Freeze dryer was produced by Termo Co., Ltd., Waltham, MA, USA; AVANCE 600 Nuclear magnetic resonance spectrometer was supplied by Bruker Co., Ltd., Karlsruhe, German; TG16A-WS Desktop high-speed centrifuge was produced by Lu Xiangyi Centrifuge Instrument Co., Ltd., Shanghai, China; BS-224 Electronic balance was produced by Eppendorf Co., Ltd., Hamburg, German; ULT178-6-V49 Ultra low temperature freezer was supplied by Revco Co., Ltd., Waltham, MA, USA; RE52-4 Rotary evaporator was produced by Huxi Analysis Instrument Co., Ltd., Shanghai, China; XW-80A Swirl mixer was supplied by Huxi Analysis Instrument Co., Ltd., Shanghai, China.

2.2. Methods

2.2.1. Preparation Method CH/LS Bio-Composite Film

LS powder (1 g) was dissolved in 50 mL of distilled water at 5000 rpm for 15 min. CH powder (1 g) was dissolved in 50 mL of distilled water at 5000 rpm for 15 min at 25 °C. Then, VC (1 g) and chitosan was added at 5000 rpm for 15 min by stirring continuously. Finally, the CH/LS film-forming solutions were degassed to remove air bubbles [27]. The 20 mL CH/LS film-forming solution was cast on a 15 cm in diameter flat glass Petri dish (B-SLPYM90, BKMAM Biological Co., Ltd., Changsha, China) and dried in an oven at 35 °C for 24 h until the surface of CH/LS film remained certain firm and viscous. Finally, the CH/LS film was stored in a small desiccator at 50% relative humidity for 20 min at 25 °C. Peel the CH/LS film from the Petri dish for analysis. The prepared films were, respectively added with 0% LS + 2% CH, 1% LS + 2% CH, 2% LS + 2% CH, so they were named as CH/LS, 1% CH/LS, 2% CH/LS, respectively [28].

2.2.2. Determination Method of CH/LS Film on the Preservation Effect of Grape Berry

Seasonal fresh grapes (Cabernet Sauvignon variety, from Ningxia, China) were tested in research packaging material experiments. Grapes of similar size and quality were tightly wrapped with the CH/LS (2%) and CH (2%) films and incubated at a constant temperature of 18 °C in an incubator with relative humidity of 68% to observe the surface changes, compared with those wrapped in polyethylene (20 cm × 20 cm) [29].

2.2.3. Determination of Decay Rates

Decay rates refer to the ratio of the weight of rotten fruit to the total weight of treated fruit [30]. The main observations related to whether the appearance of the fruit is full, whether it has edible value, and whether it is corrupted bacterial infection and deterioration. The decay rates of grapes were then calculated according to Equation (1):

Decay rates (%) = number of rotten fruits (N)/total number of fruits (N0) × 100%   (1)

2.2.4. Hardness

The hardness was checked using a fruit hardness tester (GY-1, JC Group Co., Ltd., Qingdao, China). Before measurement, the hardness tester was placed perpendicular to the surface of the grapes (without breaking the skin), the indenter was pressed into the fruit evenly, and the reading indicated when the pointer stops moving was the hardness value of the fruit [31].

2.2.5. Weight Loss Rate

The weight loss rate was measured using the weighing method [32]. The weight of the fruit on the 0th to the 16th day was measured and the average value was taken to calculate the weight loss rate.

The formula for calculating the weight loss rate content is as Equation (2):

$$\text{Weight loss rate (\%)} = \text{fruit weight (m) g/fruit original weight (m0) g} \times 100\% \text{ in a single measurement.} \quad (2)$$

2.2.6. Soluble Solids

Soluble solids refer to sugars, acids, vitamins and minerals that are soluble in water in fruit juice. After the grapes were broken and homogenized, soluble solids were determined with a PAL-1 hand-held sugar calorimeter [33]; the unit was %. Each group was tested 3 times in parallel.

2.2.7. Titratable Acid Content

The content of titratable acid was determined by acid-base titration. First, 0.5 g of a grapes sample was weighed, ground thoroughly, and then transferred to a 100 mL volumetric flask, where distilled water was added to the mark and thixotropic. Then it was filtered with filter paper and we accurately drew 20 mL of the filtrate into a 100 mL conical flask and added 2 drops of 1% phenolphthalein indicator. Using calibrated 0.01 m mol/L sodium hydroxide to titrate to pH 8.0 as the end point, we recorded the sodium hydroxide consumption. We repeated this 3 times to obtain the average value [34]. The formula for calculating the titratable acid content is as follows:

$$\text{Total acidity (\%)} = V \times C \times N \times \text{Conversion factor} \times 100/(W \times V1) \quad (3)$$

V: The total volume of sample dilution (mL); V1: the volume of the sampling solution during titration (mL);
C: The number of milliliters of sodium hydroxide standard solution consumed;
N: The molar concentration of sodium hydroxide standard solution;
W: Sample weight (g);
Conversion factor: Tartaric acid—0.075.

2.3. Determination of Metabolites in Wine

The Wine Sample

The grapes and wines used in this experiment were provided by Ningxia (Yinchuan) Helan Mountain Grape Wine Co., Ltd. (Yinchuan, China) and all the single varieties of wine in 2020 were made by standard brewing techniques; the same brewer's yeast (Lalvin CY 3079) was added during the brewing process for fermentation. The physical and chemical indicators of product quality are in compliance with the national standard GB15037-2006 [35]. The samples were stored at −4 °C for later use.

2.4. NMR Spectroscopic Analysis

2.4.1. Pre-Treatment of Wine Samples

After freeze-drying, the sample was treated with buffer solution: First, 10 mL of wine, was centrifuged at 4000 rpm for 10 min at −4 °C; 3 mL supernatant was then placed into a 20 mL lyophilized bottle and frozen overnight at −70 °C. After, it was then frozen in a freeze dryer for 48 h. Next, 400 μL of 0.2 mol/L oxalate buffer was added with pH = 4 prepared by D_2O, 140 μL of D_2O and 60 μL of 0.75% DSS, mixed well and centrifuged at 13,000 rpm for 10 min. Finally, 500 μL of the supernatant was taken and loaded into a 5 mm nuclear magnet tube; NMR experiments were carried out.

2.4.2. NMR Experimental Data Collection

^{1}H-NMR spectra of wine samples were collected by AVANCE 600 NMR spectrometer. The NMR experiment was set to a constant temperature of 298 K, the ^{1}H-NMR operating frequency was 600.23 MHz, and the spectral width was 7183.9 Hz. The Noesygpprld sequence was used to suppress the water peak signal, and all the samples were scanned 256 times.

2.5. *Statistical Analysis*

Statistical analysis was performed by Microsoft Excel 2011 and one-way analysis of variance (ANOVA) using SPSS Statistics 23.0 software. The confidence level was 95% and $p < 0.05$ was considered statistically significant. The experimental data shown in all of the results were repeated at least three times.

NMR Spectral Data Processing

After the sampling was completed, Fourier transform was performed, and the phase adjustment and baseline correction of the spectrum were performed; the spectral peaks were assigned according to the chemical shift. Using AMIX software, the spectrum was integrated into the chemical shift interval δ 0.5–10.0 ppm in the 0.005 ppm integration section; −0.5~0.5 ppm, 1.74~1.84 ppm and 2.90~2.95 ppm DSS peaks, 1.18~1.22 ppm and 3.57–3.72 ppm residual ethanol peaks, 4.8~4.96 ppm residual water peaks were not integrated. After the NMR data was normalized, it was imported into SIMCA-P 12.0 software for pattern recognition analysis.

3. Results and Discussion

3.1. The Effect of CH/LS Film on the Preservation of Grapes

3.1.1. Determination of Decay Rates

Grapes still have vigorous life activities after picking, and a series of physiological and biochemical changes occur in the fruit tissue after picking, resulting in fruit rot. The effects of different preservation treatments on the decay rates of grapes are shown in Figure 1. It can be seen that the control, CH and 2% CH/LS film packaging had no significant effect on the decay rate of grapes at the early stage of storage. After 2% CH/LS film, CH, and PE (control) packaging, the decay rates of grapes were 18.36%, 21.63%, 31.71%, respectively. However, with the extension of storage time, the decay rates of grapes demonstrated an increase, and the difference in preservation effect of the three groups was also completely different. After CH and 2% CH/LS coating film packaging, the decay rates of grapes decreased by 10.08% and 13.35%, respectively, compared with the control group. The grapes packaged in control film showed severe spoilage after 10 days of storage at 18 °C, while the grapes packaged with chitosan film and 2% CH/LS film suffered severe spoilage on the 14th and 16th days, respectively.

3.1.2. Hardness

Fruit hardness is an important indicator for measuring fruit quality and reflecting fruit ripening and senescence [36]. The effects of different treatments on the firmness of grapes are shown in Figure 2. The hardness of the grapes in each treatment group showed a downward trend, and the fruit hardness of the control group decreased most significantly: the hardness decreased by 13.05, during the 16th day of the experimental period. The fruit hardness of the experimental group decreased slowly and was significantly different from the control group ($p < 0.05$). Among them, the 2% CH/LS treatment group had the slowest decline: during the 16-day experimental period, the hardness decreased by only 6. The addition of lignosulfonate improved the coating effect of chitosan to a certain extent [37].The plump appearance of the fruit was maintained; the hardness of grapes treated with CH or 2% CH/LS composite film coating was higher than that of the control group, and the effect of maintaining the hardness was obvious, and there was no significant difference between the two ($p > 0.05$), indicating that the composite film solution is beneficial for grapes to maintain fruit firmness, delay fruit softening and rot, and prolong the storage period.

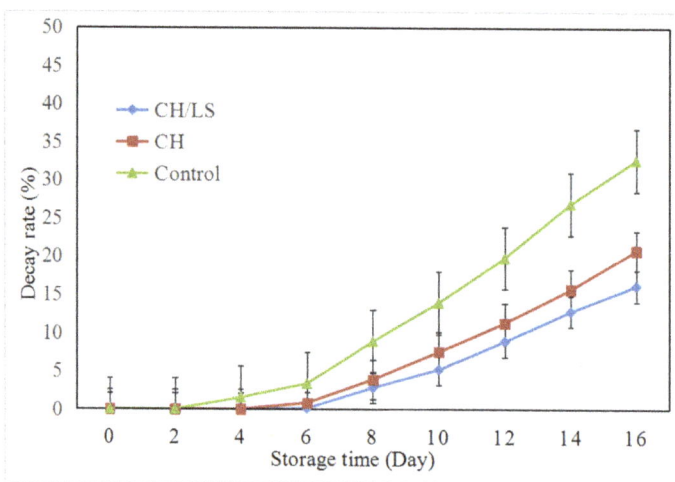

Figure 1. Decay rates of grapes during preservation with packaging films.

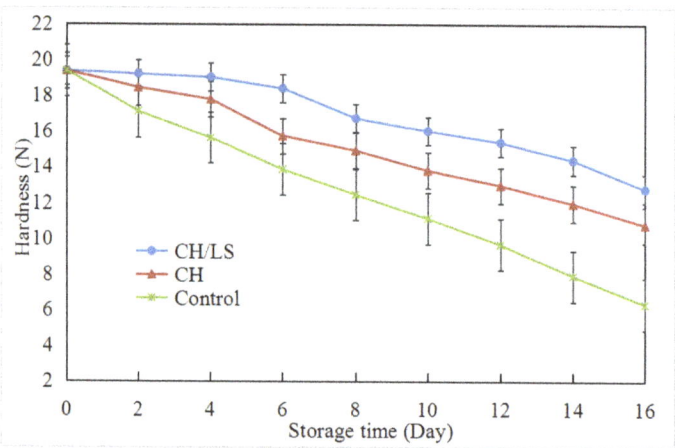

Figure 2. Hardness of grapes during preservation with different packaging films.

3.1.3. Weight Loss Rate

It is generally believed that respiration and transpiration are the main reasons for fruit weight loss. The effect of different coating treatments on the weight loss rate of grapes is shown in Figure 3. With the prolongation of storage time, the weight loss rates of grapes in CH, 2% CH/LS and control groups reached 25.49%, 21.37% and 23.24%, respectively. The weight loss rate of the CH group was higher than the control. The reason for this may be that CH film packaging has a large water vapor transmission coefficient, leading to a high water loss rate [38]. Compared with CH, 2% CH/LS film packaging can more effectively reduce the rapid loss of moisture and organic matter in grapes. The reason may be that

LS is hygroscopic, forming a barrier between the fruit epidermis and the surrounding environment to prevent gas exchange, inhibited the respiration and transpiration of grapes, reduced the water loss of grapes, and the nutrient decomposition was slower, so as to reduce the weight loss rate [39]. Although the weight loss rate of different treatment groups increased, 2% CH/LS was significantly lower than that of the rest of the two groups, the water loss of grapes was reduced, and the nutrient decomposition was slower than that of the control group.

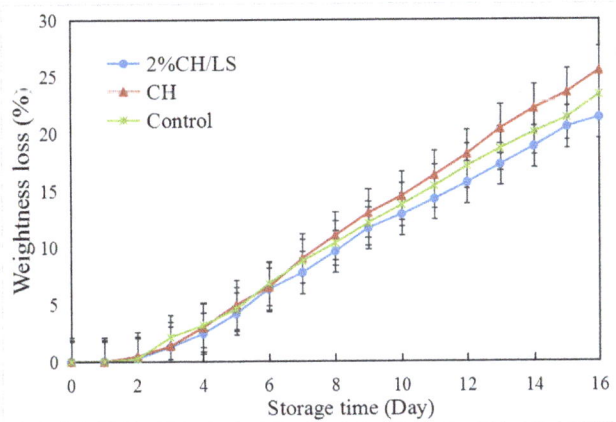

Figure 3. Weight loss rate of grapes during preservation with different packaging films.

3.1.4. Soluble Solids

The effects of different coating treatments on the soluble solids content in grapes are shown in Figure 4. As grapes respiration proceeds, there is no external source of nutrients, resulting in a decrease in the soluble solids content of grapes [40]. The changing trend of soluble solid content among different components is the same. The soluble solid content in the fruit of the control group decreased from 5.92% on 0 d to 2.14% on the 16th day. At the end of storage for 16 days, the 2% CH/LS coating film packaging group and the CH coating film packaging group had a better preservation effect, and the soluble solid content was about 3.05% and 3.28%, respectively. The loss of soluble solid content was the highest in the control group (3.78%), while the loss was 2.87% and 2.64% after CH and 2% CH/LS coating treatments. On the 6th day of storage, the soluble solids in the control group decreased rapidly. The reason may be that PE film group had poor water permeability, accumulated more water on the surface, and had a large number of microbial breeding [41], which accelerated the spoilage and deterioration of grapes. Therefore, the content of PE film group decreased steadily in the early stage, and rapidly decreased to 2.1% in the later stage. The soluble solid content of CH coating group is slightly higher than that of 2% CH/LS packaging, which may be due to that the 2% CH/LS composite film has good water retention and bacteriostasis, which can not only reduce the water loss of grapes but also inhibit the reproduction of spoilage bacteria [42]. However, the CH film has a large water vapor transmission coefficient and a fast water loss rate, leading to a slightly higher soluble solid content [43]. It indicated that CH coating and 2% CH/LS coating could effectively reduce the loss of soluble solids in grapes.

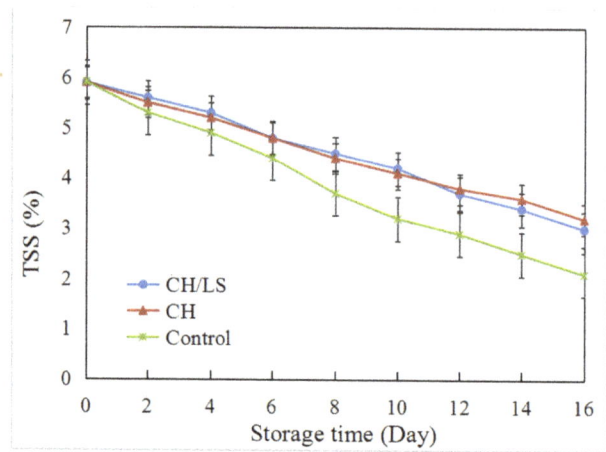

Figure 4. The Soluble solids of grapes during preservation with different packaging films.

3.1.5. Titratable Acid Content

The content of titratable acid is related to the color, aroma, taste and stability of the fruit [44]. With the prolongation of storage time, the titratable acid content of fruits showed a downward trend. Generally speaking, grapes with lower maturity will continue their physiological activities after picking, and the titratable acid may rise first and then fall [45]. However, mature grapes were selected in this study, so the titratable acid showed a continuous downward trend [46]. Figure 5 shows that the reduction rate of titratable acid of grapes coated with 2% CH/LS decreased slowly than the other two groups; during the 16th day of the experimental period, the titratable acid content decreased by only 1.39, and its content was 3.32 g/kg. In comparison, the control group decreased by 2.3, and its content was only 2.41 g/kg. CH and 2% CH/LS film-coated packaging groups decreased significantly lower than the control group ($p < 0.05$).

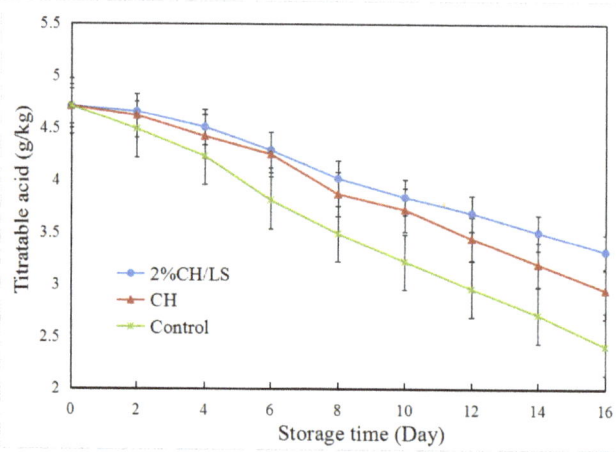

Figure 5. Titratable acid of grapes during preservation with different packaging films.

3.2. Identification of Wine Metabolites from Helan Mountain in Ningxia

Figure 6 shows the ^{1}H-NMR spectra of Cabernet Sauvignon dry red wine, Chardonnay and Italian Riesling dry white wine of Helan Mountain in Ningxia. The metabolite displacement information in wine is combined with Figure 6 and references the relevant literature [47,48]. The results are shown in Table 1.

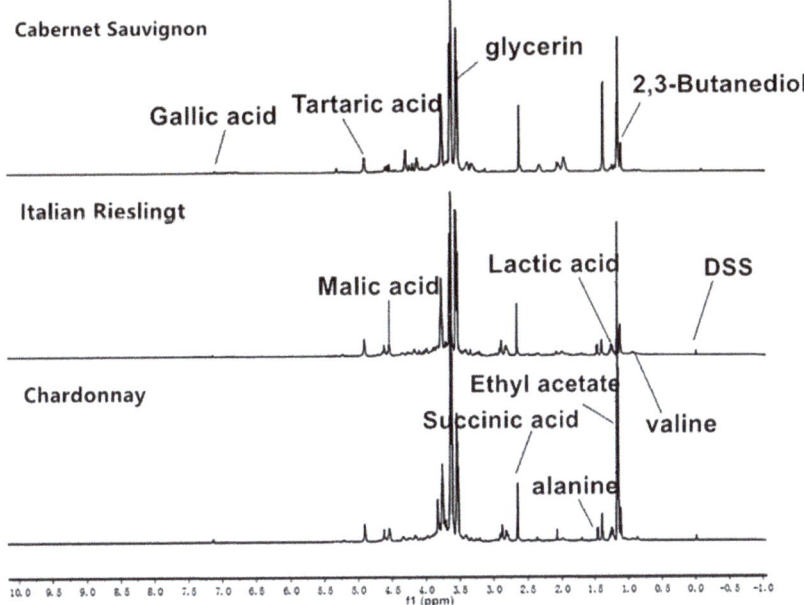

Figure 6. ^{1}H-NMR spectra of wine metabolites from Helan Mountain in Ningxia.

Table 1. ^{1}H-NMR assignment of metabolites in wines.

Keys	Metabolites	^{1}H-NMR Chemical Shift
1	Valine	0.87 (d, C4H$_3$), 1.03 (d, C5H$_3$)
2	Ethanol	1.18 (t, C2H$_3$), 3.64 (q, C1H$_2$)
3	2,3-Butanediol	1.16 (d, C1H$_3$ + C4H$_3$)
4	Succinic acid	2.64 (s, C2H$_2$ + C3H$_2$)
5	Proline	2.00 (m, u, γ-CH$_2$), 2.07 (m, u, β-CH), 2.35 (m, u, β'-CH), 3.35 (m, u, δ-CH), 3.42 (m, u, δ-CH), 4.16 (m, u, α-CH)
6	Ethyl acetate	1.26 (t, C4H$_3$), 4.16 (q, C3H$_2$)
7	Tartaric acid	4.53 (s, C2H + C3H)
8	α-Glucose	5.23 (d, αC1H)
9	β-Glucose	4.61 (d, βC1H)
10	Lactic acid	1.36 (d, C3H$_3$), 4.28 (m, C2H)
11	Gallic acid	7.14 (s, C2H + C6H)
12	Glycerol	3.56 (q, C2H$_2$), 3.65 (q, C3H$_2$), 3.81 (m, C1H)
13	α-D-Glucuronic acid	5.34 (d, C1H)
14	γ-Aminobutyric acid	2.50 (t, α-CH$_2$), 1.96 (m, β-CH$_2$), 3.05 (t, γ-CH$_2$)
15	Malic acid	2.73 (dd, βCH$_2$), 2.86 (dd, β'CH$_2$), 4.46 (q, CH)
16	Alanine	1.51 (d, βCH$_3$)
17	D-Sucrose	5.43 (d, C1H), 3.55 (dd, C2H), 3.72 (dd, C3H), 3.90 (dd, C4H), 4.215 (d, C1'H), 4.05 (dd, C2'H), 3.88 (dd, C3'H)

The characters in brackets refer to peak information: s, singlet; d, doublet; t, triplet; q, quartet; dd, doublet of doublets; m, multiple.

Because ^{1}H-NMR detection hardly needs sample pre-treatment, the inherent properties of the sample are well preserved. Using pattern recognition analysis combined with the chemical shift of nuclear magnetic spectrum, the characteristic variables with a large contribution to the difference between samples can be obtained, so as to identify the metabolites causing the difference between samples. The metabolites in the ^{1}H-NMR spectra of wine were identified, and these substances mainly included amino acids, organic acids, sugars, phenols and so on. It can be found that the composition of metabolites in these three wines is basically the same, which means that the composition of metabolites in wines is relatively stable, but there are differences in the content of metabolites between different types of wines. Each wine has its own fingerprint, and different types of wine metabolic profiles describe their physiological and biochemical states, which need to be processed to find their markers [49]. The positions of the characteristic peaks in the NMR spectra correspond to different types of metabolites in wine, and the peak intensity (such as area) represents the relative content of the corresponding metabolites.

3.3. Differences of Metabolites in Different Wine Varieties from Helan Mountain

In this experiment, the metabolite map data of three wines in the Helan Mountain production area of Ningxia were compared with the PLS-DA model to determine the main metabolites causing the difference between wine varieties. Firstly, Chardonnay and Italian Riesling dry white wine were analyzed by PLS-DA. Then, the pair PLS-DA comparison of dry white and dry red wine was carried out to determine the metabolites causing the difference between dry red and dry white wine varieties.

3.3.1. Analysis of Metabolites of Chardonnay and Italian Riesling Dry White

Two pairs of PLS-DA of dry white wine were compared to determine the main metabolites causing the difference between dry white wine varieties. The PLS-DA model of the 2020 Chardonnay and Italian Riesling Dry white wine is shown in Figure 7. In the score graph, Chardonnay and Italian Riesling dry white wine are clearly distinguished on the PC1 axis, and the cumulative contribution rate is $R^2X = 0.965$, $R^2Y = 0.994$ and $Q^2 = 0.955$, indicating that this model is effective. The validation diagram of the model in the permutation experiment further demonstrates the reliability and predictability of the model. It can be seen from the load diagram that compared with Chardonnay dry white wine, the 2,3-butanediol, lactic acid, succinic acid, glycerin, choline, tartaric acid, D-sucrose, and γ-aminobutyric acid content is relatively high in Italian Riesling Dry white wine, while the content of gallic acid, ethyl acetate, proline, malic acid, alanine, α-glucose and β-gluconic acid is relatively low.

In order to ensure the unique taste of dry white wine, the fermentation process of malic acid and lactic acid is properly controlled during the brewing of dry white wine. Malic acid plays an important physiological role in the human body. It can effectively improve the body's exercise ability, resist fatigue, accelerate the metabolism of carboxylate, protect the heart, improve memory, etc. [50]. From the perspective of organic acids, it can be considered that dry white wine has a relatively high protective effect on the body.

3.3.2. Analysis of Differences of Metabolites between Cabernet Sauvignon Dry Red Wine and Italian Riesling, Chardonnay Dry White Wine

Pairwise comparison PLS-DA were compared between dry white and dry red wines to determine the metabolites that caused the difference between the two wine varieties. Figure 8 shows the PLS-DA model of Cabernet Sauvignon dry red wine and Italian Riesling Dry white wine in 2020. In the score chart, the two wines are clearly distinguished on the PC1 axis, where the cumulative contribution rate is $R^2X = 0.76$, $R^2Y = 0.987$ and $Q^2 = 0.969$, indicating that the quality of this model is good. The validation diagram of the permutation experiment of this model once again shows the reliability and predictability of this model. It can be seen from the load diagram that compared with Cabernet Sauvignon dry white wine, Italian Riesling dry white wine has higher alanine and malic acid content, while 2,3-butanediol, glycerol, choline, lactic acid, valine, proline, ethyl acetate, succinic acid, tartaric acid, gallic acid, α-D-Glucuronic acid is low.

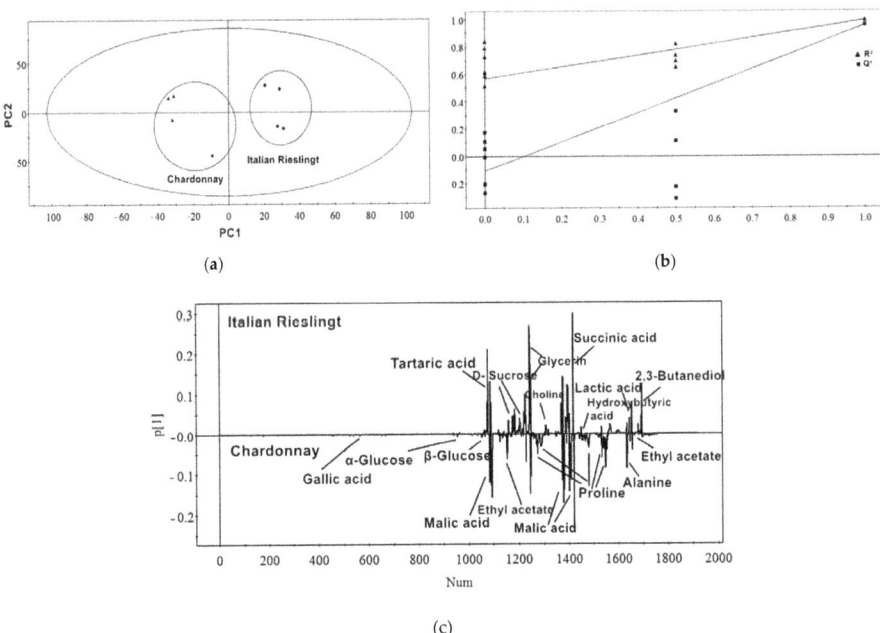

Figure 7. (a) PLS-DA model derived from the ^{1}H-NMR spectra of Chardonnay and Italian Riesling dry white wine. PLS-DA scores plot; (b) PLS-DA cross-validation plot; (c) PLS-DA loading plot.

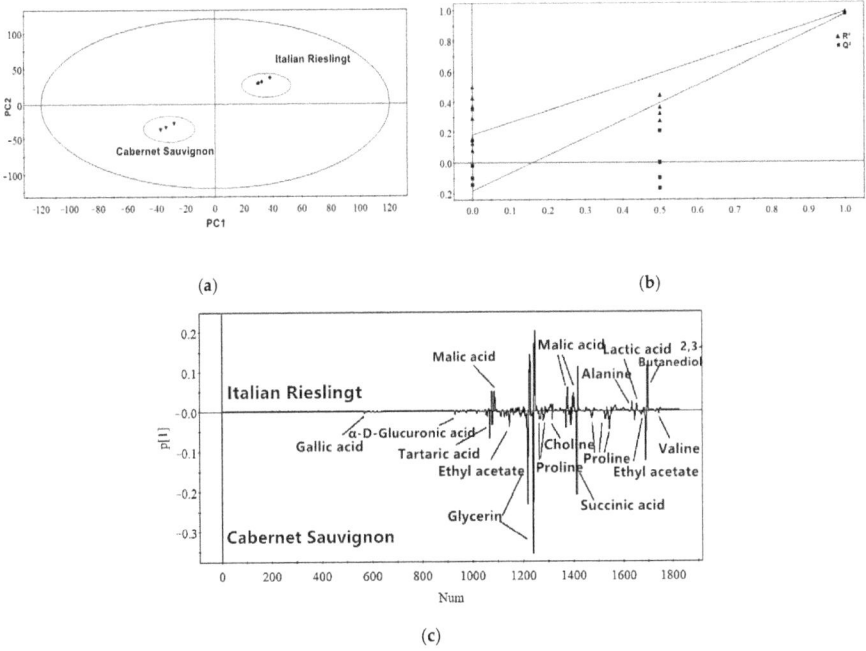

Figure 8. (a) PLS-DA model derived from the ^{1}H-NMR spectra of Cabernet Sauvignon dry red wine and Italian Riesling dry white wine. PLS-DA scores plot; (b) PLS-DA cross-validation plot; (c) PLS-DA loading plot.

The metabolites of amino acid characteristic differences among wines detected in this experiment are valine, alanine and proline. Among them, valine and alanine contribute less to the difference between wines, while proline contributes more to the difference between different varieties of wines. Song et al. [51] also recognized that the content of proline in wine is affected by environmental factors and different varieties of wine grape berries.

The PLS-DA model of 2020 Cabernet Sauvignon and Chardonnay is shown in Figure 9. In the score graph, the two wines are significantly different on the PC1 axis, where the cumulative contribution rate $R^2X = 0.798$, $R^2Y = 0.987$ and $Q^2 = 0.979$ are relatively high, which also indicates that the model established is effective. The validation diagram of the permutation experiment further demonstrates the reliability and predictability of the model. As can be seen from the load diagram, Chardonnay dry white wine compared with Cabernet sauvignon dry red wine, Chardonnay dry white wine of alanine, malic acid content is higher, and 2,3-butanediol, valine, choline, glycerin, tartaric acid, lactic acid, valine, proline, ethyl acetate, succinic acid, gallic acid, α-D-Glucuronic acid content is low.

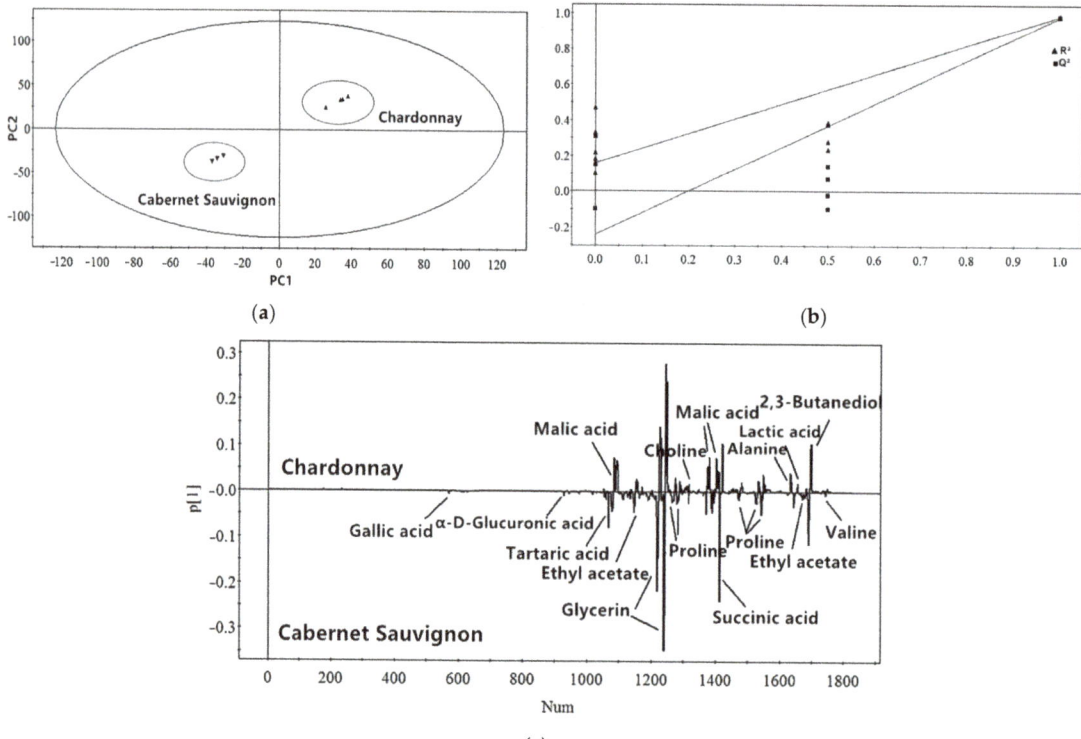

Figure 9. (a) PLS-DA model derived from the ^{1}H-NMR spectra of Cabernet Sauvignon dry red wine and Chardonnay dry white wine. PLS-DA scores plot; (b) PLS-DA cross-validation plot; (c) PLS-DA loading plot.

As an important flavor substance of wine, organic acids not only determine the quality of wine [52], but also regulate the acid-base balance in the body, enabling the physiological activities of enzymes to be realized [53]. In this experiment, the different metabolites of organic acids were tartaric acid, malic acid, succinic acid and lactic acid. Tartaric acid and malic acid are derived from grape berries, while lactic acid and succinic acid are derived from wine fermentation [54].

Based on PLS-DA analysis of Cabernet Sauvignon, Chardonnay and Italian Riesling of the Helan Mountain region of Ningxia in 2020, it was found that the components of metabolites of different varieties of wine had little difference, but the content of metabolites had great difference. The contents of 2,3-butanediol, ethyl acetate, proline, succinic acid, tartaric acid, lactic acid, glycerin, gallic acid, choline, valine and α-D-Glucuronic acid in Cabernet Sauvignon dry red wine were higher than dry white wine. Ethyl acetate is an aromatic substance abundant in wine, which brings rich aroma to wine. Its content is also related to fermentation technology, grape varieties, fermentation temperature, etc. [55]. Ethyl acetate for wine varieties in this experiment and region identification provides a larger contribution, three different varieties of wine, the ethyl acetate content of Cabernet sauvignon is highest. The contents of malic acid, alanine and γ-aminobutyric acid in Chardonnay and Italian Riesling Dry white wine are higher than Cabernet Sauvignon dry red wine, and the contents of other nutritional metabolites are relatively low, indicating that there are significant differences in nutritional metabolites among different grape varieties. Thus, a quantitative analysis of the main metabolites was carried out, as shown in Table 2. The content of metabolites can be converted by the ratio of the peak area caused by protons on a specified group of the substance to be measured in the ^{1}H-NMR spectrum to the peak area caused by protons on the specified group of the added internal standard DSS. The results were consistent with the results of PLS-DA analysis by comparing the contents of major metabolites by means of multiple samples (SNK method). The highest content of gallic acid was found in Cabernet Sauvignon. The content of alanine and malic acid is the highest in Chardonnay wine, which makes the wine made with different styles, guiding consumers to choose the appropriate wine according to their own needs and improving their health.

Table 2. Content of main metabolites in Cabernet Sauvignon dry red wine, Chardonnay, Italian Riesling dry white wine (g/L).

Metabolites	Cabernet Sauvignon	Chardonnay	Italian Riesling
Ethyl acetate	1.52 ± 0.03 b	0.92 ± 0.01 c	0.32 ± 0.01 d
Lactic acid	0.51 ± 0.03 b	0.22 ± 0.02 e	0.32 ± 0.02 d
Alanine	0.04 ± 0.01 d	0.25 ± 0.02 a	0.14 ± 0.01 b
Succinic acid	1.31 ± 0.02 b	0.46 ± 0.02 d	0.47 ± 0.01 d
Proline	3.23 ± 0.05 c	1.50 ± 0.08 d	0.49 ± 0.02 e
Malic acid	4.95 ± 0.17 c	6.08 ± 0.28 a	5.37 ± 0.02 b
Choline	0.06 ± 0.01 a	0.04 ± 0.01 b	0.05 ± 0.01 b
Glycerol	13.99 ± 0.09 b	8.92 ± 0.16 e	9.89 ± 0.19 d
Gallic acid	0.14 ± 0.01 a	0.03 ± 0.00 c	0.01 ± 0.00 c

Means followed with different letters are statistically different at the 0.05 probability level with an AVOVA-protected SNK 0.05 test.

4. Conclusions

In this study, lignosulfonate (an underutilized renewable biomass resource) was added to chitosan solution, and a separate 2% chitosan film (CH), chitosan—1% sodium lignosulfonate film (1% CH/LS) and chitosan—2% sodium lignosulfonate film (2% CH/LS) biomass composite films were developed by the classical casting method, to further study the preservation effect of 2% CH/LS film and CH film on grapes. From the test results, it can be concluded that 2% CH/LS films showed the best preservation performance. Compared with the control, CH film and 2% CH/LS coating film packaging not only effectively inhibit the evaporation of water and slow down the weight loss rate of grapes during storage, effectively alleviating the hardness, soluble solids, and titratable acid of grapes and causing the decrease of other nutrients, but they also effectively extend the shelf life of grapes. Therefore, compared with ordinary plastic packaging, 2% CH/LS film packaging is one of the promising strategies for preservation, as well as the Chitosan-lignosulfonate composite coating film. Both will be further investigated for various fruit and vegetable preservation. At the same time, this paper, based on the method of ^{1}H-NMR combined with pattern recognition analysis, analyzed the differences in metabolites in Cabernet Sauvignon dry red wine, Chardonnay and Italian Riesling dry white wine of Helan Mountain in Ningxia. As far as the research results are concerned, the types of metabolites of the three wines are similar, but their content is significantly different. Among them, Chardonnay wine has a more refreshing and delicate taste and more complete flavor, while Cabernet Sauvignon wine has the highest biological activity and health function. This provides theoretical and technical support for the quality control and evaluation of the origin of wine, the identification and protection of varieties, and provides a scientific guide for consumers to aid in the prevention of diseases and the maintenance of human health.

Author Contributions: Writing—original draft preparation, L.L.; writing—review and editing, B.H. and X.Z.; designed research, Y.F. and L.L.; validation, M.Z.; analyzed data, M.Z. and L.L.; supervision, X.Z. All authors have read and agreed to the published version of the manuscript.

Funding: This research was funded by National Natural Science Foundation of China "Wine Metabolomics and NMR Fingerprint Stud, grant number NO. 31271857".

Institutional Review Board Statement: Not applicable.

Informed Consent Statement: Not applicable.

Data Availability Statement: All raw data are available at the corresponding author.

Acknowledgments: We acknowledge financial support by the National Natural Science Foundation of China and President Zhou's Laboratory for its help.

Conflicts of Interest: The authors declare no conflict of interest.

References

1. Porat, R.; Lichter, A.; Terry, L.A.; Harker, R.; Buzby, J. Postharvest losses of fruit and vegetables during retail and in consumers' homes: Quantifications, causes, and means of prevention. *Postharvest Biol. Technol.* **2018**, *139*, 135–149. [CrossRef]
2. Del Nobile, M.A.; Sinigaglia, M.; Conte, A.; Speranza, B.; Scrocco, C.; Brescia, I.; Bevilacqua, A.; Laverse, J.; La Notte, E.; Antonacci, D. Influence of postharvest treatments and film permeability on quality decay kinetics of minimally processed grapes. *Postharvest Biol. Technol.* **2008**, *47*, 389–396. [CrossRef]
3. Fang, Y.; Wakisaka, M. A Review on the Modified Atmosphere Preservation of Fruits and Vegetables with Cutting-Edge Technologies. *Agriculture* **2021**, *11*, 992. [CrossRef]
4. Morata, A.; Loira, I.; Vejarano, R.; González, C.; Callejo, M.J.; Suárez-Lepe, J.A. Emerging preservation technologies in grapes for winemaking. *Trends Food Sci. Technol.* **2017**, *67*, 36–43. [CrossRef]
5. Auras, R.; Singh, S.P.; Singh, J. Performance evaluation of PLA against existing PET and PS containers. *J. Test. Eval.* **2006**, *34*, 530–536. [CrossRef]
6. Masmoudi, F.; Alix, S.; Buet, S.; Mehri, A.; Bessadok, A.; Jaziri, M.; Ammar, E. Design and Characterization of a New Food Packaging Material by Recycling Blends Virgin and Recovered polyethylene terephthalate. *Polym. Eng. Sci.* **2020**, *60*, 250–256. [CrossRef]
7. Aquino, A.D.; Blank, A.F.; Cristina, L.D.A.S.L. Impact of edible chitosan-cassava starch coatings enriched with Lippia gracilis Schauer genotype mixtures on the shelf life of guavas (*Psidium guajava* L.) during storage at room temperature. *Food Chem.* **2015**, *171*, 108–116. [CrossRef]
8. Baris Karsli, Emre Caglak, Witoon Prinyawiwatkul, Effect of high molecular weight chitosan coating on quality and shelf life of refrigerated channel catfish fillets. *LWT* **2021**, *142*, 111034. [CrossRef]
9. Călinoiu, L.-F.; Ştefănescu, B.E.; Pop, I.D.; Muntean, L.; Vodnar, D.C. Chitosan Coating Applications in Probiotic Microencapsulation. *Coatings* **2019**, *9*, 194. [CrossRef]
10. Youwei, Y.; Yinzhe, R. Grape Preservation Using Chitosan Combined with β-Cyclodextrin. *Int. J. Agron.* **2013**, *2013*, 1–8. [CrossRef]
11. Bilbao-Sainz, C.; Chiou, B.S.; Williams, T.; Wood, D.; Du, W.X.; Sedej, I.; Ban, Z.; Rodov, V.; Poverenov, E.; Vinokur, Y.; et al. Vitamin D-fortified chitosan films from mushroom waste. *Carbohydr. Polym.* **2017**, *167*, 97–104. [CrossRef] [PubMed]
12. Yousuf, B.; Qadri, O.S.; Srivastava, A.K. Recent developments in shelflife extension of fresh-cut fruits and vegetables by application of different edible coatings: A review. *LWT-Food Sci. Technol.* **2018**, *89*, 198–209. [CrossRef]
13. Sameni, J.; Krigstin, S.; Sain, M. Solubility of Lignin and Acetylated Lignin in Organic Solvents. *BioResources* **2017**, *12*, 1548–1565. [CrossRef]
14. Atmajayanti, A.T.; Hung, C.C.; Yuen, T.Y.P.; Shih, R.C. Influences of Sodium Lignosulfonate and High-Volume Fly Ash on Setting Time and Hardened State Properties of Engineered Cementitious Composites. *Materials* **2021**, *14*, 4779. [CrossRef]
15. OIV. *International Code of Oenological Practices*; MAPA: Paris, France, 2006.

16. Mudnic, I.; Modun, D.; Rastija, V.; Vukovic, J.; Brizic, I.; Katalinic, V.; Kozina, B.; Medic-Saric, M.; Boban, M. Antioxidative and vasodilatory effects of phenolic acids in wine. *Food Chem.* **2009**, *119*, 1205–1210. [CrossRef]
17. Hubner, A.; Sobreira, F.; Vetore Neto, A.; Pinto, C.A.; Dario, M.F.; Díaz, I.E.; Lourenço, F.R.; Rosado, C.; Baby, A.R.; Bacchi, E.M. The Synergistic Behavior of Antioxidant Phenolic Compounds Obtained from Winemaking Waste's Valorization, Increased the Efficacy of a Sunscreen System. *Antioxidants* **2019**, *8*, 530. [CrossRef]
18. Puškaš, V.; Jović, S.; Antov, M.; Tumbas, V. Antioxidative activity of red wine with the in-creased share of phenolic compounds from solid parts of grape. *Chem. Ind. Chem. Eng. Q.* **2010**, *16*, 65–71. [CrossRef]
19. Jin, L.; Huang, Z.; Li, Y.; Sun, Z.; Li, H.; Zhang, J. On Modified Multi-Output Chebyshev-Polynomial Feed-Forward Neural Network for Pattern Classification of Wine Regions. *IEEE Access.* **2018**, *7*, 1973–1980. [CrossRef]
20. Williamson, P.O.; Robichaud, J.; Francis, I.L. Comparison of Chinese and Australian consumers' liking responses for red wines. *Aust. J. Grape Wine Res.* **2012**, *18*, 256–267. [CrossRef]
21. Ricardo-Rodrigues, S.; Laranjo, M.; Coelho, R.; Martins, P.; Rato, A.E.; Vaz, M.; Valverde, P.; Shahidian, S.; Véstia, J.; Agulheiro-Santos, A.C. Terroir Influence on Quality of 'Crimson' Table Grapes. *Sci. Hortic.* **2019**, *245*, 244–249. [CrossRef]
22. Hu, B.; Gao, J.; Xu, S.; Zhu, J.; Fan, X.; Zhou, X. Quality Evaluation of Different Varieties of Dry Red Wine Based on Nuclear Magnetic Resonance Metabolomics. *Appl. Biol. Chem.* **2020**, *64*, 24. [CrossRef]
23. Shimizu, H.; Akamatsu, F.; Kamada, A.; Koyama, K.; Okuda, M.; Fukuda, H.; Iwashita, K.; Goto-Yamamoto, N. Discrimination of wine from grape cultivated in Japan, imported wine, and others by multi-elemental analysis. *J. Biosci. Bioeng.* **2018**, *125*, 413–418. [CrossRef] [PubMed]
24. Da Costa, G.; Gougeon, L.; Guyon, F.; Richard, T. ^{1}H NMR metabolomics applied to Bordeaux red wines. *Food Chem.* **2019**, *301*, 125257.
25. Trygg, J.; Holmes, E.; Lundstedt, T. Chemometrics in metabonomics. *J. Proteome Res.* **2007**, *6*, 469–479. [CrossRef]
26. Godelmann, R.; Fang, F.; Humpfer, E.; Schutz, B.; Bansbach, M.; Schafer, H.; Spraul, M. Targeted and Nontargeted Wine Analysis by ^{1}H NMR Spectroscopy Combined with Multivariate Statistical Analysis. Differentiation of Important Parameters: Grape Variety, Geographical Origin, Year of Vintage. *J. Agric. Food Chem.* **2013**, *61*, 5610–5619. [CrossRef]
27. Tan, Y.M.; Lim, S.H.; Tay, B.Y.; Lee, M.W.; Thian, E.S. Functional chitosan-based grapefruit seed extract composite films for applications in food packaging technology. *Mater. Res. Bull.* **2015**, *69*, 142–146. [CrossRef]
28. Yao, X.; Liu, J.; Hu, H.; Yun, D.; Liu, J. Development and comparison of different polysaccharide/PVA-based active/intelligent packaging films containing red pitaya betacyanins. *Food Hydrocoll.* **2022**, *124*, 107305. [CrossRef]
29. Li, C.; Tao, J.; Zhang, H. Peach Gum Polysaccharides-Based Edible Coatings Extend Shelf Life of Cherry Tomatoes. *Biotech* **2017**, *7*, 168. [CrossRef]
30. Wang, S.; Zhou, Q.; Zhou, X.; Wei, B.; Ji, S. The Effect of Ethylene Absorbent Treatment on the Softening of Blueberry Fruit. *Food Chem.* **2018**, *246*, 286–294. [CrossRef]
31. Shariatinia, Z.; Fazli, M. Mechanical properties and antibacterial activities of novel nanobiocomposite films of chitosan and starch. *Food Hydrocoll.* **2015**, *46*, 112–124. [CrossRef]
32. Zhang, J.; Li, D.; Xu, W.; Fu, Y. Preservation of Kyoho grapes stored in active, slow-releasing pasteurizing packaging at room temperature. *LWT-Food Sci. Technol.* **2014**, *56*, 440–444. [CrossRef]
33. Yu, Y.M. *The Starch and Chitosan Film Containing Antibacterial Agent Natamycins Giant Peak Grape Preservation Effect Evaluation*; Heilongjiang University: Harbin, China, 2013.
34. Francisco, C.B.; Pellá, M.G.; Silva, O.A.; Raimundo, K.F.; Caetano, J.; Linde, G.A.; Colauto, N.B.; Dragunski, D.C. Shelf-life of guavas coated with biodegradable starch and cellulose-based films. *Int. J. Biol. Macromol.* **2020**, *152*, 272–279. [CrossRef] [PubMed]
35. China National Standardization Administration Committee. *GB15037-2006*; National Wine Standard. China Standards Press: Beijing, China, 2006.
36. Zhang, M.; Huan, Y.; Tao, Q.; Wang, H.; Li, C.L. Studies on Preservation of Two Cultivars of Grapes at Controlled Temperature. *LWT-Food Sci. Technol.* **2001**, *34*, 502–506. [CrossRef]
37. García, M.A.; Pinotti, A.; Martino, M.N.; Zaritzky, N.E. Characterization of starch and composite edible films and coatings. In *Edible Films and Coatings for Food Applications*; Springer: New York, NY, USA, 2009; pp. 169–209.
38. Zhou, X.; Liu, L.; Li, J.; Wang, L.; Song, X. Extraction and Characterization of Pectin from Jerusalem ArtiChoke Residue and Its Application in Blueberry Preservation. *Coatings* **2022**, *12*, 385. [CrossRef]
39. Han, C.; Zhao, Y.; Leonard, S.W.; Traber, M.G. Edible coatings to improve storability and enhance nutritional value of fresh and frozen strawberries (*Fragaria × ananassa*) and raspberries (*Rubus idaeus*). *Postharvest Biol. Technol.* **2004**, *33*, 67–78. [CrossRef]
40. Youssef, K.; Roberto, S.R. Chitosan/silica nanocomposite-based formulation alleviated gray mold through stimulation of the antioxidant system in table grapes. *Int. J. Biol. Macromol.* **2021**, *168*, 242–250. [CrossRef]
41. Wu, Z.; Zhou, W.; Pang, C.; Deng, W.; Xu, C.; Wang, X. Multifunctional chitosan-based coating with liposomes containing laurel essential oils and nanosilver for pork preservation. *Food Chem.* **2019**, *295*, 16–25. [CrossRef]
42. Zhang, X.; Wu, H.; Zhang, L.A.; Sun, Q.J. Horseradish peroxidase-mediated synthesis of an antioxidant gallic acid-g-chitosan derivative and its preservation application in cherry tomatoes. *RSC Adv.* **2018**, *8*, 20363–20371. [CrossRef]
43. Kan, J.; Liu, J.; Yong, H.; Liu, Y.; Qin, Y.; Liu, J. Development of active packaging based on chitosan-gelatin blend films functionalized with Chinese hawthorn (*Crataegus pinnatifida*) fruit extract. *Int. J. Biol. Macromol.* **2019**, *140*, 384–392. [CrossRef]

44. Opara, U.L.; Al-Ani, M.R.; Al-Rahbi, N.M. Effect of Fruit Ripening Stage on Physico-Chemical Properties, Nutritional Composition and Antioxidant Components of Tomato (*Lycopersicum esculentum*) Cultivars. *Food Bioprocess Technol.* **2012**, *5*, 3236–3243. [CrossRef]
45. Jiang, Y.R.; Fu, Y.B.; Li, D.L.; Xu, W.C. Effects of 1-MCP and Controllable-Release SO_2 Packaging on Cold Preservation of Grapes (C.V. Muscat Hamburg). *Adv. Mater. Res.* **2013**, *750–752*, 2335–2339. [CrossRef]
46. Verheul, M.J.; Slimestad, R.; Tjøstheim, I.H. From Producer to Consumer: Greenhouse Tomato Quality As Affected by Variety, Maturity Stage at Harvest, Transport Conditions, and Supermarket Storage. *J. Agric. Food Chem.* **2015**, *63*, 5026. [CrossRef] [PubMed]
47. Son, H.S.; Kim, K.M.; Van Den Berg, F.; Hwang, G.S.; Park, W.M.; Lee, C.H.; Hong, Y.S. ^{1}H Nuclear Magnetic Resonance-Based Metabolomic Characterization of Wines by Grape Varieties and Production Areas. *J. Agric. Food Chem.* **2008**, *56*, 8007–8016. [CrossRef]
48. Gougeon, L.; Da Costa, G.; Le Mao, I.; Ma, W.; Teissedre, P.L.; Guyon, F.; Richard, T. Wine Analysis and Authenticity Using H-1-NMR Metabolomics Data: Application to Chinese Wines. *Food Anal. Methods* **2018**, *11*, 3425–3434. [CrossRef]
49. Goodacre, R.; Broadhurst, D.; Smilde, A.K.; Kristal, B.S.; Baker, J.D.; Beger, R.; Bessant, C.; Connor, S.; Capuani, G.; Craig, A.; et al. Proposed minimum reporting standards for data analysis in metabolomics. *Metabolomics* **2007**, *3*, 231–241. [CrossRef]
50. Cassino, C.; Tsolakis, C.; Bonello, F.; Gianotti, V.; Osella, D. Effects of area, year and climatic factors on Barbera wine characteristics studied by the combination of 1H-NMR metabolomics and chemometrics. *J. Wine Res.* **2017**, *28*, 259–277. [CrossRef]
51. Nho-Eul, S.; Chan-Mi, L.; Sang-Ho, B. Isolation and molecular identification for autochthonous starter Saccharomyces cerevisiae with low biogenic amine synthesis for black raspberry (*Rubus coreanus* Miquel) wine fermentation. *J. Gen. Appl. Microbiol.* **2019**, *65*, 188–196.
52. Wojdyo, A.; Samoticha, J.; Chmielewska, J. The influence of different strains of Oenococcus oeni malolactic bacteria on profile of organic acids and phenolic compounds of red wine cultivars Rondo and Regent growing in a cold region. *J. Food Sci.* **2020**, *85*, 1070–1081. [CrossRef]
53. Ferreira, A.M.; Mendes-Faia, A. The Role of Yeasts and Lactic Acid Bacteria on the Metabolism of Organic Acids during Winemaking. *Foods* **2020**, *9*, 1231. [CrossRef]
54. Zheng, Y.J.; Duan, Y.T.; Zhang, Y.F.; Pan, Q.H.; Li, J.M.; Huang, W.D. Determination of Organic Acids in Red Wine and Must on Only One RP-LC-Column Directly After Sample Dilution and Filtration. *Chromatographia* **2009**, *69*, 1391–1395. [CrossRef]
55. Lasik, M. Influence of malolactic bacteria inoculation scenarios on the efficiency of the vinification process and the quality of grape wine from the Central European region. *Eur. Food Res. Technol.* **2017**, *243*, 2163–2173. [CrossRef]

Article

Assessing the Use of *Aloe vera* Gel Alone and in Combination with Lemongrass Essential Oil as a Coating Material for Strawberry Fruits: HPLC and EDX Analyses

Hanaa S. Hassan [1], Mervat EL-Hefny [2,*], Ibrahim M. Ghoneim [1], Mina S. R. Abd El-Lahot [3], Mohammad Akrami [4], Asma A. Al-Huqail [5], Hayssam M. Ali [5] and Doaa Y. Abd-Elkader [1]

1. Department of Vegetable, Faculty of Agriculture (El-Shatby), Alexandria University, Alexandria 21545, Egypt; hanaa.saad@alexu.edu.eg (H.S.H.); ibrahimghonim85@yahoo.com (I.M.G.); doaa.abdelkader@alexu.edu.eg (D.Y.A.-E.)
2. Department of Floriculture, Ornamental Horticulture and Garden Design, Faculty of Agriculture (El-Shatby), Alexandria University, Alexandria 21545, Egypt
3. Department of Food Science & Technology, Faculty of Agriculture (El-Shatby), Alexandria University, Alexandria 21545, Egypt; mina.ragheb@alexu.edu.eg
4. Department of Engineering, University of Exeter, Exeter EX4 4QF, UK; m.akrami@exeter.ac.uk
5. Botany and Microbiology Department, College of Science, King Saud University, P.O. Box. 2455, Riyadh 11451, Saudi Arabia; aalhuqail@ksu.edu.sa (A.A.A.-H.); hayhassan@ksu.edu.sa (H.M.A.)
* Correspondence: mervat.mohamed@alexu.edu.eg

Abstract: Strawberry is a non-climacteric fruit but exhibits a limited postharvest life due to rapid softening and decay. A strawberry coating that is natural and safe for human consumption can be used to improve the appearance and safeguard the fruits. In this study, 20% and 40% *Aloe vera* gel alone or in combination with 1% lemongrass essential oil (EO) was used as an edible coating for strawberries. After application of all the treatments, the strawberry fruits were stored at a temperature of 5 ± 1 °C at a relative humidity (RH) of 90%–95% for up to 16 days and all the parameters were analyzed and compared to control (uncoated fruits). The results show that *A. vera* gel alone or with lemongrass EO reduced the deterioration and increased the shelf life of the fruit. Treatment with *A. vera* gel and lemongrass EO decreased acidity and total anthocyanins and maintained fruit firmness. Treatment with *A. vera* gel 40% + lemongrass EO 1% led to the lowest weight loss, retained firmness and acidity, but increased the total soluble solids and total anthocyanins compared to uncoated fruits during storage of up to 16 days. The phenolic compounds of *A. vera* gel were analyzed by HPLC, and the most abundant compounds were found to be caffeic (30.77 mg/mL), coumaric (22.4 mg/mL), syringic (15.12 mg/mL), sinapic (14.05 mg/mL), ferulic (8.22 mg/mL), and cinnamic acids (7.14 mg/mL). Lemongrass EO was analyzed by GC–MS, and the most abundant compounds were identified as α-citral (neral) (40.10%) β-citral (geranial) (30.71%), γ-dodecalactone (10.24%), isoneral (6.67%), neryl acetal (5.64%), and linalool (1.77%). When the fruits were treated with 20% or 40% *A. vera* gel along with 1% lemongrass, their total phenolic content was maintained during the storage period (from 4 to 8 days). The antioxidant activity was relatively stable during the 8 days of cold storage of the fruits coated with *A. vera* gel combined with lemongrass EO because the activity of both 20% and 40% gel was greater than that for the other treatments after 12 days of storage in both experiments. Moreover, all the treatments resulted in lower numbers of total microbes at the end of the storage period compared with the control treatment. This study indicates that the use of *Aloe vera* gel with lemongrass EO as an edible coating considerably enhances the productivity of strawberry fruits and the treatment could be used on a commercial scale.

Keywords: edible coatings; *Aloe vera* gel; lemongrass essential oil; shelf life; strawberry fruit

Citation: Hassan, H.S.; EL-Hefny, M.; Ghoneim, I.M.; El-Lahot, M.S.R.A.; Akrami, M.; Al-Huqail, A.A.; Ali, H.M.; Abd-Elkader, D.Y. Assessing the Use of *Aloe vera* Gel Alone and in Combination with Lemongrass Essential Oil as a Coating Material for Strawberry Fruits: HPLC and EDX Analyses. *Coatings* **2022**, *12*, 489. https://doi.org/10.3390/coatings12040489

Academic Editor: Elena Torrieri

Received: 18 February 2022
Accepted: 1 April 2022
Published: 6 April 2022

Publisher's Note: MDPI stays neutral with regard to jurisdictional claims in published maps and institutional affiliations.

Copyright: © 2022 by the authors. Licensee MDPI, Basel, Switzerland. This article is an open access article distributed under the terms and conditions of the Creative Commons Attribution (CC BY) license (https://creativecommons.org/licenses/by/4.0/).

1. Introduction

Strawberry (*Fragaria × ananassa*) is an economically important crop worldwide [1,2]. It has a great nutritional value because it contains minerals, vitamins, flavonoids, and phenolic compounds with beneficial biological properties, for instance antioxidant, anticancer, and anti-inflammatory activities [3,4]. Strawberry has tremendous prospects for commercial use, e.g., for the extraction of natural color with great potential for diverse value-added processed products [5]. However, the physiological characteristics of strawberry fruits deteriorate easily, as their softening reduces their postharvest shelf life during cold storage [6,7].

Postharvest losses in produce are a constant struggle for modern agriculture, which makes it urgent to develop new alternatives to reduce the waste [8,9]. To prolong the storage life of fresh and minimally processed fruits and vegetables, several physical, chemical, and biological alternatives and treatments have been proposed [10]. Recently, methods of ozone, electrolyzed water, modified/controlled atmospheric packaging, natural compounds, antifungal edible coatings, and biocontrol agents have emerged as safe alternatives and efficient preservation methods in the fresh produce industry [11–13].

Edible coatings are made up of natural polymers, such as carbohydrates, proteins, waxes, and their composites, that separate fruits from the surrounding atmosphere [14,15]. Coatings with edible films and essential oils (EOs) can also help to maintain the postharvest quality of fruits by reducing transpiration and respiration [16,17]. They also protect fruits and vegetables from deterioration by reducing the microbial growth and enhancing the textural quality [18–20].

Aloe vera (*Aloe barbadensis* Miller) is a succulent plant belonging to the family Asphodelaceae [21]. *A. vera* leaves have been used for many centuries for their therapeutic properties, and over 75 active ingredients have been identified in its gel [22].

A. vera gel is rich in soluble sugars and polysaccharides but has low properties of hydrophobic and lipid levels with gas barrier efficacy, making it an ideal edible coating material [23,24]. Moreover, *A. vera* gel coatings act as a barrier to moisture and O_2, reducing the respiration rate, thereby preventing anaerobic conditions and conserving fruit quality [25].

As a coating material, *A. vera* gel maintains the texture, color, and shelf life of fruits and vegetables [26,27]. It is edible, invisible, odorless, and does not affect the quality of the fruit and vegetables, moreover, it is safe for human health and ecofriendly [28]. Furthermore, it reduces respiration rate, moisture loss, softening of tissues, oxidative browning, and proliferation of microorganisms in fruits, such as strawberry, cherry laurel fruit, and grapes [27,29,30]. Using *A. vera* gel dip coating reduced weight loss, changes in the physicochemical parameters, and decay, extending the shelf life of figs and litchi fruits [30,31].

EOs play an important role in the protection of the plants as they are antimicrobials and insecticides [32]. One of the advantages of plant EOs is their bioactivity in the vapor phase, which makes them possible fumigants to control postharvest rotting fungi in fruits and grains [33]. For example, lemongrass (*Cymbopogon citratus*), one of the important medicinal herbs, belonging to family Poaceae, is known to have strong antimicrobial and insecticidal properties [34–38]. It shows a strong fungicidal effect against microorganisms in fruit juices [39]. Lemongrass EOs are composed of terpenes and phenylpropenes compounds [40]. In addition, they contain other chemical groups like ketones, alcohols, esters, aldehyde, and flavonoid compounds [41]. The major components of Lemongrass EOs are nerol, α-citral, citronellal, β-citral, geraniol, terpinolene, myrcene, geranyl acetate, terpinol, and methylheptenone [34–37,42,43]. Mixtures of *A. vera* gel and EOs are widely studied as edible coatings for fresh-cut and whole fruit [14].

The present research assessed the effect of *A. vera* gel and lemongrass EO to enhance the postharvest quality, bioactive constituents, and shelf life of strawberry fruit.

2. Materials and Methods

2.1. Plant Material

In two successive experiments during 2020, fruits of strawberry (*Fragaria* × *ananassa* Duch.) cv. Winterstar, a short-day genotype adapted to an annual plastic culture growing system, were harvested at commercial maturity (red color on 80% of the fruit surface) at 30°35′34.5″ N, 30°42′58.4″ E, Behira Governorate, Egypt. The plant is compact, upright, and with long pedicels, making the fruit easy to harvest. This variety produces conical and firm fruit that is uniform in shape throughout the season and has low sourness. The mature fruit has red color on about 90% of its surface [44]. Those fruits were chosen that had red color on over 80% of their surface and were free from mechanical damage, blemishes, and disease [27]. On the same day of harvesting, the fruits were delivered to the laboratory of Alex Postharvest Center (APHC), Faculty of Agriculture, Alexandria University Then, they were washed with fresh water, air dried, and used in the post-harvest treatments.

2.2. A. vera Gel Extraction and HPLC Analysis of Phenolic Compounds

A. vera mature leaves were obtained from the Nursery of Floriculture, Ornamental Horticulture and Garden Design Department, Faculty of Agriculture, Alexandria University (Alexandria, Egypt). The leaves were washed in tap water and then the gel was separated and blended to obtain a homogeneous mixture. The mixture was filtered using a muslin cloth and then centrifuged at $10000\times g$ for 25 min to remove the fibers [45]. Then, concentrations were prepared for HPLC analysis as follows: 0.5 g of powdered *A. vera* gel was extracted by ultrasound for 30 min at 25 °C using methanol/water (80%, *v/v*) and filtered.

Phenolic compounds were identified by high-performance liquid chromatography equipment (Agilent 1100, pump PU-1580; UV detector UV-1570; injector equipped with a 20 µL loop) (Agilent Technologies, Santa Clara, CA, USA). The samples were separated using a 250 mm × 4.6 mm stainless-steel column Discovery-C18 4 µm (Agilent Technologies, Santa Clara, CA, USA). The flow rate of the mobile phase was kept at 1 mL/min. Solvent A was water containing 0.05% formic acid, and solvent B was acetonitrile/methanol (80%:20%, *v/v*). The gradient conditions were as follows: 0–5 min, 10% B; 5–15 min, 10%–18% B; 15–25 min, 18% B; 25–30 min, 18%–25% B; 30–35 min, 25% B; 35–40 min, 25%–35% B; 40–45 min, 35%–60% B; 45–50 min, 60%–10% B; and 50–55 min, 10% B. The temperature of the column was controlled at 25 °C.

2.3. Extraction and Chemical Analysis of Lemongrass EO

Lemongrass leaves were obtained from the Nursery of Floriculture, Ornamental Horticulture and Garden Design Department, Faculty of Agriculture, Alexandria University. About 100 g of fresh leaves were chopped and put in a 2 L flask and the essential oil (EO) was hydrodistillated using a Clevenger-type apparatus for 3 h. The collected EO was kept in brown bottles at 4 °C until use [46].

The EO chemical composition was determined using a Trace GC Ultra-ISQ mass spectrometer (Thermo Scientific, Austin, TX, USA) with a direct capillary column TG–5MS (30 m × 0.25 mm × 0.25 µm film thickness). To prepare the EO for GC–MS, 5 µL from the pure lemongrass EO was dissolved in 1.5 mL of hexane. Then, 1 µL was injected into GC–MS. The temperatures of column oven, chemical separation and identification conditions can be found in a previous study [47]. The match factor (MF) between the mass spectrum obtained for each compound and the library mass spectra for each compound was measured and reported, where it was accepted if its value ≥ 650 [35].

2.4. Preparation of A. vera–Lemongrass EO Coating

A. vera gel solution (20%–40%) + lemongrass EO 1% was mixed by dissolving lemongrass EO in distilled water with a few drops of Tween-80 (0.01% *w/v*) for 2 min, then the gel was added under vigorous shaking for approximately 2 min [12].

2.5. Treatment Application and Analysis

The fruits were separated into five groups (150 fruits per group and 3 replicates/treat-ment). Each group was treated by immersing fruits with the treatments mentioned in Table 1, for 1 min. Then, the fruits were left to air-dry at room temperature for 1 h so that their surfaces were dry [27,48]. The treated fruits were drained, packed in perforated polystyrene bags (1 L), and stored at 5 ± 1 °C under 90%–95% relative humidity for 16 days. The parameters were recorded every 4 days for each treatment.

Table 1. Fruit coating treatments used in the present study.

Treatments	Concentration
1	Control
2	A. vera gel 20% (v/v)
3	A. vera gel 40% (v/v)
4	A. vera gel 20% + lemongrass EO 1%
5	A. vera gel 40% + lemongrass EO 1%

2.6. Physical Parameters of Strawberry

2.6.1. Weight Loss (%)

The fresh weight of fruit of each replicate was measured on the treatment day and at 4, 8, 12, and 16 days of sampling time. The cumulative weight loss was expressed as a percentage loss of the original fresh weight: weight loss (%) = $(F_0 - F_1)/F_0 \times 100$, where F_0 is the initial fresh weight and F_1 is the measured weight on each sampling day.

2.6.2. Fruit Firmness

The strawberry fruit firmness was determined using a texture analyzer for each treatment and storage period using FT011 Fruit Firmness Tester (Wagner Instruments, Greenwich, CT, USA). This instrument consists of penetrating cylinder (1 mm in diameter) to penetrate inside the pulp of fruits up to a constant distance of 5 mm at a speed 2 mm/s. The firmness per Newton (N) was measured.

2.7. Color Fruit Samples

HunterLab Colormeter (HunterLab Labscan 600 spectrocolorimeter, version 3.0; Hunter Associates Laboratory Inc., Reston, VA, USA) was used according to the Granato and Masson [49]. Strawberry fruits from each treatment were measured at three equidistant points for L* (lightness), a* (redness), b* (yellowness), while hue angle (h°) was measured [(h° = $\tan^{-1}(b^*/a^*)$] for each sample.

2.8. Physicochemical and Bioactive Constituents of Strawberry

2.8.1. Soluble Solid Content (SSC)

Firstly, in a mortar, the fruit samples were crushed then squeezed to acquire juice by hand. A digital refractometer (model PR101, Co. Ltd., Atago, Tokyo, Japan) was used to measure the soluble solid content (SSC) in the fruit juice.

2.8.2. Titratable Acidity (TA), pH and Total Anthocyanin Measurements

Fruits titratable acidity (TA) was analyzed using the Association of Official Analytical Chemists' (AOAC) standard [50]. pH was measure by a digital pH meter (Martini, Temperature Laboratory Bench meter Mi 150 pH, Da Nang City, Vietnam) according to the AOAC standard [50]. All parameters of fruit samples were determined after 4, 8, 12, and 16 days. Total anthocyanins were extracted and then calculated (mg of cyanidin chloride g^{-1}) [51].

2.8.3. Total Phenolic Content

Phenolic compounds were extracted with methanol (containing 0.1% HCl) as a solvent. One gram of the sample was individually blended with the solvent at a ratio of 1:20 (w/v) and the extraction was carried out twice at the room temperature. Then the extract was stored at −18 °C until use. The total phenolic content was measured at 750 nm by Optizin UV–Vis spectrophotometer model (Thermo Electron Corporation, Waltham, MA, USA) [50] followed by Folin–Ciocalteu reagent and gallic acid (GA) as the standard. The results were expressed as mg of GA equivalent (GAE)/100 g of FW.

2.8.4. Antioxidant Activity

Strawberry fruit samples (10 g) were soaked in 80% ethanol (50 mL) for 1 week at room temperature, filtered through Whatman paper No. 1, and stored at 4 °C in a refrigerator until use [52]. The antioxidant activity was assessed by evaluating the free-radical-scavenging activity of the 2,2-diphenyl-1-picryl-hydrazyl (DPPH) radical according to a modified method as described previously [53] using Optizin UV–Vis spectrophotometer model (Thermo Electron Corporation, Waltham, MA, USA). The radical-scavenging activity was calculated as a percentage of DPPH discoloration using the following equation: scavenging activity (%) = $((A_{Control} - A_{Sample})/A_{Control}) \times 100$, where A_{Sample} is the absorbance of the tested sample and $A_{Control}$ is the absorbance of the control (DPPH solution).

2.8.5. Fruit Extraction and HPLC Analysis of Flavonoid Compounds

From each treatment, about 50 gm of strawberry fruits were soaked in 60 mL of ethanol for 1 week to acquire the extract. The extracts were filtered using filter paper (Whatman No. 1), concentrated, and in brown vials were stored for further analysis. The flavonoid compounds from the extracts were identified by HPLC (Agilent 1100, Santa Clara, CA, USA), composed of two LC pumps, a UV/Vis detector, and a C18 column (250 mm × 4.6 mm, 5 µm) [54].

2.9. Elemental Analysis of Strawberry Fruit EDX Analysis

Elemental analysis of strawberry fruits was performed by scanning electron microscopy (SEM), attached with energy dispersive spectrometry (EDX), and a JFC-1100E ion sputtering device (model JEOL/MP, JSMIT200 Series, Tokyo, Japan) with an acceleration voltage of 20.00 kV (SEM–EDX) [55] to measure the changes in the elemental chemical composition of strawberry fruits due to different treatments.

2.10. Microbiological Analysis

Samples of a specific weight were pacified to decimal serial dilutions in Ringer's solution (Sigma-Aldrich, Milan, Italy), and 25 g of the fruit samples were homogenized in a flask containing 225 mL of Ringer's solution (Sigma-Aldrich, Milan, Italy) using the Bag-Mixer 400 stomacher (Interscience, Saint Nom, France) for 2 min at the highest speed (blending power 4). All test tubes of serial dilutions used contained Ringer's solution (9 mL). Peptone Dextrose agar (PDA) was used for all plate media. A total of 1 mL of the bacterial suspension was pipetted into a dilution tube containing 9 mL of Ringer's solution. This tube was vortexed, and 1 mL of this volume was removed and placed into a second dilution tube containing 9 mL of Ringer's solution. This process was repeated until the sample was sufficiently diluted. These PDA plates are prepared in three replicates, and 100 mL of this suspension was added to the plates and repeated for every tube in the dilution series. The different microbial groups were investigated as follows: total mesophilic microorganisms (TMM) on plate count agar (PCA) incubated at 37 °C for 48 h and total yeasts and molds (TYM) on PDA supplemented with chloramphenicol (0.1 g/L) to avoid the growth of bacteria then incubated for 48 h at 25 °C. Plate counts were achieved by the spread plate method [56] by inoculating 100 µL from each sample's suspension of appropriate dilution. All media were supplied from Oxoid (Milan, Italy). At each collection time, the microbiological counts were performed in triplicate.

2.11. Statistical Analysis

Data were subjected to statistical analysis for calculation of means, variance and standard error using CoStat Software Program Version 6.303 (CoHort Software, Monterey, CA, USA) using one-factor analysis of variance (ANOVA, general linear model), followed by Duncan multiple range test for $p < 0.05$ [57] was used to test the differences among treatments.

3. Results and Discussion

3.1. Phenolic Profile of A. vera Gel by HPLC

The data in Table 2 and Figure 1 show the six identified phenolic compounds of *A. vera* gel by HPLC. Caffeic acid is the most abundant, followed by coumaric, syringic, sinapic, ferulic, and cinnamic acids, and their percentages are displayed in Table 2. These data were matched with the studies by López et al. [58], who found that *A. vera* gel contains catechin, sinapic acid, and quercitrin; Elbandy et al. [59], who observed that gallic protocatecuic, vanillic, ferulic acids, cinnamic, p-coumaric acids, hesperidin, rosmarinicrutin, quercitrin, narengenin, hesperitin, kampferol, and apigenin are the main components of *A. vera* gel; and Numan [60], who confirmed that the gel contains the phenolic compounds quercetin, catechin, aloe emodin, sinapic acid, and aloin [58,60].

Table 2. Identification of the phenolic compounds in *A. vera* gel by HPLC.

R.T. (min)	Compound	Concentration (mg/mL)
5.00	Coumaric acid	22.4
7.01	Ferulic acid	8.22
8.00	Caffeic acid	30.77
9.00	Syringic acid	15.12
11.10	Sinapic acid	14.05
15.00	Cinnamic acid	7.14

R.T.: retention time.

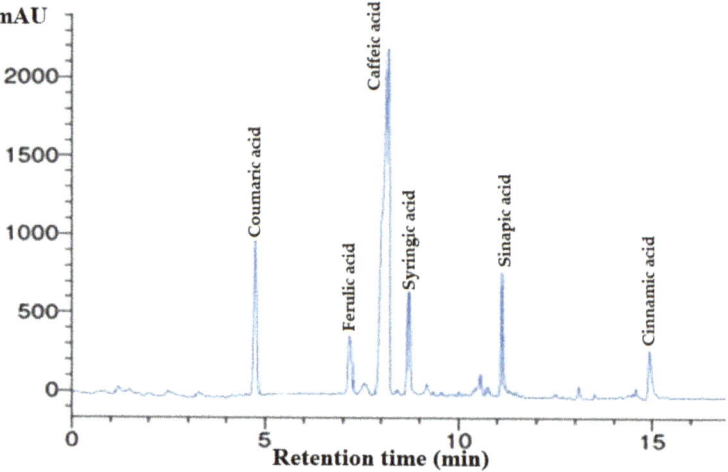

Figure 1. Phenolic profile of *A. vera* gel by HPLC.

3.2. Chemical Constituents of Lemongrass Oil

Table 3 shows the chemical constituents of the EO from lemongrass fresh leaves, where the main components were α-citral (neral) (40.10%), β-citral (geranial) (30.71%), γ-dodecalactone (10.24%), isoneral (6.67%), neryl acetal (5.64%), linalool (1.77%), citral (1.22%), isocitral (0.97%), and geraniol acetate (0.91%). Previous studies have reported that the major constituents of EO are α-citral, β-citral, geraniol, nerol, terpinolene, geranyl acetate, citronellal, myrcene, and terpinol methylheptenone [37]. In one study, α-citral (44.97%), α-citral (33.06%), and myrcene (7.68%) were identified as the major compounds of lemongrass EO [61].

Table 3. Phytochemical constituents of lemongrass EO identified by GC–MS.

Chemical Compound	Percentage (%)	MF
Linalool	1.77	861
Isocitral	0.97	853
Isoneral	6.67	943
α-Citral (Neral)	40.10	930
β-Citral (Geranial)	30.71	916
Citral	1.22	931
Neryl acetal	5.64	876
γ-Dodecalactone	10.24	912
Geraniol acetate	0.91	897
2-Tridecanone	0.70	867
Nizatidine	0.32	979
β-Caryophyllene epoxide	0.36	917
Selin-6-en-4β-ol	0.40	892

MF: match factor.

In general, lemongrass EO contains greater than 45% of α- or β-citral but the amount can vary widely depending on the factors the plants are exposed to, such as genetic diversity, weather, and extraction techniques [62–64].

3.3. Physical Parameters of Strawberry fruits

3.3.1. Weight Loss (%)

The effects of different treatments on the weight loss (%) of strawberry fruits during the 16 days of storage period in both experiments are shown in Figure 2. During the time in all treatments, the weight loss (%) was significantly ($p \leq 0.05$) increased, but all the edible coating treatments helped to reduce the weight loss of strawberry fruit in the two experiments. At the end of storage (day 16), the highest weight loss (6.11%) was observed for the control sample, while the lowest value (11.87%) was obtained for fruits treated with *A. vera* gel 40% + lemongrass EO 1% coating, followed by fruits treated with *A. vera* gel 20% + EO 1% coating (7.63%), in the two experiments.

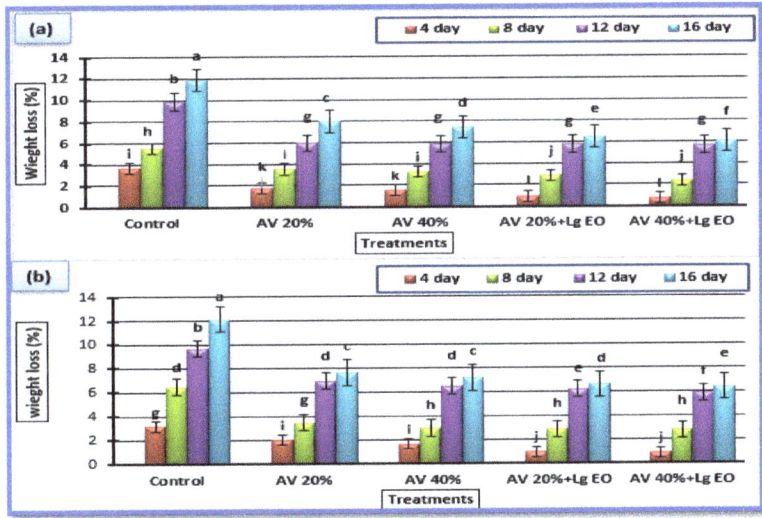

Figure 2. Weight loss (%) (mean ± S.E.) of strawberry fruits stored at 5 °C as affected by coating treatments when stored for different lengths of time in both experiments. The mean ± S.E. of treatments in the figures with the same letter/s shows a nonsignificant difference according to Duncan multiple range test for $p < 0.05$. AV: *A. vera* gel. (**a**) First experiment; (**b**) second experiment.

The reduction in the weight loss in the fruits treated with *A. vera* with EO could be due to composition of polysaccharides in *A. vera* gel, which act as an effective moisture barrier [23,24]. The water loss reduction mechanism is based on the hygroscopic water pressure between the fruit and environment, whereas *A. vera* gel can form a film on the strawberry surface with a thin layer, sealing small wounds and reducing moisture loss [27,65,66].

3.3.2. Fruit Firmness

The firmness of strawberry fruits significantly reduced after all treatments during storage of 4, 8, 12, and 16 days in the two experiments, as shown in Figure 3. The coating treatments showed a higher firmness value than the control strawberry fruits. The firmness of strawberries treated with *A. vera* gel at 20% and 40% was found to be lower than that of other coated samples with *A. vera* gel combined with EO during different days of storage, showing that the treated and control strawberry fruit become less firm due to reaping (Figure 3). Compared to that of the control, the highest firmness value was that of strawberry fruits treated with *A. vera* gel 40%, followed by the fruits treated with *A. vera* gel 20% and those treated with *A. vera* gel with EO 1%. *A. vera* gel coatings retarded the postharvest ripening process and reduced the firmness of table grape and sour cherry [67,68]. Lower water vapor from fruits subjected to *A. vera* gel coating results in maintained turgor pressure of the cell wall [69,70]. The treated fruits with *A. vera* with EO showed slightly higher firmness, that could be due to the higher hydrophobic properties in this treatment. In agreement with this work, the coated strawberry fruits with *A. vera* gel alone and combined with basil EO had a lower softening rate compared to the control treatment [71].

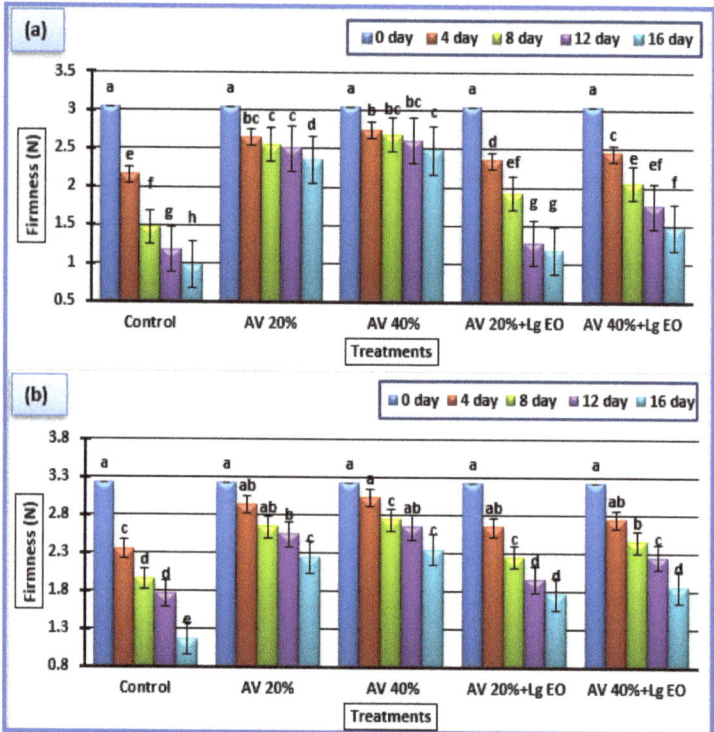

Figure 3. Firmness (N) (mean ± S.E.) of strawberry fruits stored at 5 °C as affected by coating treatments when stored for different lengths of time in both experiments. The mean ± S.E. of treatments in the figures with the same letters shows a nonsignificant difference according to Duncan multiple range test for $p \leq 0.05$. AV: *A. vera* gel. (**a**) First experiment; (**b**) second experiment.

3.4. Color Value

The data in Tables 4 and 5 show that the color value of strawberry fruits was influenced by the different coating treatments. The L* of the strawberry fruits increased in lightness in both coating treatments during the shelf-life study. In other words, a coating of *A. vera* gel or *A. vera* gel with lemongrass EO helped to maintain the lightness of the fruit compared to the control samples (Tables 4 and 5). The chromaticity coordinate a* value, representing the red-green color, was slightly affected by the coatings and remained stable in all treatments during the shelf life. The chromaticity coordinate b* value of strawberry fruits slightly reduced over storage time and coating treatment by *A. vera* gel only. However, *A. vera* gel with lemongrass EO increased this parameter by day 16 in the two experiments. Coated and control fruits showed a significant decrease in hue during the first 8 days of storage and the hue increased as the storage period was increased to 16 days (Tables 4 and 5).

Table 4. Color values of strawberry fruits as affected by coating treatments after 0, 8, and 16 days of storage at 5 °C in Experiment 1.

Treatments	Days	L*	a*	b*	h°
			Experiment 1		
Control	0	43.83 ± 1.21 [b]	43.64 ± 1.89 [b]	21.05 + 2.87 [b]	1.97 + 1.12 [ab]
	8	35.66 ± 2.24 [d]	44.93 ± 2.13 [b]	20.87 + 1.87 [b]	1.57 + 0.87 [d]
	16	33.70 ± 1.54 [e]	45.45 ± 1.97 [ab]	19.16 + 1.22 [cd]	1.90 + 0.58 [b]
A. vera gel 20%	0	43.83 ± 1.21 [b]	43.64 ± 1.89 [b]	21.05 ± 2.87 [b]	1.97 ± 1.12 [ab]
	8	39.40 ± 1.32 [cd]	49.47 ± 2.54 [a]	19.48 ± 2.01 [cd]	1.37 ± 1.02 [ef]
	16	35.61 ± 1.12 [d]	44.63 ± 2.33 [b]	17.07 ± 1.33 [d]	1.57 ± 0.98 [d]
A. vera gel 40%	0	43.83 ± 1.21 [b]	43.64 ± 1.89 [b]	21.05 ± 2.87 [b]	1.97 ± 1.12 [ab]
	8	38.40 ± 2.12 [cd]	44.05 ± 2.01 [b]	18.49 ± 1.21 [cd]	1.45 ± 1.03 [e]
	16	36.84 ± 1.34 [d]	44.87 ± 2.44 [b]	16.28 ± 1.32 [d]	1.15 ± 0.45 [f]
A. vera gel 20% + lemongrass EO 1%	0	43.83 ± 1.21 [b]	43.64 ± 1.89 [b]	21.05 ± 2.87 [b]	1.97 ± 1.12 [ab]
	8	37.80 ± 2.65 [cd]	36.43 ± 2.12 [c]	25.25 ± 1.56 [ab]	2.03 ± 1.11 [a]
	16	49.90 ± 1.67 [a]	36.97 ± 2.03 [c]	28.59 ± 1.12 [a]	2.03 ± 1.01 [a]
A. vera gel 20% + lemongrass EO 1%	0	43.83 ± 1.21 [b]	43.64 ± 1.89 [b]	21.05 ± 2.87 [b]	1.97 ± 1.12 [ab]
	8	44.05 ± 1.78 [b]	35.14 ± 2.43 [c]	22.42 ± 1.65 [b]	1.84 ± 0.55 [cd]
	16	46.80 ± 2.03 [a]	34.57 ± 2.63 [c]	25.93 ± 1.78 [ab]	1.81 ± 0.98 [cd]

The mean values with the same superscript letter/s in the same column show a nonsignificant difference according to Duncan multiple range test for $p \leq 0.05$.

Table 5. Color values of strawberry fruits as affected by coating treatments after 0, 8, and 16 days of storage at 5 °C in Experiment 2.

Treatments	Days	L*	a*	b*	h°
			Experimental 2		
Control	0	43.49 ± 2.44 [c]	42.07 ± 2.61 [b]	20.39 ± 1.77 [b]	1.89 ± 0.31 [c]
	8	34.71 ± 2.03 [e]	43.87 ± 2.11 [b]	19.83 ± 1.34 [c]	1.27 ± 0.61 [e]
	16	33.23 ± 1.65 [f]	42.69 ± 1.34 [b]	18.66 ± 1.01 [c]	2.07 ± 0.54 [b]
A. vera gel 20%	0	43.49 ± 2.44 [c]	42.07 ± 2.61 [b]	20.39 ± 1.77 [b]	1.89 ± 0.31 [c]
	8	40.52 ± 2.23 [d]	47.48 ± 2.43 [a]	18.73 ± 1.22 [c]	1.96 ± 1.21 [bc]
	16	34.67 ± 2.01 [e]	43.25 ± 1.87 [b]	17.16 ± 1.02 [d]	2.75 ± 1.32 [a]
A. vera gel 40%	0	43.49 ± 2.44 [c]	42.07 ± 2.61 [b]	20.39 ± 1.77 [b]	1.89 ± 0.31 [c]
	8	39.94 ± 2.46 [d]	46.71 ± 2.21 [a]	18.49 ± 1.67 [d]	1.54 ± 0.64 [d]
	16	35.17 ± 1.78 [e]	43.22 ± 1.44 [b]	20.10 ± 1.25 [b]	1.37 ± 0.67 [e]
A. vera gel 20% + lemongrass EO 1%	0	43.49 ± 2.44 [c]	42.07 ± 2.61 [b]	20.39 ± 1.77 [b]	1.89 ± 0.31 [c]
	8	47.18 ± 2.67 [a]	36.20 ± 1.98 [c]	26.88 ± 1.06 [ab]	1.08 ± 0.23 [g]
	16	48.50 ± 2.43 [a]	37.30 ± 1.66 [c]	29.73 ± 1.33 [a]	1.95 ± 0.35 [bc]
A. vera gel 20% + lemongrass EO 1%	0	43.49 ± 2.44 [c]	42.07 ± 2.61 [b]	20.39 ± 1.77 [b]	1.89 ± 0.31 [c]
	8	45.37 ± 1.56 [b]	35.86 ± 1.78 [d]	26.99 ± 1.89 [ab]	1.16 ± 0.33 [f]
	16	47.08 ± 1.32 [d]	34.19 ± 1.56 [d]	27.19 ± 1.65 [a]	1.86 ± 0.54 [c]

The mean values with the same superscript letter/s in the same column show a nonsignificant difference according to Duncan multiple range test for $p \leq 0.05$.

The color of strawberry fruits is an important property for product reception by the consumer; although the coating did not change the fruit initially color [72], and with the increased storage time, the fruit became redder and darker. This increase was probably due to a reduction in both the respiration rate and some enzymatic processes, maintaining the quality of the fruit and preventing its browning [73]. At the end of storage time, control fruits and those treated with gluten plus $CaCl_2$ had a low L* value (darker color) [74].

3.5. Physicochemical and Bioactive Constituents of Strawberry

3.5.1. Soluble Solid Content (SSC)

The effect of *A. vera* gel and lemongrass EO coating on the SSC content of strawberry fruits during the storage period of 16 days is shown in Figure 4. The SSC content in treated strawberry fruits increased gradually until day 16 of the storage period in the two experiments (Figure 4). Until the end of the trial in the treatments with control and *A. vera* gel 20%, a gradual increase in the SSC was found, which indicates that *A. vera* gel 20% and 40% + lemongrass EO 1% treatments had slowed the respiration rate of strawberry fruits during the storage period. The same result was reported previously, indicating a link between SSC and respiration rate [27]. In both experiments, the maximum increase in the SSC content was found in the control on day 16. In the case of treatment with *A. vera* gel 40% + lemongrass EO 1%, the least SSC content was observed on days 4 and 8 in the second experiment (Figure 4b). The hydrolysis of starch into sugar might cause the initial increase in the SSC and subsequently the decline in SSC could be due to the decreased respiration rate and the metabolism of sugars into organic acids [48]. A lower SSC could be related to the hydrolysis of carbohydrates into sugar [65].

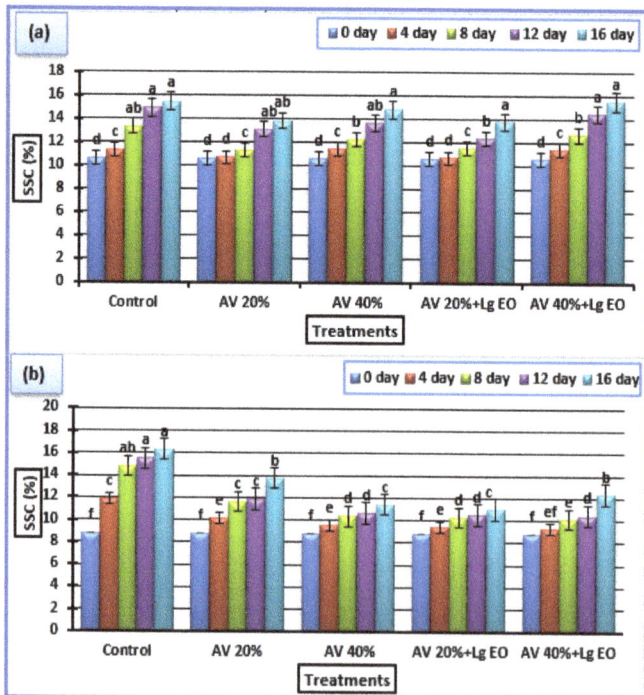

Figure 4. SSC (mean ± S.E.) of strawberry fruits stored at 5 °C as affected by coating treatments when stored for different lengths of time in both experiments. The mean ± S.E. of treatments in the figures with the same letters shows a nonsignificant difference according to Duncan multiple range test for $p \leq 0.05$. AV: *A. vera* gel. (**a**) First experiment; (**b**) second experiment.

3.5.2. Titratable Acidity (TA) and pH

The changes in the Titratable acidity (TA) amount and pH of fruit strawberry in the two experiments during storage are shown in Figures 5 and 6, respectively. The pH of strawberry juices increased in all treatments during the storage period until day 8 in both experiments (Figure 6). Moreover, the coated fruits in both experiments had steadied pH around 3.5. However, coating treatments slowed down the titratable acidity (TA) change in the strawberries during the shelf-life study compared with uncoated fruits (control).

The TA amount in strawberry is directly correlated to fruit organic acids content [48]. The content of fruit acid tends to decrease over time, that could due to the organic acids oxidation as the fruit ripens [75]. The edible coatings of the fruits reduce the respiration rate, decreasing the consumption of organic acids in the respiratory metabolic activities of the fruits [48,76].

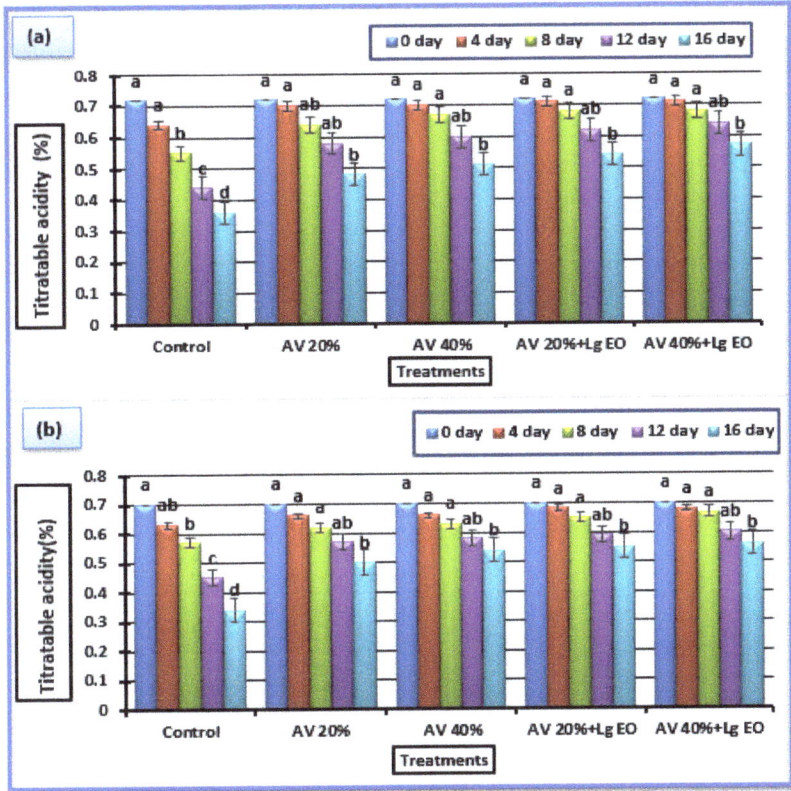

Figure 5. Titratable acidity (%) (mean ± S.E.) of strawberry fruits stored at 5 °C as affected by coating treatments when stored for different lengths of time in both experiments. The mean ± S.E. of treatments in the figures with the same letters shows a nonsignificant difference according to Duncan multiple range test for $p \leq 0.05$. AV: *A. vera* gel. (**a**) First experiment; (**b**) second experiment.

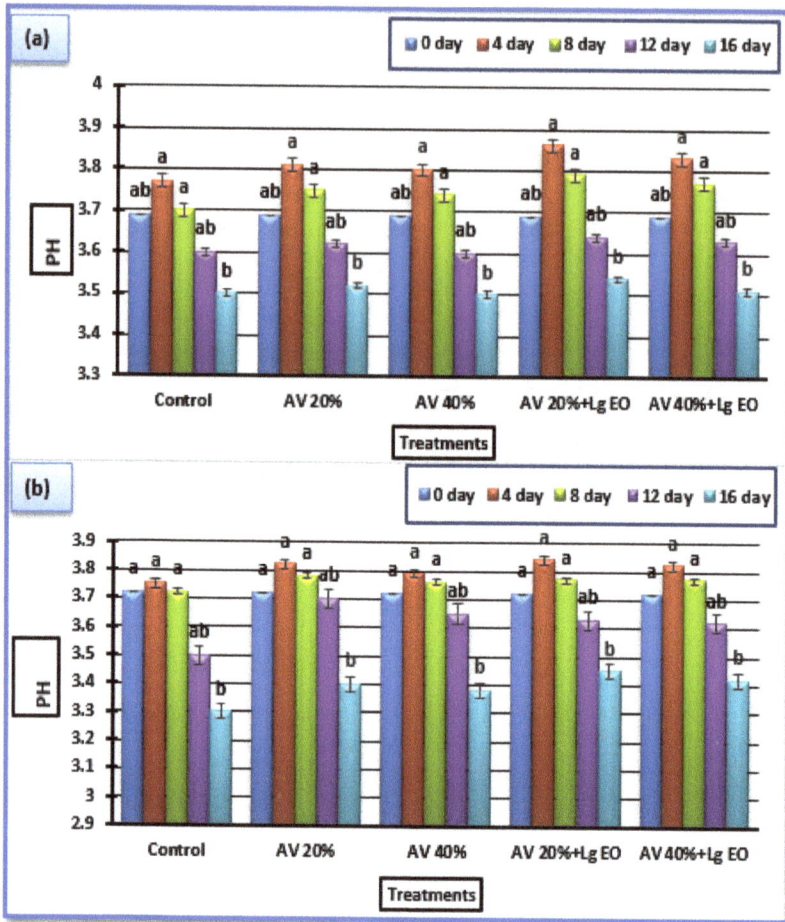

Figure 6. pH (mean ± S.E.) of strawberry fruits stored at 5 °C as affected by coating treatments when stored for different lengths of time in both experiments. The mean ± S.E. of treatments in the figures with the same letters shows a nonsignificant difference according to Duncan multiple range test for $p \leq 0.05$. AV: *A. vera* gel. (**a**) First experiment; (**b**) second experiment.

3.5.3. Total Anthocyanins

The change in the total anthocyanins of strawberry fruits coated with *A. vera* gel and *A. vera* gel with lemongrass EO is shown in Figure 7. The total anthocyanin content in all the treatments increased for the first 12 days of storage in both experiments. Thereafter, it decreased gradually for the remainder of storage. The untreated fruit showed the maximum anthocyanin concentration (277 mg·kg^{-1}) on day 12 of the storage, followed by fruits treated with *A. vera* gel 40% + lemongrass EO 1% (246 mg·kg^{-1}) and those treated with *A. vera* gel 20% + lemongrass EO 1% (221 mg·kg^{-1}) at the end of the storage period, compared to 163 mg·kg^{-1} when the fruits were initially stored.

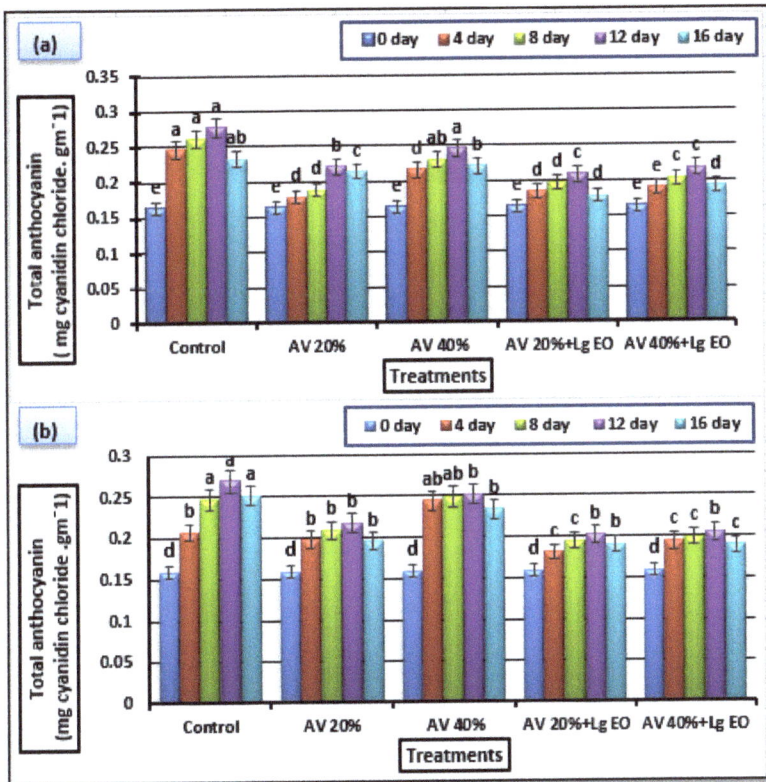

Figure 7. Total anthocyanin (mg of cyanidin chloride.gm^{-1}) (mean ± S.E.) of strawberry fruits stored at 5 °C as affected by coating treatments when stored for different lengths of time in both experiments. The mean ± S.E. of treatments in the figures with the same letters shows a nonsignificant difference according to Duncan multiple range test for $p \leq 0.05$. A. vera gel. (**a**) First experiment; (**b**) second experiment.

The significant increase in anthocyanin in control treatment could probably be related to the natural process during fruit ripening. However, the fruits treated with EOs showed a lower concentration of anthocyanin than the untreated ones. During cold storage the anthocyanin of treated fruits was increased, similar to those reported previously [77], which may be due to the continued biosynthesis of these compounds after harvest. Furthermore, total anthocyanin showed significant differences among fruits coated with a lemongrass EO and alginate-based edible coating [78].

3.5.4. Total Phenolic Content

It is clear from Figure 8 that all examined postharvest treatments decreased the total phenolic content (TPC) in both experiments. However, the highest TPC was recorded by untreated fruits, followed by fruits treated with *A. vera* gel 20%. The lowest values of TPC were scored by the treated fruits with *A. vera* gel 40% + lemongrass EO 1% and those treated with *A. vera* gel 20% + lemongrass EO 1% during both experiments. Figure 8 also indicates that regardless of the initial reading, the TPC was increased from day 4 to day 8 of storage.

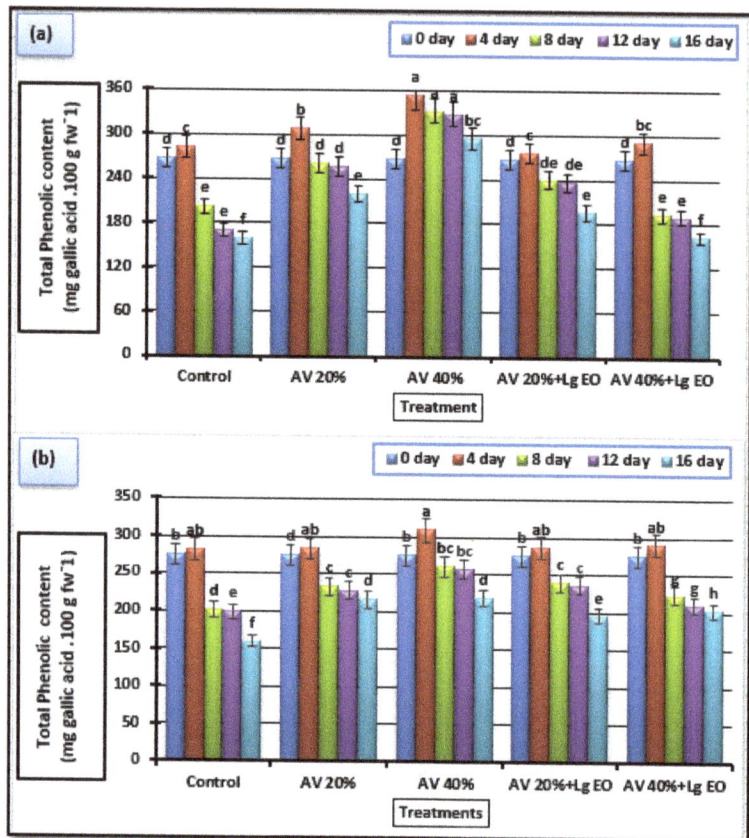

Figure 8. Total phenolic content (mg of gallic acid.100 g fw^{-1}) (mean ± S.E.) of strawberry fruits stored at 5 °C as affected by coating treatments when stored for different lengths of time in both experiments. The mean ± S.E. of treatments in the figures with the same letters shows a nonsignificant difference according to Duncan multiple range test for $p \leq 0.05$. AV: *A. vera*. (**a**) First experiment; (**b**) second experiment.

Anthocyanins are a group of phenolic compounds responsible for the red-blue color of many fruits and are important for human health [79]. The TPC and anthocyanin may be one of their most significant biological properties [80]. In the current study, the TPC decreased while anthocyanin increased in the untreated fruit. It is important for fruits to retain high levels of these compounds during storage and over their shelf life. The anthocyanin and TPC of the treated fruit increased during cold storage (Figure 8), similar to those reported previously [77], which may be due to the continued biosynthesis of these compounds after harvest. The evolution of the TPC of fruits during storage could be different depending on the species, temperature, cultivar, and climactic and environmental conditions during the growth period [48]. The findings indicate that both TPC and anthocyanin content in fruits treated with *A. vera* + ascorbic acid were higher than those in either untreated fruits or fruits treated with *A. vera* alone. Similarly, the use of ascorbic acid as a reducing agent prevented a decrease in the TPC in fresh-cut fruits [14,66].

3.5.5. Antioxidant Activity

The free-radical-scavenging activity (% inhibition) of strawberry fruits' ethanolic extracts was assessed by the DPPH test (Figure 9). Treatments with *A. vera* gel at 20%

and 40% were more effective than treatment with *A. vera* gel with lemongrass EO, since the radical-scavenging activity was 77.04%, 74.58% and 58.22%, 54.29% for *A. vera* gel at 20% and 40% and *A. vera* gel with lemongrass EO, respectively, while it decreased in untreated extract to 64.24% at the end of the storage period. However, antioxidant activity was relatively stable during the 8 days of cold storage in fruits treated with *A. vera* gel and lemongrass EO, and the activity in fruits treated with *A. vera* gel at 20% or 40% was greater than the activity in fruits that underwent other treatments after 12 days of storage in both experiments. Moreover, the antioxidant activity decreased in untreated fruits and fruits treated with *A. vera* gel alone or combined with lemongrass EO. This means that *A. vera* gel at 20% and 40% has powerful potential antioxidant activity and increased the quality and stability of strawberry fruits.

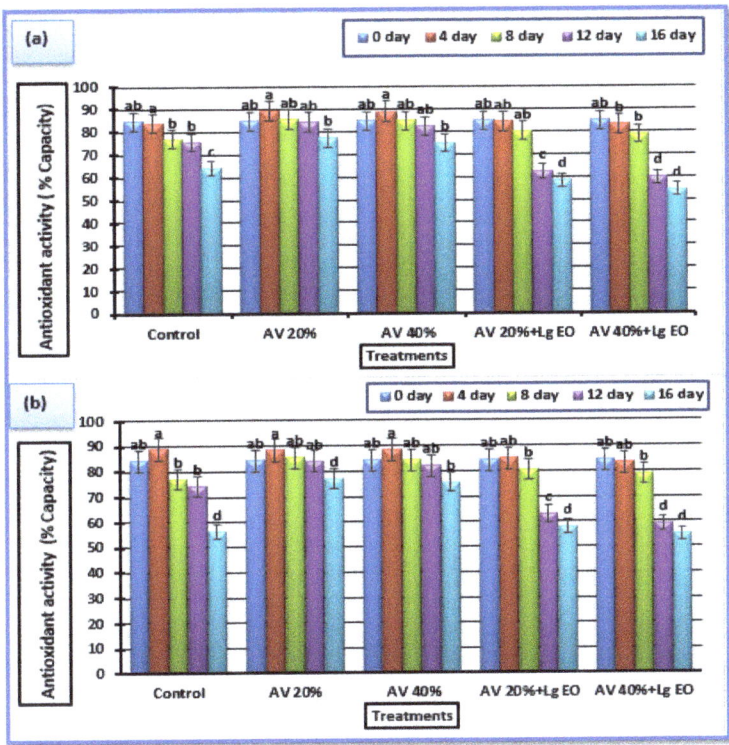

Figure 9. Antioxidant activity (% capacity) (mean ± S.E.) of strawberry fruits stored at 5 °C as affected by coating treatments after being stored for different lengths of time in both experiments. The mean ± S.E. of treatments in the figures with the same letters shows a nonsignificant difference according to Duncan multiple range test for $p \leq 0.05$. AV: *A. vera* gel. (**a**) First experiment; (**b**) second experiment.

Several studies have shown that strawberry is a good source of natural antioxidants [27]. It has been reported that fruits treated with *A. vera* had higher antioxidant capacity than the sample in the case of mango [81], raspberry [77], and table grapes (*Vitis vinifera* L. cv. Yaghouti) [82]. *A. vera* may also increase tissue resistance to decay by enhancing their antioxidant system and free-radical-scavenging capability [83]. Hidayati et al. [84] stated that antioxidant activity can be affected by the phenolic compounds and pigment content. Phenolic compounds and flavonoids as primary antioxidants can play an important role in absorbing and neutralizing free radicals, preventing the progress of diseases such as cancer [85].

3.5.6. Fruit Extraction and HPLC Analysis of the Fruit's Flavonoids

As presented in Table 6, the flavonoid concentration (µg/mL) of strawberry fruit was affected on treatment with *A. vera* gel and lemongrass EO. The highest value of rutin was obtained on treatment with *A. vera* gel 40% + lemongrass EO 1% compared with the initial value in the fruits and the value in the control fruit sample (16.25, 6.14, and 9.14 µg/mL, respectively). Naringin and hesperidin values were the best in the control, with concentrations of 8.16 and 14.56 µg/mL, respectively. Isorhamnetin and genistein were not detected in any treatment except for fruits treated with *A. vera* gel 40% + lemongrass EO 1%, with concentrations of 10.23 and 3.52 µg/mL, respectively. The highest concentration of quercetin was identified in strawberry fruits treated with *A. vera* gel 20% (15.36 µg/mL), the highest value of kaempferol was obtained on treatment with *A. vera* gel 40% (20.47 µg/mL), the highest values for luteolin and catechin were observed in fruits treated with *A. vera* gel 20% (14.66 and 20.56 µg/mL, respectively), and the highest value of 7-hydroxyflavone was obtained in fruits treated with *A. vera* gel 40% (14.16 µg/mL). The best value of chrysoeriol was observed in the initial sample, followed by fruits treated with *A. vera* gel 40%, with concentrations of 25.08 and 17.44 µg/mL. The compound myricetin was not detected in any treatment except in fruits treated with *A. vera* 40%, with a concentration of 2.25 µg/mL. These results are in good agreement with the studies of Hannum [86] and Co and Markakis [87].

Table 6. Flavonoid concentration in strawberry fruit as affected by different treatments of *A. vera* gel and lemongrass EO.

R.T. (min)	Compound	Initial (0 day)	Control	*A. vera* 20%	*A. vera* 40%	*A. vera* 20% + Lemongrass EO 1%	*A. vera* 40% + Lemongrass EO 1%
4.6	Rutin	6.14	9.14	ND	7.13	7.14	16.25
5.2	Naringin	5.16	8.16	ND	6.19	ND	5.66
6.0	Isorhamnetin	ND	ND	ND	ND	ND	10.23
6.9	Quercetin	10.41	15.23	15.36	8.47	9.56	9.66
8.1	Kaempferol	5.17	6.17	6.15	20.47	22.17	11.43
9.0	Luteolin	7.13	7.46	14.66	ND	8.15	ND
10.0	Hesperidin	13.45	14.56	8.12	ND	ND	22.15
11.0	7-Hydroxyflavone	ND	ND	ND	14.16	12.02	8.14
12.01	Catechin	8.14	9.52	20.56	11.78	16.11	1.13
14.6	Genistein	ND	ND	ND	ND	ND	3.52
15.0	Chrysoeriol	25.08	ND	4.21	17.44	7.66	15.04
15.2	Myricetin	ND	ND	ND	2.25	ND	ND

ND: not detected; R.T.: retention time.

3.6. EDX Analysis for Elemental Composition of Strawberry Fruits

Table 7 and Figure 10 present the EDX analysis to measure the changes in the element composition of strawberry fruits due to different treatments. There was a significant effect of various treatments on element O percentage ($p < 0.05$), the highest value obtained on treatment with *A. vera* gel 20% (56.1%), followed by control (55.61%). There was a significant effect of treatments on element Ca percentage ($p < 0.0001$), with the highest value observed in strawberry fruits treated with *A. vera* gel 40% + lemongrass EO 1% (1.23%), followed by fruits treated with *A. vera* gel 20% + lemongrass EO 1% (0.48%). The rest of the treatments were not significant. However, the highest values of elements C, P, and K in strawberry fruits were obtained on treatment with *A. vera* gel 40% + lemongrass EO 1%, with percentages of 45.06%, 0.17%, and 1.8%, respectively, compared with the other treatments. The highest value of Mg in strawberry fruits was obtained on treatment with *A. vera* gel 40% (0.19%), whereas N was identified only in strawberry fruits treated with *A. vera* gel 40% alone.

Table 7. Elemental analysis of strawberry fruits coated by *A. vera* gel and lemongrass EO.

Treatment	Element (Atom %)							
	C	O	Si	P	K	Ca	Mg	N
Control	42.36 ± 0.26 [ad]	55.61 ± 0.21 [ab]	0.25 ± 0.12 [a]	0.14 ± 0.02 [a]	1.33 ± 0.12 [b]	0.31 ± 0.05 [b]	nd	nd
A. vera gel 20%	41.30 ± 1.35 [b]	56.10 ± 1.11 [a]	0.49 ± 0.49 [a]	0.17 ± 0.05 [a]	1.54 ± 0.16 [ab]	0.25 ± 0.02 [b]	0.15 ± 0.08 [ab]	nd
A. vera gel 40%	41.75 ± 1.12 [b]	55.01 ± 0.17 [ab]	1.04 ± 0.76 [ab]	0.11 ± 0.01 [ab]	1.10 ± 0.06 [b]	0.38 ± 0.13 [b]	0.19 ± 0.04 [a]	0.40 ± 0.40
A. vera gel 20% + lemongrass EO 1%	43.61 ± 0.80 [ad]	53.74 ± 0.56 [b]	0.63 ± 0.41 [b]	0.04 ± 0.01 [ab]	1.48 ± 0.20 [ab]	0.48 ± 0.03 [b]	0.03 ± 0.03 [bc]	nd
A. vera gel 40% + lemongrass EO 1%	45.06 ± 0.66 [a]	51.54 ± 0.56 [c]	0.15 ± 0.15 [a]	0.17 ± 0.02 [a]	1.80 ± 0.07 [a]	1.23 ± 0.14 [a]	0.05 ± 0.03 [abc]	nd
p-value	0.0908 [ns]	0.0030	0.6712 [ns]	0.0652 [ns]	0.0516 [ns]	0.0001	0.0612 [ns]	-

ns: not significant; the mean values with the same superscript letter/s within the same column show a nonsignificant difference according to LSD at 0.05 level of probability. nd: not detected.

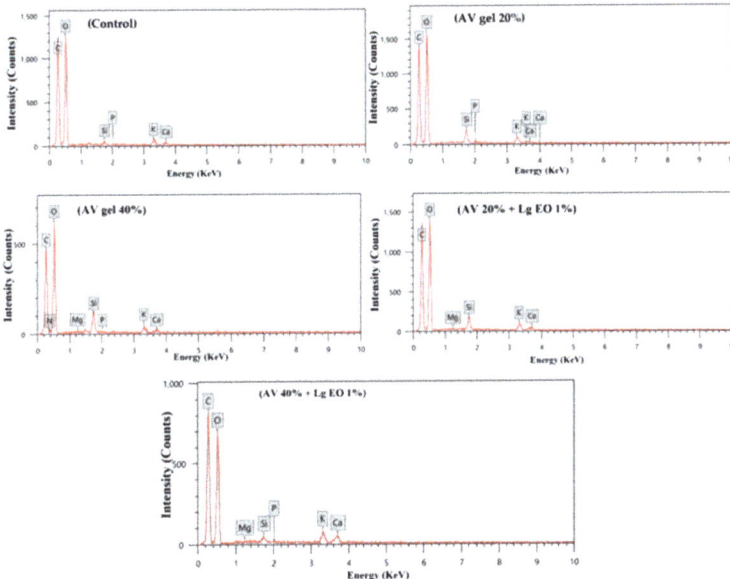

Figure 10. EDX analysis for the elemental composition of strawberry fruits as affected by different treatments; each treatment was measured at three points.

3.7. Microbiological Analysis

Strawberry shows high metabolic activities and sensitivity against pathogens. The bioactive compounds and phytochemicals of the fruit rapidly decrease during storage [88]. Increased soluble sugars and sweetness and decreased acidity and defense metabolites, such as phenolic and antioxidants, make the fruit more susceptible to pathogen attack and postharvest losses [77].

The initial populations of total aerobic mesophilic bacteria and yeasts + molds in the fruit were 80 and 6 CFU g^{-1}, respectively, which increased in the untreated and treated fruit during 16 days of cold storage (Figures 11 and 12), but the samples treated with *A. vera* gel alone or with lemongrass EO showed a strong effect on the total count of microbes in terms of preservation during the storage period, and the counts remained lower than in untreated fruits. These results are comparable with the results of coating with *A. vera* gel and cinnamon EO in modified atmosphere packaging of strawberry. A reduction in microbial populations during storage was observed but there was no change observed in the mold and yeast counts until day 10 of storage; however, on day 15, a decrease in the microbial load was noticed [89].

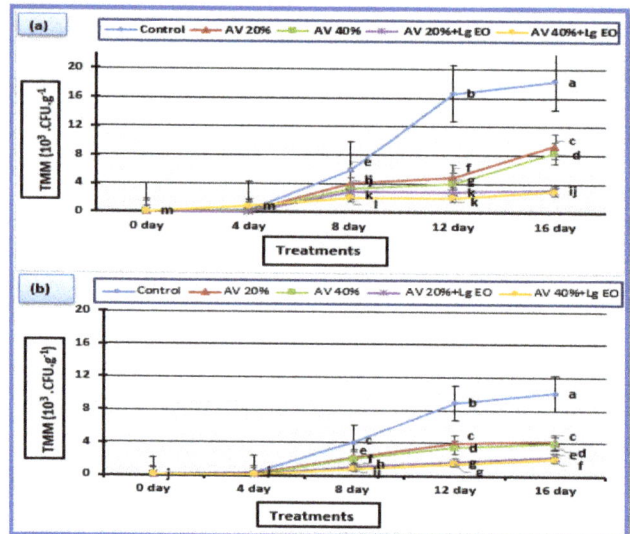

Figure 11. TMM (10^3 CFU.g^{-1}) (mean ± S.E.) of strawberry fruits stored at 5 °C as affected by coating treatments when store for different lengths of time in both experiments. The mean ± S.E. of treatments in the figures with the same letters shows a nonsignificant difference according to Duncan multiple range test for $p \leq 0.05$. AV: *A. vera* gel. (**a**) First experiment; (**b**) second experiment.

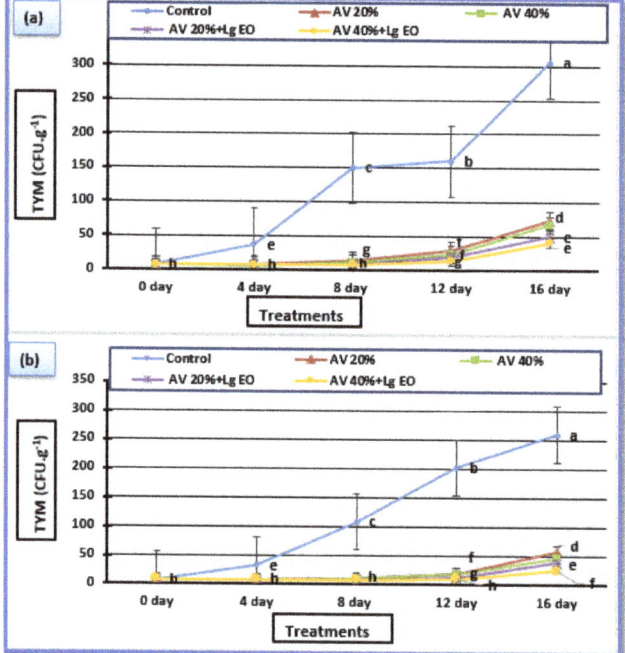

Figure 12. TYM (CFU g^{-1}) (mean ± S.E.) of strawberry fruits stored at 5 °C as affected by coating treatments when stored for different lengths of time in both experiments. The mean ± S.E. of treatments in the figures with the same letters shows a nonsignificant difference according to Duncan multiple range test for $p \leq 0.05$. AV: *A. vera* gel. (**a**) First experiment; (**b**) second experiment.

At the end storage, the fruits treated with *A. vera* gel 40% and lemongrass EO recorded a lower microbial count than the fruits that underwent all other treatments. The control sample had the highest microbial count at the end of storage. Therefore, fruits treated with *A. vera* gel at 20% and 40% with lemongrass EO were able to resist fungal growth better than fruits treated with *A. vera* gel alone, and *A. vera* also remarkably reduced aerobic bacteria and yeast and mold counts during the 16 days of storage. In our study, a combination of *A. vera* gel and lemongrass EO seemed to have a synergistic effect on controlling microbial growth in strawberry during storage in a concentration-dependent manner. In a similar study, the effect of coating using *A. vera* gel 20% with 3% starch + 0.1% mandarin EO on physical and mechanical properties of blackberry indicated that this coating is suitable due to its thickness and shows the best mechanical properties observed, providing the fruit with greater thickness and improving its resistance to possible damage [90]. A bioactive coating combined with cinnamon EO significantly reduced mesophilic bacteria and yeast and molds in apple slices during a storage time of 25 days [91].

Coatings of *A. vera* gel with lemongrass EO effectively controlled or inhibited microbial populations (Figures 11 and 12). The present results are comparable with the results when sweet cherries and table grapes were coated with *A. vera* gel, which showed a reduction in the populations of mesophilic aerobic bacteria and yeast and mold during storage. *A. vera* gel compounds such as saponins, acemannan, and anthraquinones derivatives are reported to be responsible for antibacterial activity [92].

Rasouli et al. [93] reported that the inhibition effect of *A. vera* gel on microbial load arises from the presence of ingredients such as aleonin and aloeemodin, which is a possible rationale for the diminishing of germination and mycelial growth of fungi. Antara et al. [94] stated that the compounds responsible for the antimicrobial mechanism of lemongrass EO are a group of terpenoids, e.g., geranial (β-citral) and neral (α-citral).

Certain phenolic compounds are reported to be associated with antioxidant activity, such as radical-scavenging activity [95]. As shown in Figure 8, extracts contain polyphenols and flavonoids, which exhibit not only antioxidant activity but also antimicrobial activity (e.g., ferulic acid, caffeic acid, *p*-coumaric acid, syringic acid, sinapic acid, and cinnamic acid). Therefore, the antioxidant and antibacterial activities of extracts are linked to the activity of individual phenolic and flavonoid compounds. Extracts of *A. vera* gel 40% with lemongrass EO presented higher antimicrobial activity due to their high contents of total phenol content, total flavonoid content, and terpenoids. These results thus suggest that *A. vera* gel with lemongrass EO can be used as a natural antimicrobial.

Strawberries coated with *A. vera* gel 40% + lemongrass EO 1% had increased storage time because this treatment contributed to a decrease in the decay rate. Therefore, *A. vera* gel with lemongrass EO helps maintain the quality of strawberries during storage.

4. Conclusions

The results obtained from this study show that an *Aloe vera* gel coating with lemongrass EO on strawberry fruit has a positive influence on the quality and biochemical properties of the fruit and reduces the microbial growth on the fruit. It was observed that between the two best treatments, treatment with *A. vera* gel 40% followed by *A. vera* gel 20% + lemongrass EO 1% gives better results as compared to treatment with *A. vera* gel (20% or 40%); additionally, for both treatments, significantly higher results were observed as compared to control. Treatment with *A. vera* gel 40% + lemongrass EO 1% enhanced the shelf life of the fruit at 5 °C by maintaining its quality and reducing the spoilage by postharvest pathogens. Thus, this treatment has the potential to be practiced on other different types of strawberry fruits. The present results may show an economical and natural way to improve fruit quality as well as resistance to a wide range of microorganisms.

Author Contributions: Conceptualization, H.S.H., M.E.-H. and D.Y.A.-E.; methodology, H.S.H., M.E.-H., M.S.R.A.E.-L., A.A.A.-H. and D.Y.A.-E.; software, H.S.H., M.S.R.A.E.-L., I.M.G., M.E.-H., A.A.A.-H., M.A. and D.Y.A.-E.; validation, H.S.H., M.E.-H., I.M.G., M.S.R.A.E.-L. and D.Y.A.-E.; formal analysis, H.S.H., M.E.-H. and D.Y.A.-E.; investigation, H.S.H., M.S.R.A.E.-L., I.M.G., M.E.-H., A.A.A.-H., M.A. and D.Y.A.-E.; resources, H.S.H., M.S.R.A.E.-L., I.M.G., M.E.-H., A.A.A.-H., M.A. and D.Y.A.-E.; data curation, H.S.H., M.S.R.A.E.-L., M.E.-H. and D.Y.A.-E.; writing—original draft preparation, H.S.H., M.S.R.A.E.-L., I.M.G., M.E.-H., H.M.A., M.A. and D.Y.A.-E.; writing—review and editing, H.S.H., M.S.R.A.E.-L., I.M.G., M.E.-H., A.A.A.-H., M.A. and D.Y.A.-E.; visualization, H.S.H., M.S.R.A.E.-L., I.M.G., M.E.-H., A.A.A.-H., M.A. and D.Y.A.-E.; supervision, H.S.H. and D.Y.A.-E.; project administration, D.Y.A.-E.; funding acquisition, A.A.A.-H. and H.M.A. Article revision, figure amendment, and proofreading of the revised article, H.S.H., M.E.-H., D.Y.A.-E. and H.M.A. All authors have read and agreed to the published version of the manuscript.

Funding: This research was funded by the Researchers Supporting Project number (RSP-2021/186) King Saud University, Riyadh, Saudi Arabia.

Institutional Review Board Statement: Not applicable.

Informed Consent Statement: Not applicable.

Data Availability Statement: Not applicable.

Acknowledgments: Authors would like to extend their sincere appreciation to the Researchers Supporting Project number (RSP-2021/186), King Saud University, Riyadh, Saudi Arabia. Authors also acknowledge Alexandria University, Egypt for facilitating performing this research work as well.

Conflicts of Interest: The authors declare no conflict of interest.

References

1. Ahn, M.G.; Kim, D.S.; Ahn, S.R.; Sim, H.S.; Kim, S.; Kim, S.K. Characteristics and Trends of Strawberry Cultivars throughout the Cultivation Season in a Greenhouse. *Horticulturae* **2021**, *7*, 30. [CrossRef]
2. Ilari, A.; Toscano, G.; Boakye-Yiadom, K.A.; Duca, D.; Foppa Pedretti, E. Life Cycle Assessment of Protected Strawberry Productions in Central Italy. *Sustainability* **2021**, *13*, 4879. [CrossRef]
3. Mozafari, A.a.; Dedejani, S.; Ghaderi, N. Positive responses of strawberry (*Fragaria* × *ananassa* Duch.) explants to salicylic and iron nanoparticle application under salinity conditions. *Plant Cell Tissue Organ Cult.* **2018**, *134*, 267–275. [CrossRef]
4. Abu Salha, B.; Gedanken, A. Extending the Shelf Life of Strawberries by the Sonochemical Coating of their Surface with Nanoparticles of an Edible Anti-Bacterial Compound. *Appl. Nano* **2021**, *2*, 14–24. [CrossRef]
5. Kumar, R.; Bakshi, P.; Singh, M.; Singh, A.; Vikas, V.; Srivatava, J.; Kumar, V.; Gupta, V. Organic production of strawberry: A review. *Int. J. Chem. Stud.* **2018**, *6*, 1231–1236.
6. Eum, H.-L.; Han, S.-H.; Lee, E.-J. High-CO2 Treatment Prolongs the Postharvest Shelf Life of Strawberry Fruits by Reducing Decay and Cell Wall Degradation. *Foods* **2021**, *10*, 1649. [CrossRef] [PubMed]
7. Chiabrando, V.; Garavaglia, L.; Giacalone, G. The Postharvest Quality of Fresh Sweet Cherries and Strawberries with an Active Packaging System. *Foods* **2019**, *8*, 335. [CrossRef]
8. De Corato, U. Improving the shelf-life and quality of fresh and minimally-processed fruits and vegetables for a modern food industry: A comprehensive critical review from the traditional technologies into the most promising advancements. *Crit. Rev. Food Sci. Nutr.* **2020**, *60*, 940–975. [CrossRef] [PubMed]
9. Ziv, C.; Fallik, E. Postharvest Storage Techniques and Quality Evaluation of Fruits and Vegetables for Reducing Food Loss. *Agronomy* **2021**, *11*, 1133. [CrossRef]
10. Agriopoulou, S.; Stamatelopoulou, E.; Sachadyn-Król, M.; Varzakas, T. Lactic Acid Bacteria as Antibacterial Agents to Extend the Shelf Life of Fresh and Minimally Processed Fruits and Vegetables: Quality and Safety Aspects. *Microorganisms* **2020**, *8*, 952. [CrossRef] [PubMed]
11. Romanazzi, G.; Feliziani, E.; Sivakumar, D. Chitosan, a Biopolymer With Triple Action on Postharvest Decay of Fruit and Vegetables: Eliciting, Antimicrobial and Film-Forming Properties. *Front. Microbiol.* **2018**, *9*, 2745. [CrossRef] [PubMed]
12. Tzortzakis, N.; Xylia, P.; Chrysargyris, A. Sage Essential Oil Improves the Effectiveness of *Aloe vera* Gel on Postharvest Quality of Tomato Fruit. *Agronomy* **2019**, *9*, 635. [CrossRef]
13. Abd-Elkader, D.Y.; Salem, M.Z.M.; Komeil, D.A.; Al-Huqail, A.A.; Ali, H.M.; Salah, A.H.; Akrami, M.; Hassan, H.S. Post-Harvest Enhancing and *Botrytis cinerea* Control of Strawberry Fruits Using Low Cost and Eco-Friendly Natural Oils. *Agronomy* **2021**, *11*, 1246. [CrossRef]
14. Farina, V.; Passafiume, R.; Tinebra, I.; Palazzolo, E.; Sortino, G. Use of *Aloe vera* Gel-Based Edible Coating with Natural Anti-Browning and Anti-Oxidant Additives to Improve Post-Harvest Quality of Fresh-Cut 'Fuji' Apple. *Agronomy* **2020**, *10*, 515. [CrossRef]

15. Ghoora, M.D.; Srividya, N. Effect of Packaging and Coating Technique on Postharvest Quality and Shelf Life of *Raphanus sativus* L. and *Hibiscus sabdariffa* L. Microgreens. *Foods* **2020**, *9*, 653. [CrossRef] [PubMed]
16. Hosseini, S.F.; Amraie, M.; Salehi, M.; Mohseni, M.; Aloui, H. Effect of chitosan-based coatings enriched with savory and/or tarragon essential oils on postharvest maintenance of kumquat (*Fortunella* sp.) fruit. *Food Sci. Nutr.* **2019**, *7*, 155–162. [CrossRef] [PubMed]
17. Yuan, G.; Chen, X.; Li, D. Chitosan films and coatings containing essential oils: The antioxidant and antimicrobial activity, and application in food systems. *Food Res. Int.* **2016**, *89*, 117–128. [CrossRef] [PubMed]
18. Khodaei, D.; Hamidi-Esfahani, Z.; Rahmati, E. Effect of edible coatings on the shelf-life of fresh strawberries: A comparative study using TOPSIS-Shannon entropy method. *NFS J.* **2021**, *23*, 17–23. [CrossRef]
19. Kahramanoğlu, İ. Effects of lemongrass oil application and modified atmosphere packaging on the postharvest life and quality of strawberry fruits. *Sci. Hortic.* **2019**, *256*, 108527. [CrossRef]
20. Nair, M.S.; Tomar, M.; Punia, S.; Kukula-Koch, W.; Kumar, M. Enhancing the functionality of chitosan- and alginate-based active edible coatings/films for the preservation of fruits and vegetables: A review. *Int. J. Biol. Macromol.* **2020**, *164*, 304–320. [CrossRef] [PubMed]
21. Singh, A.; Verma, K.; Kumar, D.; Nilofer; Lothe, N.B.; Kumar, A.; Chaudhary, A.; Kaur, P.; Singh, K.P.; Singh, A.K.; et al. Optimized irrigation regime and planting technique improve yields and economics in *aloe vera* [*Aloe barbadensis* (Miller)]. *Ind. Crops Prod.* **2021**, *167*, 113539. [CrossRef]
22. Habeeb, F.; Shakir, E.; Bradbury, F.; Cameron, P.; Taravati, M.R.; Drummond, A.J.; Gray, A.I.; Ferro, V.A. Screening methods used to determine the anti-microbial properties of *Aloe vera* inner gel. *Methods* **2007**, *42*, 315–320. [CrossRef] [PubMed]
23. Zapata, P.J.; Navarro, D.; Guillén, F.; Castillo, S.; Martínez-Romero, D.; Valero, D.; Serrano, M. Characterisation of gels from different *Aloe* spp. as antifungal treatment: Potential crops for industrial applications. *Ind. Crop. Prod.* **2013**, *42*, 223–230. [CrossRef]
24. Morillon, V.; Debeaufort, F.; Blond, G.; Capelle, M.; Voilley, A. Factors Affecting the Moisture Permeability of Lipid-Based Edible Films: A Review. *Crit. Rev. Food Sci. Nutr.* **2002**, *42*, 67–89. [CrossRef] [PubMed]
25. Maan, A.A.; Nazir, A.; Khan, M.K.I.; Ahmad, T.; Zia, R.; Murid, M.; Abrar, M. The therapeutic properties and applications of *Aloe vera*: A review. *J. Herb. Med.* **2018**, *12*, 1–10. [CrossRef]
26. Nicolau-Lapeña, I.; Colàs-Medà, P.; Alegre, I.; Aguiló-Aguayo, I.; Muranyi, P.; Viñas, I. Aloe vera gel: An update on its use as a functional edible coating to preserve fruits and vegetables. *Prog. Org. Coat.* **2021**, *151*, 106007. [CrossRef]
27. Sogvar, O.B.; Koushesh Saba, M.; Emamifar, A. *Aloe vera* and ascorbic acid coatings maintain postharvest quality and reduce microbial load of strawberry fruit. *Postharvest Biol. Technol.* **2016**, *114*, 29–35. [CrossRef]
28. Hasan, M.U.; Riaz, R.; Malik, A.U.; Khan, A.S.; Anwar, R.; Rehman, R.N.U.; Ali, S. Potential of *Aloe vera* gel coating for storage life extension and quality conservation of fruits and vegetables: An overview. *J. Food Biochem.* **2021**, *45*, e13640. [CrossRef] [PubMed]
29. Ozturk, B.; Karakaya, O.; Yıldız, K.; Saracoglu, O. Effects of *Aloe vera* gel and MAP on bioactive compounds and quality attributes of cherry laurel fruit during cold storage. *Sci. Hortic.* **2019**, *249*, 31–37. [CrossRef]
30. Ali, S.; Khan, A.S.; Nawaz, A.; Anjum, M.A.; Naz, S.; Ejaz, S.; Hussain, S. *Aloe vera* gel coating delays postharvest browning and maintains quality of harvested litchi fruit. *Postharvest Biol. Technol.* **2019**, *157*, 110960. [CrossRef]
31. Marpudi, S.L.; Pushkala, R.; Srividya, N. *Aloe vera* gel coating for post-harvest quality maintenance of fresh fig fruits. *Res. J. Pharm. Biol. Chem. Sci.* **2013**, *4*, 878–887.
32. Ding, P.; Lee, Y. Use of essential oils for prolonging postharvest life of fresh fruits and vegetables. *Int. Food Res. J.* **2019**, *26*, 363–366.
33. Perumal, A.B.; Sellamuthu, P.S.; Nambiar, R.B.; Sadiku, E.R. Effects of Essential Oil Vapour Treatment on the Postharvest Disease Control and Different Defence Responses in Two Mango (*Mangifera indica* L.) Cultivars. *Food Bioprocess Technol.* **2017**, *10*, 1131–1141. [CrossRef]
34. Tawfeek, M.E.; Ali, H.M.; Akrami, M.; Salem, M.Z.M. Potential Insecticidal Activity of Four Essential Oils against the Rice Weevil, *Sitophilus oryzae* (L.) (Coleoptera: Curculionidae). *BioResources* **2021**, *16*, 7767–7783. [CrossRef]
35. Moustafa, M.A.M.; Awad, M.; Amer, A.; Hassan, N.N.; Ibrahim, E.-D.S.; Ali, H.M.; Akrami, M.; Salem, M.Z.M. Insecticidal Activity of Lemongrass Essential Oil as an Eco-Friendly Agent against the Black Cutworm *Agrotis ipsilon* (Lepidoptera: Noctuidae). *Insects* **2021**, *12*, 737. [CrossRef] [PubMed]
36. Mansour, M.M.A.; El-Hefny, M.; Salem, M.Z.M.; Ali, H.M. The Biofungicide Activity of Some Plant Essential Oils for the Cleaner Production of Model Linen Fibers Similar to Those Used in Ancient Egyptian Mummification. *Processes* **2020**, *8*, 79. [CrossRef]
37. Shah, G.; Shri, R.; Panchal, V.; Sharma, N.; Singh, B.; Mann, A.S. Scientific basis for the therapeutic use of *Cymbopogon citratus*, stapf (Lemon grass). *J. Adv. Pharm. Technol. Res.* **2011**, *2*, 3–8. [CrossRef] [PubMed]
38. Utama, I.M.S.; Yulianti, N.L.; Prastya, O.A.; Luther, G. Sesame and Lemon Grass Oils as Coating Materials to Reduce the Deterioration of Tomato Fruits during Storage. In Proceedings of the Indonesian Horticultural Society National Seminar, Malang, Indonesia, 6–9 November 2014; pp. 1–9.
39. Helal, G.A.; Sarhan, M.M.; Shahla, A.N.K.A.; Ei-Khair, E.K.A. Antimicrobial Activity of Some Essential Oils Against Microorganisms Deteriorating Fruit Juices. *Mycobiology* **2006**, *34*, 219–229. [CrossRef] [PubMed]

40. Alamgir, A.N.M. (Ed.) Secondary Metabolites: Secondary Metabolic Products Consisting of C and H; C, H, and O; N, S, and P Elements; and O/N Heterocycles. In *Therapeutic Use of Medicinal Plants and Their Extracts: Volume 2: Phytochemistry and Bioactive Compounds*; Springer International Publishing: Cham, Switzerland, 2018; pp. 165–309.
41. El-Gioushy, S.F.; Baiea, M.H.M. Impact of gelatin, lemongrass oil and peppermint oil on storability and fruit quality of Samany date palm under cold storage. *Bull. Natl. Res. Cent.* **2020**, *44*, 14. [CrossRef]
42. Trang, D.T.; Hoang, T.K.V.; Nguyen, T.T.M.; Van Cuong, P.; Dang, N.H.; Dang, H.D.; Nguyen Quang, T.; Dat, N.T. Essential Oils of Lemongrass *Cymbopogon citratus* Stapf Induces Apoptosis and Cell Cycle Arrest in A549 Lung Cancer Cells. *BioMed Res. Int.* **2020**, *2020*, 5924856. [CrossRef] [PubMed]
43. Tran, T.H.; Tran, T.K.N.; Ngo, T.C.Q.; Pham, T.N.; Bach, L.G.; Phan, N.Q.A.; Le, T.H.N. Color and composition of beauty products formulated with lemongrass essential oil: Cosmetics formulation with lemongrass essential oil. *Open Chem.* **2021**, *19*, 820–829. [CrossRef]
44. Whitaker, V.M.; Chandler, C.K.; Santos, B.M.; Peres, N.; Cecilia do Nascimento Nunes, M.; Plotto, A.; Sims, C.A. Winterstar™ ('FL 05-107') Strawberry. *HortScience* **2012**, *47*, 296–298. [CrossRef]
45. Kumar, S.; Tiku, A.B. Immunomodulatory potential of acemannan (polysaccharide from *Aloe vera*) against radiation induced mortality in Swiss albino mice. *Food Agric. Immunol.* **2016**, *27*, 72–86. [CrossRef]
46. El-Hefny, M.; Abo Elgat, W.A.A.; Al-Huqail, A.A.; Ali, H.M. Essential and Recovery Oils from *Matricaria chamomilla* Flowers as Environmentally Friendly Fungicides Against Four Fungi Isolated from Cultural Heritage Objects. *Processes* **2019**, *7*, 809. [CrossRef]
47. Okla, M.K.; Alamri, S.A.; Salem, M.Z.M.; Ali, H.M.; Behiry, S.I.; Nasser, R.A.; Alaraidh, I.A.; Al-Ghtani, S.M.; Soufan, W. Yield, Phytochemical Constituents, and Antibacterial Activity of Essential Oils from the Leaves/Twigs, Branches, Branch Wood, and Branch Bark of Sour Orange (*Citrus aurantium* L.). *Processes* **2019**, *7*, 363. [CrossRef]
48. Shehata, S.A.; Abdeldaym, E.A.; Ali, M.R.; Mohamed, R.M.; Bob, R.I.; Abdelgawad, K.F. Effect of Some Citrus Essential Oils on Post-Harvest Shelf Life and Physicochemical Quality of Strawberries during Cold Storage. *Agronomy* **2020**, *10*, 1466. [CrossRef]
49. Granato, D.; Masson, M.L. Instrumental color and sensory acceptance of soy-based emulsions: A response surface approach. *Food Sci. Technol.* **2010**, *30*, 1090–1096. [CrossRef]
50. The Association of Official Analytical Chemists (A.O.A.C.). *Official Methods of Analysis*, 17th ed.; The Association of Official Analytical Chemists: Gaithersburg, MD, USA, 2000.
51. Fuleki, T.; Francis, F.J. Quantitative Methods for Anthocyanins. *J. Food Sci.* **1968**, *33*, 72–77. [CrossRef]
52. El-Hefny, M.; Ashmawy, N.A.; Salem, M.Z.M.; Salem, A.Z.M. Antibacterial activities of the phytochemicals-characterized extracts of *Callistemon viminalis*, *Eucalyptus camaldulensis* and *Conyza dioscoridis* against the growth of some phytopathogenic bacteria. *Microb. Pathog.* **2017**, *113*, 348–356. [CrossRef] [PubMed]
53. Hwang, E.-S.; Thi, N.D. Effects of Extraction and Processing Methods on Antioxidant Compound Contents and Radical Scavenging Activities of Laver (*Porphyra tenera*). *Prev. Nutr. Food Sci.* **2014**, *19*, 40–48. [CrossRef] [PubMed]
54. Hassan, H.S.; Mohamed, A.A.; Feleafel, M.N.; Salem, M.Z.M.; Ali, H.M.; Akrami, M.; Abd-Elkader, D.Y. Natural Plant Extracts and Microbial Antagonists to Control Fungal Pathogens and Improve the Productivity of Zucchini (*Cucurbita pepo* L.) In Vitro and in Greenhouse. *Horticulturae* **2021**, *7*, 470. [CrossRef]
55. Salem, M.Z.M.; Ali, H.M.; Akrami, M. *Moringa oleifera* seeds-removed ripened pods as alternative for papersheet production: Antimicrobial activity and their phytoconstituents profile using HPLC. *Sci. Rep.* **2021**, *11*, 19027. [CrossRef]
56. Herigstad, B.; Hamilton, M.; Heersink, J. How to optimize the drop plate method for enumerating bacteria. *J. Microbiol. Methods* **2001**, *44*, 121–129. [CrossRef]
57. Snedecor, W.; Cochran, G. *Statistical Methods*, 8th ed.; Iowa State University Press: Ames, IA, USA, 1989.
58. López, A.; De Tangil, M.S.; Vega-Orellana, O.; Ramírez, A.S.; Rico, M. Phenolic Constituents, Antioxidant and Preliminary Antimycoplasmic Activities of Leaf Skin and Flowers of *Aloe vera* (L.) Burm. f. (syn. *A. barbadensis* Mill.) from the Canary Islands (Spain). *Molecules* **2013**, *18*, 4942–4954. [CrossRef] [PubMed]
59. Elbandy, M.A.; Abed, S.; Gad, S.; Abdel-Fadeel, M. *Aloe vera* gel as a functional ingredient and natural preservative in mango nectar. *World J. Dairy Food Sci.* **2014**, *9*, 191–203. [CrossRef]
60. Numan, I.N. Identification of Flavonoids and Phenolic Compound in *Aloe vera* gel by HPLC. *Tikrit J. Pure Sci.* **2018**, *23*, 91–94. [CrossRef]
61. Do, D.N.; Nguyen, H.T.T.; Huynh, T.H.; Nguyen, N.P.; Luu, X.C. Chemical composition, antibacterial and antioxidant activities of lemongrass (*Cymbopogon citratus*) essential oil and its fractions obtained by vacuum distillation. *IOP Conf. Ser. Mater. Sci. Eng.* **2021**, *1166*, 012051. [CrossRef]
62. Verma, R.K.; Verma, R.S.; Chauhan, A.; Bisht, A. Evaluation of essential oil yield and chemical composition of eight lemongrass (*Cymbopogon* spp.) cultivars under Himalayan region. *J. Essent. Oil Res.* **2015**, *27*, 197–203. [CrossRef]
63. Shaikh, M.N.; Suryawanshi, Y.C.; Mokat, D.N. Volatile Profiling and Essential Oil Yield of *Cymbopogon citratus* (DC.) Stapf Treated with Rhizosphere Fungi and Some Important Fertilizers. *J. Essent. Oil Bear. Plants* **2019**, *22*, 477–483. [CrossRef]
64. Tajidin, N.; Ahmad, S.; Rosenani, A.; Azimah, H.; Munirah, M. Chemical composition and citral content in lemongrass (*Cymbopogon citratus*) essential oil at three maturity stages. *Afr. J. Biotechnol.* **2012**, *11*, 2685–2693. [CrossRef]
65. Rehman, M.A.; Asi, M.R.; Hameed, A.; Bourquin, L.D. Effect of Postharvest Application of *Aloe vera* Gel on Shelf Life, Activities of Anti-Oxidative Enzymes, and Quality of 'Gola' Guava Fruit. *Foods* **2020**, *9*, 1361. [CrossRef] [PubMed]

66. Khaliq, G.; Abbas, H.T.; Ali, I.; Waseem, M. *Aloe vera* gel enriched with garlic essential oil effectively controls anthracnose disease and maintains postharvest quality of banana fruit during storage. *Hortic. Environ. Biotechnol.* **2019**, *60*, 659–669. [CrossRef]
67. Castillo, S.; Navarro, D.; Zapata, P.J.; Guillén, F.; Valero, D.; Serrano, M.; Martínez-Romero, D. Antifungal efficacy of *Aloe vera* in vitro and its use as a preharvest treatment to maintain postharvest table grape quality. *Postharvest Biol. Technol.* **2010**, *57*, 183–188. [CrossRef]
68. Ravanfar, R.; Niakousari, M.; Maftoonazad, N. Postharvest sour cherry quality and safety maintenance by exposure to Hot- water or treatment with fresh Aloe vera gel. *J. Food Sci. Technol.* **2014**, *51*, 2872–2876. [CrossRef] [PubMed]
69. Hazrati, S.; Beyraghdar Kashkooli, A.; Habibzadeh, F.; Tahmasebi-Sarvestani, Z.; Sadeghi, A.R. Evaluation of *Aloe vera* Gel as an Alternative Edible Coating for Peach Fruits During Cold Storage Period. *Gesunde Pflanz.* **2017**, *69*, 131–137. [CrossRef]
70. Pinzon, M.I.; Sanchez, L.T.; Garcia, O.R.; Gutierrez, R.; Luna, J.C.; Villa, C.C. Increasing shelf life of strawberries (*Fragaria* ssp) by using a banana starch-chitosan-*Aloe vera* gel composite edible coating. *Int. J. Food Sci. Technol.* **2020**, *55*, 92–98. [CrossRef]
71. Mohammadi, L.; Ramezanian, A.; Tanaka, F.; Tanaka, F. Impact of *Aloe vera* gel coating enriched with basil (*Ocimum basilicum* L.) essential oil on postharvest quality of strawberry fruit. *J. Food Meas. Charact.* **2021**, *15*, 353–362. [CrossRef]
72. Del-Valle, V.; Hernández-Muñoz, P.; Guarda, A.; Galotto, M.J. Development of a cactus-mucilage edible coating (*Opuntia ficus indica*) and its application to extend strawberry (*Fragaria ananassa*) shelf-life. *Food Chem.* **2005**, *91*, 751–756. [CrossRef]
73. Nadim, Z.; Ahmadi, E.; Sarikhani, H.; Amiri Chayjan, R. Effect of Methylcellulose-Based Edible Coating on Strawberry Fruit's Quality Maintenance During Storage. *J. Food Process. Preserv.* **2015**, *39*, 80–90. [CrossRef]
74. Atress, A.S.H.; El-Mogy, M.; Aboul-Anean, H.; Alsanius, B. Improving strawberry fruit storability by edible coating as a carrier of thymol or calcium chloride. *J. Hortic. Sci. Ornam. Plants* **2010**, *2*, 88–97.
75. Gol, N.B.; Patel, P.R.; Rao, T.V.R. Improvement of quality and shelf-life of strawberries with edible coatings enriched with chitosan. *Postharvest Biol. Technol.* **2013**, *85*, 185–195. [CrossRef]
76. Dhital, R.; Mora, N.B.; Watson, D.G.; Kohli, P.; Choudhary, R. Efficacy of limonene nano coatings on post-harvest shelf life of strawberries. *LWT Food Sci. Technol.* **2018**, *97*, 124–134. [CrossRef]
77. Hassanpour, H. Effect of *Aloe vera* gel coating on antioxidant capacity, antioxidant enzyme activities and decay in raspberry fruit. *LWT Food Sci. Technol.* **2015**, *60*, 495–501. [CrossRef]
78. Azarakhsh, N.; Osman, A.; Ghazali, H.M.; Tan, C.P.; Mohd Adzahan, N. Lemongrass essential oil incorporated into alginate-based edible coating for shelf-life extension and quality retention of fresh-cut pineapple. *Postharvest Biol. Technol.* **2014**, *88*, 1–7. [CrossRef]
79. García-Alonso, M.; Rimbach, G.; Rivas-Gonzalo, J.C.; de Pascual-Teresa, S. Antioxidant and Cellular Activities of Anthocyanins and Their Corresponding Vitisins A Studies in Platelets, Monocytes, and Human Endothelial Cells. *J. Agric. Food Chem.* **2004**, *52*, 3378–3384. [CrossRef] [PubMed]
80. Bhat, R.; Stamminger, R. Impact of ultraviolet radiation treatments on the physicochemical properties, antioxidants, enzyme activity and microbial load in freshly prepared hand pressed strawberry juice. *Food Sci. Technol. Int.* **2014**, *21*, 354–363. [CrossRef] [PubMed]
81. Shah, S.; Hashmi, M.S. Chitosan–*Aloe vera* gel coating delays postharvest decay of mango fruit. *Hortic. Environ. Biotechnol.* **2020**, *61*, 279–289. [CrossRef]
82. Ehtesham Nia, A.; Taghipour, S.; Siahmansour, S. Pre-harvest application of chitosan and postharvest *Aloe vera* gel coating enhances quality of table grape (*Vitis vinifera* L. cv. 'Yaghouti') during postharvest period. *Food Chem.* **2021**, *347*, 129012. [CrossRef]
83. Hu, Q.; Hu, Y.; Xu, J. Free radical-scavenging activity of *Aloe vera* (*Aloe barbadensis* Miller) extracts by supercritical carbon dioxide extraction. *Food Chem.* **2005**, *91*, 85–90. [CrossRef]
84. Hidayati, J.R.; Yudiati, E.; Pringgenies, D.; Oktaviyanti, D.T.; Kusuma, A.P. Comparative Study on Antioxidant Activities, Total Phenolic Compound and Pigment Contents of Tropical *Spirulina platensis*, *Gracilaria arcuata* and *Ulva lactuca* Extracted in Different Solvents Polarity. *E3S Web Conf.* **2020**, *147*, 03012. [CrossRef]
85. Tungmunnithum, D.; Thongboonyou, A.; Pholboon, A.; Yangsabai, A. Flavonoids and Other Phenolic Compounds from Medicinal Plants for Pharmaceutical and Medical Aspects: An Overview. *Medicines* **2018**, *5*, 93. [CrossRef]
86. Hannum, S.M. Potential Impact of Strawberries on Human Health: A Review of the Science. *Crit. Rev. Food Sci. Nutr.* **2004**, *44*, 1–17. [CrossRef]
87. Co, H.; Markakis, P. Flavonoid Compounds in the Strawberry Fruit. *J. Food Sci.* **1968**, *33*, 281–283. [CrossRef]
88. Sun, Y.; Asghari, M.; Zahedipour-Sheshgelani, P. Foliar Spray with 24-Epibrassinolide Enhanced Strawberry Fruit Quality, Phytochemical Content, and Postharvest Life. *J. Plant Growth Regul.* **2020**, *39*, 920–929. [CrossRef]
89. Esmaeili, Y.; Zamindar, N.; Paidari, S.; Ibrahim, S.A.; Mohammadi Nafchi, A. The synergistic effects of *Aloe vera* gel and modified atmosphere packaging on the quality of strawberry fruit. *J. Food Process. Preserv.* **2021**, *45*, e16003. [CrossRef]
90. Arrubla Vélez, J.P.; Guerrero Álvarez, G.E.; Vargas Soto, M.C.; Cardona Hurtado, N.; Pinzón, M.I.; Villa, C.C. *Aloe Vera* Gel Edible Coating for Shelf Life and Antioxidant Proprieties Preservation of Andean Blackberry. *Processes* **2021**, *9*, 999. [CrossRef]
91. Solís-Contreras, G.A.; Rodríguez-Guillermo, M.C.; de la Luz Reyes-Vega, M.; Aguilar, C.N.; Rebolloso-Padilla, O.N.; Corona-Flores, J.; de Abril Alexandra Soriano-Melgar, L.; Ruelas-Chacon, X. Extending Shelf-Life and Quality of Minimally Processed Golden Delicious Apples with Three Bioactive Coatings Combined with Cinnamon Essential Oil. *Foods* **2021**, *10*, 597. [CrossRef]
92. Martínez-Romero, D.; Alburquerque, N.; Valverde, J.M.; Guillén, F.; Castillo, S.; Valero, D.; Serrano, M. Postharvest sweet cherry quality and safety maintenance by *Aloe vera* treatment: A new edible coating. *Postharvest Biol. Technol.* **2006**, *39*, 93–100. [CrossRef]

93. Rasouli, M.; Koushesh Saba, M.; Ramezanian, A. Inhibitory effect of salicylic acid and *Aloe vera* gel edible coating on microbial load and chilling injury of orange fruit. *Sci. Hortic.* **2019**, *247*, 27–34. [CrossRef]
94. Antara, N.S.; Paramita, D.; Duwipayana, A.A.; Gunam, I. Inhibitory activity of lemongrass essential oil against *Eschericia coli, Staphylococcus aureus*, and *Vibrio cholera*. In Proceedings of the Seminar Nasional Patpi 2013, Jember, Indonesia, 26–29 August 2013.
95. Velkov, Z.; Balabanova, E.; Tadjer, A. Radical scavenging activity prediction of o-coumaric acid thioamide. *J. Mol. Struct. THEOCHEM* **2007**, *821*, 133–138. [CrossRef]

Article

Effects of *Pleurotus ostreatus* on Physicochemical Properties and Residual Nitrite of the Pork Sausage

Xiaoguang Wu [1], Peiren Wang [2], Qiyao Xu [1], Bin Jiang [1], Liangyu Li [1], Lili Ren [3], Xiuyi Li [1] and Liyan Wang [1,*]

[1] College of Food Science and Engineering, Jilin Agricultural University, 2888 Xincheng Street, Changchun 130118, China; xiaoguangw@jlau.edu.cn (X.W.); xqy19970326@163.com (Q.X.); jiangbin1117@126.com (B.J.); lly2115513336@163.com (L.L.); lxy18104304692@163.com (X.L.)

[2] Institute for Materials Discovery, University College London, London WC1E 7JE, UK; peiren.wang.20@alumni.ucl.ac.uk

[3] Key Laboratory of Bionic Engineering (Ministry of Education), College of Biological and Agricultural Engineering, Jilin University, 5988 Renmin Street, Changchun 130022, China; liliren@jlu.edu.cn

* Correspondence: wangliyan@jlau.edu.cn; Tel.: +86-8453 3312

Abstract: In this work, a novel sausage incorporated with the *Pleurotus ostreatus* (PO) puree was successfully developed to reduce the residual nitrite and lipid oxidation during refrigerated storage (4 ± 1 °C) for 20 days. Five recipes with the supplement proportion of 0 wt.%, 10 wt.%, 20 wt.%, 30 wt.%, and 40 wt.% PO were produced and their physicochemical properties, nitrite residue, and sensory characteristics were measured. The results show that the content of moisture and all the essential amino acids (especially lysine and leucine) and the non-essential amino acids (especially aspartic and glutamic), lightness, springiness, and water holding capacity of the sausages were increased. However, the content of protein, fat, ash, pH, redness, hardness, gumminess, and chewiness of the sausages was decreased. For the sensory evaluation, the sausage with 20 wt.% PO had better sensory performance including flavor, aroma, and acceptability compared with other experimental groups and the control group. Moreover, the sausages with PO reduced the residual nitrite and inhibited lipid oxidation during storage. All of these results indicate that adding PO puree into pork sausage is a realizable and effective way to obtain nutritional and healthy pork sausages.

Keywords: mushroom; sausage; nitrite residue; meat product

Citation: Wu, X.; Wang, P.; Xu, Q.; Jiang, B.; Li, L.; Ren, L.; Li, X.; Wang, L. Effects of *Pleurotus ostreatus* on Physicochemical Properties and Residual Nitrite of the Pork Sausage. *Coatings* **2022**, *12*, 484. https://doi.org/10.3390/coatings12040484

Academic Editor: Maria Jose Fabra

Received: 25 February 2022
Accepted: 2 April 2022
Published: 4 April 2022

Publisher's Note: MDPI stays neutral with regard to jurisdictional claims in published maps and institutional affiliations.

Copyright: © 2022 by the authors. Licensee MDPI, Basel, Switzerland. This article is an open access article distributed under the terms and conditions of the Creative Commons Attribution (CC BY) license (https://creativecommons.org/licenses/by/4.0/).

1. Introduction

Sausage is a popular meat product manufactured from different meat species such as pork, beef, chicken, fish, and buffalo [1–3]. This kind of meat product has important economic value for the meat-packing industry and is relished by consumers around the world for its delicious taste and high nutrition. In order to produce the sausages with high quality, nitrite is widely used as a preservative that can control foodborne pathogens [4]. Moreover, the additional function of the nitrites is to prevent lipid oxidation and rancidity, facilitate stabilization of the bright red color, and guarantee a typical "cured" flavor [5–7]. To fight against lipid oxidation, the nitrites could be associated with the binding of heme and prevent the release of the catalytic iron [8]. For keeping the bright red color of meat products, the nitrites can bind to myoglobin, forming the heat-stable NO-myoglobin. However, due to the reaction between nitrites and protein components in meat resulting in the carcinogen nitrosamine formation, the nitrites have been classified as potentially carcinogenic agents by the International Agency for Cancer Research of World Health Organizations (WHO) [9,10]. Consequently, it is important to obtain healthier meat products with a low content of residual nitrite without compromising the quality of the sausage.

Recently, researchers have focused on finding ways to decrease the additional content of nitrite in meat products, especially replacing it with natural resources such as vegetables, mushroom, and their extracts [11–14]. The reports showed that a part of plant essential oils exhibited strong antioxidative, antimicrobial, anticarcinogenic, and antimutagenic properties, which could be a compressive solution to substitute the nitrite [12]. For instance, Tang et al. (2021) reported that the combination of *Flos Sophorae* and chilli pepper can improve redness and reduce lipid oxidation of the meat product to replace nitrite in processed meat [15]. Vegetable powder extracted from radish and beetroot can substitute nitrite in sausage and increase the weight loss of sausages [16]. However, the studies on mushroom sausages have mainly focused on the nutritional components and sensory evaluation of the sausages and less concentrated on reducing the content of nitrite by adding edible mushrooms to sausage [17–19].

The mushroom is not only popular for its taste and flavor, but also for its high nutritional value and bioactive compounds. The addition of edible mushrooms to meat products can be used as a substitute for salt and phosphates in the formula of meat products and improve the quality of meat products, protein, dietary fiber, and ash content [20]. For example, the addition of shiitake mushrooms to sausage can improve the antioxidant and antibacterial properties of the product [21]. Replacing pork lean meat with *Lentinus edodes* can enhance total dietary fiber content, total phenol content, and the 2,2-diphenyl-1-picrylhydrazyl (DPPH) scavenging ability of sausage [22].

Pleurotus ostreatus (PO) is rich in amino acids (glutamic acid, aspartic acid, and arginine), polysaccharides (PSPO-1a, PSPO-4a, and branched β-glucans and α-glucans), vitamins (riboflavin and ascorbic acid), and dietary fiber with the function of antioxidant, antitumor, hypoglycemic, lipid-lowering, anti-inflammatory, sterilization, liver protection, improving immunity, and other significant effects [23–27]. Herein, a novel sausage incorporated with the *Pleurotus ostreatus* (PO) puree was developed to reduce the residual nitrite content in low-meat sausages during storage. Five recipes with the supplement proportion of 0 wt.%, 10 wt.%, 20 wt.%, 30 wt.%, and 40 wt.% PO were produced and their compositions, water activities, textures, colors, water holding capacities (WHC), amino acid compositions, and sensory evaluation were characterized. During storage at 4 °C, several characterizations including residual nitrite analysis, lipid oxidation analysis, and microbiological analysis were employed to characterize the quality and shelf-life of the products on days 0, 5, 10, 15, and 20.

2. Materials and Methods

2.1. Materials

The lean pork meat and back fat were purchased from the local market (WalMart Inc., Bentonville, Arkansas, USA) in China and stored at −20 °C. The fresh *Pleurotus ostreatus* (PO), salt, sugar, white pepper, cinnamon, and dry starch were also purchased from the local market (WalMart Inc., Bentonville, Arkansas, USA) in China. All other additives and chemical reagents shown in Table 1 were purchased from Sichuan Jinshan Pharmaceutical Co., Ltd. (Chengdu, China) and Beijing Beihua Co., Ltd., (Beijing, China).

Table 1. Formulations of *Pleurotus ostreatus* (PO) pork sausages.

Ingredients (g)	Control	PO10	PO20	PO30	PO40
Pork lean meat	400	400	400	400	400
Pork back fat	100	100	100	100	100
PO	0	50	100	150	200
Salt	10	10	10	10	10
Sugar	5	5	5	5	5
Sodium tripolyphosphate	0.400	0.400	0.400	0.400	0.400
Sodium nitrite	0.050	0.050	0.050	0.050	0.050
White pepper	1	1	1	1	1
Cinnamon (refined)	0.500	0.500	0.500	0.500	0.500
Carrageenan	2	2	2	2	2
Isolated soy protein	20	20	20	20	20
Dry starch	30	30	30	30	30
Monascus	0.075	0.075	0.075	0.075	0.075
Ice	125	125	125	125	125

2.2. Formulations and Processing of Sausage

Table 1 shows the formulations of the PO pork sausages with different PO contents. The sausages with 0 wt.%, 10 wt.%, 20 wt.%, 30 wt.%, and 40 wt.% PO were named as the Control, PO10, PO20, PO30, and PO40, respectively. The content of PO supplement used to process the sausages was based on the total content of pork lean meat and pork back fat.

To process the sausages, the pork lean meat was first sliced into sections and pickled at $4 \pm 1\ °C$ for 24 h with salt, sugar, sodium tripolyphosphate, and sodium nitrite. Then, the pork back fat was cut into cubes with a side length of 1 cm. The PO was cleaned with water, cut into small pieces, and mashed with a chopping machine into the puree. According to the formulation of each group, the pickled pork lean meat, PO puree, white pepper, cinnamon, carrageenan, isolated soy protein, dry starch, *Monascus*, ice, and back fat were mixed homogeneously in a mixer (Busch, Marburg, Germany) for 140 s. After that, the mixture was filled into the artificial collagen casings with a diameter of 2.6 cm by a sausage filler (Guangdong Shunde Fangzhan Electric Industrial Co., Ltd., Foshan, China). The methods of cooking, cooling, and preserving sausages were the same as the method reported by Wang et al. (2013) [28]. Finally, the sausage was stored after vacuum packing (Shandong Xiaokang Machinery Co., Ltd., Weifang, China) at 4 °C, which was counted as day 0.

2.3. Proximate Composition

The proximate compositions including protein, fat, ash, and moisture content were analyzed according to the Association of Official Analytical Chemists (AOAC) [29]. Each measurement was replicated three times for repeatability.

2.4. Water Activity (a_w) and pH

The water activity (a_w) of sausages was measured by a four-channel desktop water activity meter (Hygro Lab 2, Rotronic, Bassesdorf, Switzerland) at 25 °C. The sausage (10 g) was chopped and put into the water activity meter with 75% RH at 25 °C for 20 min, and then the data were recorded.

The pH values of sausages were characterized by a portable digital potentiometer (pH meter, Mettler Toledo, Zurich, Switzerland). The sausage (10 g) was crushed in 90 mL distilled water by a homogenate (Changzhou Magnetar Instrument Co., Ltd., Guangzhou, China). Then, we filled the mixture and measured the pH value of the solution at 25 °C in triplicate.

2.5. Color

The L^* (lightness), a^* (redness), and b^* (yellowness) values of sausages were determined by a colorimeter (HunterLab ColorFlex, Reston, VA, USA) with a standard illuminant D65 light source (Xinlian Creation Electronic Co., Ltd., Shanghai, China) and a standard plate ($L^* = 94.52$, $a^* = -0.86$, $b^* = 0.68$). The color values were determined at room temperature by five different areas on the cross-section of sausages. In particular, the area of fat in the sausage slice was not selected. The color differences (ΔE^*) can be calculated by the following equations [28,30]:

$$\begin{cases} \Delta E* = \left(\Delta a^2 + \Delta b^2 + \Delta L^2\right)^{0.5} \\ \Delta a = a*_{control} - a*_{PO} \\ \Delta b = b*_{control} - b*_{PO} \\ \Delta L = L*_{control} - L*_{PO} \end{cases} \quad (1)$$

where ΔL, Δa, and Δb are the differences between the L^*, a^*, and b^* values of the control groups and the groups with *PO*, respectively.

2.6. Textural Profile Analysis (TPA)

Texture profile analysis was employed by a texture analyzer (CT3-50kg, Brookfield Engineering Labs, New York City, NY, USA) to calculate the hardness, springiness, gumminess, chewiness, and cohesiveness of the sausages. The sausage was cut into cylinders with a diameter of 2 cm and a height of 1 cm. Then, the cylinder was axially compressed twice until reaching 80% of its initial height with a 20 s pause time between the descent, 30 mm probe retraction, 6 cm/min detection speed, and 100 N force induction. The analysis was repeated at 25 °C in triplicate.

2.7. Cooking Loss and Water Holding Capacity (WHC)

Cooking loss was characterized by the method reported by Fu et al. (2016) [30]. Sausage (50 g) was cooked at 80 °C for 50 min then cooled at room temperature to determine the weight difference before and after cooking. The cooking loss can be calculated by:

$$Cooking\ loss = [(m_1 - m_2)/m_1] \times 100\% \tag{2}$$

where m_1 and m_2 are the weight before and after cooking, respectively.

The WHC of the sample was measured by the method reported by Jridi et al. (2015) with slight modifications [31]. Sausage (10 g) was centrifuged at 12,000 rpm for 30 min at 4 °C. Finally, the WHC can be calculated by:

$$WHC = (W_2/W_1) \times 100\% \tag{3}$$

where W_1 and W_2 are the weight of the sample before and after centrifugation, respectively.

2.8. Amino Acids Content

The amino acid content was analyzed according to the method reported by Serrano et al. (2005), which separated the amino acids by cation exchange chromatography via an automatic amino acid analyzer (L8900, Hitachi High-Technologies Corporation, Tokyo, Japan) for measurement [32]. Ninhydrin derivative reagents were used to measure the amino acid content at 570 nm.

2.9. Sensory Evaluation

The sensory evaluation of sausages was carried out according to the research reported by Hu et al. (2014) with modification [33]. The cooked sausages (20 min at 80 °C) were assessed by a panel of 30 members (15 males and 15 females). The team members were chosen from students and the faculties of Jilin Agricultural University (Changchun, China). The samples were cut into 5 mm thick slices at room temperature and identified with a three-digit random code. In addition, the sensory panel were provided with water and salt-free biscuits to clean their taste buds. The appearance, texture, flavor, aroma, and overall acceptability were evaluated using a 9-point Hedonic scale (1 = dislike very much, 9 = like very much).

2.10. Residual Nitrite Analysis

GB5009.33-2016 was used to measure the residual nitrite in sausage samples on days 0, 5, 10, 15, and 20 [34]. The sausage (5 g) was mixed in 12.5 mL saturated borax solution (50 g/L) and 150 mL water. Then, the solution was heated in a boiling water bath for 15 min. After cooling, potassium ferrocyanide solution (5 mL, 106 g/L) and zinc acetate solution (5 mL, 220 g/L) were mixed homogeneously and stood for 30 min. Finally, the pink dye formed by the coupling of sulfonamide and naphthalene ethylenediamine hydrochloride was determined by spectrophotometry to obtain the residual nitrite content. Each sample was analyzed in triplicate.

2.11. Thiobarbituric Acid Reactive Species (TBARs)

TBARs of the sausage was analyzed by spectrophotometry according to GB5009.181-2016 on days 0, 5, 10, 15, and 20 [35]. The sausage (5 g) was mixed into a solution with trichloroacetic acid (75 g/L) and disodium EDTA (1 g/L) by a thermostatic oscillator at 50 °C for 30 min. Then, the filtrate (5 mL) was added into 2.88 g/L thiobarbituric acid solution and mixed at 90 °C for 30 min. Then, the solution was cooled down to room temperature and the absorbance was measured at 532 nm.

2.12. Microbiological Analysis

The total number of bacterial colonies in sausage was determined by the method in GB4789.2-2016 on days 0, 5, 10, 15, and 20. The total number of germs, yeasts, and molds was analyzed and identified with the same method in GB4789.2-2016 [36].

2.13. Statistical Analysis

All determinations were designed three times and the values were shown as means ± standard deviations. The difference between factors and levels were submitted to the analysis of variance (ANOVA). Duncan's multiple range tests were used to determine the differences among mean values ($p < 0.05$). The analysis was taken by SPSS software version 19.

3. Results

3.1. Proximate Composition

The proximate composition contents of sausage are presented in Table 2. According to the reference, the chemical composition of the PO included 20~25% protein, 2.5~2.9% fat, 5.9~6.7% ash, and 88.0~90% moisture [37]. Obviously, the protein and fat of the PO were lower than those of the meat. For the protein content, the control group was at the highest level (13.23%) and significantly different from other groups ($p < 0.05$). This result was comparable with that of the sausages reported by Lee et al. (2016) and Silva et al. (2019) [38,39]. With the increase in PO content, the protein level of PO40 decreased from 13.08% (the control group) to 12.73% (PO40), which was because the fresh PO contained less protein compared to the meat [40]. The fat content in the sausages decreased from 17.60% to 14.82% with the increase in PO content, which was also because the PO had less fat than the meat.

Table 2. Proximate composition (%), water activity, and pH of *Pleurotus ostreatus* (PO) pork sausages.

Parameters	Control	PO10	PO20	PO30	PO40
Protein	13.23 ± 0.08 [a]	13.08 ± 0.02 [b]	12.98 ± 0.02 [b]	12.92 ± 0.06 [b]	12.73 ± 0.05 [c]
Fat	17.60 ± 0.12 [a]	17.22 ± 0.04 [b]	16.53 ± 0.06 [c]	15.79 ± 0.02 [d]	14.82 ± 0.07 [e]
Ash	3.23 ± 0.02 [a]	3.01 ± 0.12 [b]	2.95 ± 0.01 [b]	2.75 ± 0.01 [c]	2.64 ± 0.02 [d]
Moisture	56.56 ± 0.01 [d]	63.83 ± 0.12 [c]	63.90 ± 0.11 [c]	64.78 ± 0.21 [b]	66.26 ± 0.62 [a]
a_w	0.98 ± 0.00 [a]	0.98 ± 0.00 [a]	0.98 ± 0.00 [a]	0.98 ± 0.00 [a]	0.98 ± 0.00 [a]
pH	6.86 ± 0.01 [a]	6.74 ± 0.03 [b]	6.55 ± 0.04 [c]	6.21 ± 0.02 [d]	6.12 ± 0.01 [d]

[a–e] Means within the same row with different letters differ significantly among the treatments ($p < 0.05$). Values are given as mean ± standard deviations. Control, PO10, PO20, PO30, and PO40 were 0 wt.%, 10 wt.%, 20 wt.%, 30 wt.%, and 40 wt.% addition of PO puree, respectively.

The ash content was significantly reduced from 3.23% to 2.64%, which was comparable with that of the beetroot sausages and Toscana sausages reported by Sucu et al. (2018) and Monteiro et al. (2017) [41,42]. The moisture was a significant difference between the control group and the experimental groups ($p < 0.05$). This is because the moisture content of PO (>80%) was much richer than the meat.

3.2. Water Activity and pH

Among all samples, there was no significant difference in the water activity (a_w) of sausages ($p > 0.05$) (Table 2). The results were close to bologna sausages as reported by Câmara et al. (2021), in which the a_w was 0.97~0.98 [43]. The pH value of the control group was the highest and the pH value gradually decreased with the increase in PO puree content significantly ($p < 0.05$). The result of pH in this research was close to the research reported by Riel et al. (2017) (pH = 6~7) [44].

3.3. Color

Color is an important parameter for consumers to accept processed meat products. The effects of PO puree on the color traits of sausages are evaluated in Table 3. With the increase in PO content, lightness (L^*) values of the sausages significantly ($p < 0.05$) increased. This could be explained by the higher lightness of the PO, which enhanced that of the sausages. For redness (a^*), the decreasing trend may be due to the decrease in lean meat content, which contributed to the pink color of the sausage product. For the yellowness (b^*) and color difference (ΔE^*), there were no significant differences among all groups ($p < 0.05$).

Table 3. Color comparison of *Pleurotus ostreatus* (PO) pork sausages.

Parameters	Control	PO10	PO20	PO30	PO40
L^*	47.46 ± 0.12 [e]	48.08 ± 0.17 [d]	51.17 ± 0.24 [c]	52.50 ± 0.21 [b]	54.62 ± 0.31 [a]
a^*	17.96 ± 0.01 [a]	16.60 ± 0.03 [b]	15.94 ± 0.13 [c]	14.56 ± 0.31 [d]	13.41 ± 0.57 [e]
b^*	16.35 ± 0.02 [a]	16.39 ± 0.16 [a]	16.45 ± 0.11 [a]	16.29 ± 0.04 [a]	16.41 ± 0.11 [a]
ΔE^*	-	44.20 ± 0.04 [a]	44.37 ± 0.13 [a]	44.47 ± 0.31 [a]	44.31 ± 0.59 [a]

[a–e] Means within the same row with different letters differed significantly among the treatments ($p < 0.05$). Values are given as mean ± standard deviations. Control, PO10, PO20, PO30, and PO40 were 0 wt.%, 10 wt.%, 20 wt.%, 30 wt.%, and 40 wt.% addition of PO puree, respectively.

3.4. Textural Profile Analysis (TPA)

The addition of the PO affected the apparent texture properties of the sausages (Table 4). The experimental groups had higher hardness (11.71~47.63%), cohesiveness (7.27~23.63%), gumminess (14.90~51.07%), and chewiness (6.44~21.81%) values, while it had a lower springiness (−45.71~−49.71%) value than those of the control group significantly ($p < 0.05$), showing that the experimental groups were more conducive and convenient for chewing. In the experimental groups, the PO with high water content prevented the gel production of the pork myofibril protein, which caused lower cohesiveness of sausage fillings and the appearance of more inside gaps. Furthermore, high amounts of dietary fiber in PO was another factor leading PO sausages to show lower degrees of hardness, gumminess, and chewiness [45,46].

Table 4. Texture parameter analysis (TPA), cooking loss (%), and water holding capacity (WHC) of *Pleurotus ostreatus* (PO) pork sausages.

Parameters	Control	PO10	PO20	PO30	PO40
Hardness (N)	155.63 ± 15.15 [a]	137.40 ± 4.12 [b]	111.93 ± 8.70 [c]	111.83 ± 6.91 [c]	81.50 ± 7.04 [d]
Cohesiveness	0.55 ± 0.04 [a]	0.51 ± 0.06 [a]	0.42 ± 0.13 [a]	0.44 ± 0.08 [a]	0.48 ± 0.01 [a]
Springiness (mm)	3.50 ± 0.22 [b]	5.24 ± 0.93 [a]	5.16 ± 0.36 [a]	5.10 ± 0.70 [a]	5.15 ± 0.94 [a]
Gumminess (N)	85.23 ± 13.32 [a]	72.53 ± 8.83 [a]	46.70 ± 11.61 [b]	45.77 ± 13.74 [b]	41.70 ± 2.41 [b]
Chewiness (N)	296.80 ± 14.51 [a]	277.68 ± 15.02 [b]	251.94 ± 11.14 [c]	232.05 ± 81.70 [c]	241.14 ± 29.23 [c]
Cooking loss (%)	4.39 ± 0.20 [e]	16.54 ± 0.21 [d]	18.44 ± 0.26 [c]	19.30 ± 0.26 [b]	21.25 ± 0.26 [a]
WHC (%)	69.25 ± 0.44 [e]	73.90 ± 0.56 [d]	75.81 ± 0.46 [c]	77.01 ± 0.40 [b]	78.66 ± 0.42 [a]

[a–e] Means within the same row with different letters differ significantly among the treatments ($p < 0.05$). Values are given as mean ± standard deviations. Control, PO10, PO20, PO30, and PO40 were 0 wt.%, 10 wt.%, 20 wt.%, 30 wt.%, and 40 wt.% addition PO puree, respectively.

3.5. Cooking Loss and Water Holding Capacity (WHC)

The cooking loss and WHC of sausages in each group are shown in Table 4. With the increase in PO content, the cooking loss and WHC were significantly ($p < 0.05$) increased. The higher cooking loss might be due to the high water content of fresh PO, which evaporated during the cooking process. WHC of the supplement group increased 4.7~13.55% compared with that of the control group. This phenomenon was probably due to the higher water retention ability of the fibers in PO [47]. These results were similar to the research about adding shiitake mushrooms to replace lean pork in sausages reported by Wang et al. (2019), in which the shiitake mushroom enhanced the cooking loss and WHC of the sausages [48].

3.6. Amino Acids Profile

Amino acids have many health benefits. For example, lysine can enhance immunity and improve the central nervous function of the human body [49]. Moreover, glutamic and aspartic are umami amino acids that could contribute to the enhancement of the flavor of meat products [50]. Therefore, the amino acid content is another key factor of the meat products' nutrients [51]. As shown in Table 5, the addition of the PO enhanced the contents of all the essential amino acids and non-essential amino acids compared with the control group ($p < 0.05$). It could be seen that the essential amino acids of the control sausage were mainly lysine, leucine, and valine; the non-essential amino acids were mainly glutamic, aspartic, and arginine. Comparing PO40 with the control group, the contents of lysine, leucine, valine, glutamic, aspartic, and arginine in sausage increased by 39.26%, 40.00%, 35.53%, 34.65%, 32.26%, and 40.00%, respectively. All of these increases might be contributed by the abundance of amino acids in PO [27].

Table 5. Amino acid profile (%) of *Pleurotus ostreatus* (PO) sausages during storage.

Parameters	Control	PO10	PO20	PO30	PO40
		Essential amino acid			
Threonine	0.67 ± 0.04 [d]	0.67 ± 0.03 [d]	0.81 ± 0.07 [c]	0.86 ± 0.01 [b]	0.87 ± 0.01 [a]
Valine	0.76 ± 0.05 [e]	0.81 ± 0.04 [d]	0.93 ± 0.02 [c]	0.98 ± 0.05 [b]	1.03 ± 0.05 [a]
Methionine	0.12 ± 0.01 [d]	0.12 ± 0.03 [d]	0.21 ± 0.01 [c]	0.25 ± 0.01 [b]	0.29 ± 0.01 [a]
Isoleucine	0.72 ± 0.01 [e]	0.77 ± 0.04 [d]	0.92 ± 0.03 [c]	0.96 ± 0.03 [b]	1.01 ± 0.03 [a]
Phenylalanine	0.69 ± 0.02 [e]	0.71 ± 0.03 [d]	0.84 ± 0.02 [c]	0.89 ± 0.04 [b]	0.93 ± 0.01 [a]
Lysine	1.35 ± 0.03 [e]	1.36 ± 0.01 [d]	1.69 ± 0.03 [c]	1.80 ± 0.02 [b]	1.88 ± 0.05 [a]
Leucine	1.30 ± 0.02 [e]	1.34 ± 0.03 [d]	1.60 ± 0.03 [c]	1.72 ± 0.03 [b]	1.82 ± 0.02 [a]
		Non-essential amino acids			
Histidine	0.53 ± 0.03 [e]	0.64 ± 0.03 [d]	0.65 ± 0.06 [c]	0.69 ± 0.01 [b]	0.70 ± 0.07 [a]
Glycine	0.76 ± 0.03 [e]	0.81 ± 0.02 [d]	0.85 ± 0.06 [c]	0.90 ± 0.04 [b]	0.97 ± 0.02 [a]
Aspartic	1.55 ± 0.01 [e]	1.70 ± 0.05 [d]	1.85 ± 0.03 [c]	2.00 ± 0.03 [b]	2.05 ± 0.04 [a]
Arginine	1.00 ± 0.03 [e]	1.05 ± 0.03 [d]	1.24 ± 0.01 [c]	1.30 ± 0.03 [b]	1.40 ± 0.01 [a]
Alanine	0.88 ± 0.03 [e]	0.91 ± 0.01 [d]	1.05 ± 0.03 [c]	1.12 ± 0.02 [b]	1.19 ± 0.06 [a]
Tyrosine	0.39 ± 0.07 [e]	0.47 ± 0.03 [d]	0.65 ± 0.06 [c]	0.66 ± 0.06 [b]	0.72 ± 0.05 [a]
Cysteine	0.03 ± 0.04 [e]	0.04 ± 0.07 [d]	0.05 ± 0.03 [c]	0.06 ± 0.02 [b]	0.11 ± 0.09 [a]
Serine	0.68 ± 0.06 [e]	0.70 ± 0.02 [d]	0.80 ± 0.02 [c]	0.87 ± 0.04 [b]	0.90 ± 0.03 [a]
Glutamic	2.54 ± 0.07 [e]	2.60 ± 0.03 [d]	3.08 ± 0.07 [c]	3.28 ± 0.03 [b]	3.42 ± 0.04 [a]
Proline	0.68 ± 0.03 [e]	0.71 ± 0.03 [d]	0.74 ± 0.03 [c]	0.76 ± 0.06 [b]	0.96 ± 0.03 [a]

[a–e] Means within the same row with different letters differ significantly among the treatments ($p < 0.05$). Values are given as mean ± standard deviations. Control, PO10, PO20, PO30, and PO40 were 0 wt.%, 10 wt.%, 20 wt.%, 30 wt.%, and 40 wt.% addition of PO puree, respectively.

3.7. Sensory Evaluation

The sensory evaluation results of each group are shown in Figure 1. The appearance of all groups had excellent performance (>8). In the PO10 and PO20 groups, an appropriate increase in brightness of the sausages might be more loved by the group members. For the

PO30 and PO40 groups, the decrease in the redness might reduce group acceptance of the sausages in appearance. In terms of aroma and flavor, all scores exceeded eight and PO20 had the best aroma and flavor. This might be because PO was rich in free amino acids, especially glutamic and aspartic, which were conducive to enhance the aroma and flavor of the sausage. The texture of the sausages decreased with the increase in PO content. This result might be caused by the soft texture of the sausage and more gaps inside the sausages (Section 3.2.) which were disliked by the group member. Although a difference was found between each group, all of these samples were judged as acceptable (>7), suggesting that PO had an opportunity to in sausages. From what has been discussed above, PO20 can be regarded as the best group in sensory evaluation through comprehensive consideration.

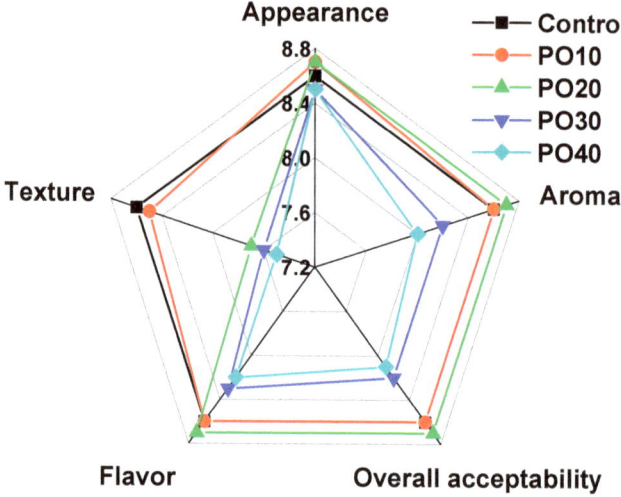

Figure 1. Sensory evaluations of each group of sausages.

3.8. Residual Nitrite Analysis

The residual nitrate in meat products can convert to nitrosamine compounds in the human body and cause cancer in the lungs, stomach, esophagus, liver, and bladder [52,53]. Therefore, the residual nitrite content in meat products should be as low as possible. The results of residual nitrite analysis of all groups on days 0, 5, 10, 15, and 20 are shown in Tables 6 and 7. Because the PO was the supplement in the sausage and did not replace the meat in the sausage, the residual nitrite concentrations were not convincing to clarify the true content of nitrite in each group shown in Table 1. To solve this problem, the data in Table 7 were calculated to show the weight of the residual nitrite of each group. With the passage of storage time and increase in PO content, the residual nitrite content of sausage showed a gradually decreasing trend before 15 days. On the 15th day of the storage period, the residual amount of nitrite in the supplement group decreased 24.7~37.35% compared with the control group. After 15 days, the nitrite content in the sausages decreased sharply. On the last day (day 20) of the storage period, the residual amount of nitrite in the supplement group could decrease 45.06~81.97% compared with the control group, which was probably due to the microbial proliferation affecting nitrogen compounds. The low residual nitrite of the supplement group can be explained by the decrease in initial nitrite (day 0). The residual nitrite could react with active substances such as phenolics and terpenes into nitrite acid or nitric oxide. This might be because PO was rich in strong antioxidant activities such as Vitamin C (Vc), which could improve the activity of antioxidant enzymes, interrupt the chain reaction of free radicals, maintain the integrity of cell membrane, and convert methemoglobin to hemoglobin and thereby reduce the nitrite content [54–58]. In a word, the nitrite residue of sausage with PO was lower than that of the control group over the entire storage period.

Table 6. Residual nitrite analysis of *Pleurotus ostreatus* (PO) pork sausages during storage (mg/kg).

Days	Control	PO10	PO20	PO30	PO40
0	11.75 ± 0.25 [a]	10.31 ± 0.01 [b]	9.74 ± 0.03 [c]	9.33 ± 0.02 [d]	8.77 ± 0.01 [e]
5	11.24 ± 0.17 [a]	9.90 ± 0.17 [b]	9.35 ± 0.01 [c]	9.01 ± 0.02 [d]	8.52 ± 0.03 [e]
10	10.95 ± 0.11 [a]	8.26 ± 0.12 [b]	7.94 ± 0.13 [c]	6.95 ± 0.13 [d]	6.87 ± 0.11 [d]
15	10.79 ± 0.25 [a]	8.15 ± 0.01 [b]	7.37 ± 0.10 [c]	6.81 ± 0.10 [d]	6.76 ± 0.10 [d]
20	2.33 ± 0.03 [a]	1.28 ± 0.01 [b]	0.70 ± 0.07 [c]	0.58 ± 0.02 [d]	0.42 ± 0.01 [e]

[a–e] Means within the same row with different letters differ significantly among the treatments ($p < 0.05$). Values are given as mean ± standard deviations. Control, PO10, PO20, PO30, and PO40 were 0 wt.%, 10 wt.%, 20 wt.%, 30 wt.%, and 40 wt.% addition of PO puree, respectively.

Table 7. Residual nitrite of *Pleurotus ostreatus* (PO) pork sausages during storage for every group (mg).

Days	Control	PO10	PO20	PO30	PO40
0	7.80 ± 0.17 [a]	6.40 ± 0.01 [b]	6.31 ± 0.02 [c]	6.35 ± 0.01 [bc]	6.17 ± 0.01 [d]
5	7.46 ± 0.11 [a]	6.15 ± 0.11 [b]	6.06 ± 0.01 [c]	6.14 ± 0.01 [b]	6.00 ± 0.02 [d]
10	7.27 ± 0.07 [a]	5.13 ± 0.07 [b]	5.14 ± 0.08 [b]	4.73 ± 0.09 [c]	4.84 ± 0.08 [c]
15	7.16 ± 0.17 [a]	5.06 ± 0.01 [b]	4.77 ± 0.06 [c]	4.64 ± 0.07 [c]	4.76 ± 0.07 [c]
20	1.55 ± 0.02 [a]	0.79 ± 0.01 [b]	0.45 ± 0.05 [c]	0.40 ± 0.01 [d]	0.30 ± 0.01 [e]

[a–e] Means within the same row with different letters differ significantly among the treatments ($p < 0.05$). Values are given as mean ± standard deviations. Control, PO10, PO20, PO30, and PO40 were 0 wt.%, 10 wt.%, 20 wt.%, 30 wt.%, and 40 wt.% addition of PO puree, respectively.

3.9. TBARs

Lipid oxidation is considered as the main mechanism leading to the rancidity, low quality, and low shelf life of meat products [42,59]. In this work, TBARs was employed for the evaluation of the lipid oxidation of the sausages (Table 8). With the increase in PO content, the TBARs values of the sausages showed a decreasing trend. The TBARs value of the experimental group could decrease 3.92~31.37% and 4.76~32.14% compared with that of the control group on the initial day and 15th day, respectively. This phenomenon could be explained by the good antioxidant biological chemical composition of the PO such as phenols, ascorbic acid, α-tocopherol, β-carotene, and flavonoid compounds (rutin and chrysin), which could effectively delay the lipid oxidation of meat products [55,60,61]. Moreover, Selli et al. (2021) reported that the phenolic substance content of heat-treated mushrooms increased significantly, which could also enhance the antioxidant activity directly [62].

Table 8. The TBARs value of *Pleurotus ostreatus* (PO) pork sausages during storage (mg/kg).

Days	Control	PO10	PO20	PO30	PO40
0	0.51 ± 0.03 [a]	0.49 ± 0.01 [ab]	0.46 ± 0.01 [b]	0.39 ± 0.02 [c]	0.35 ± 0.01 [d]
5	0.61 ± 0.01 [a]	0.58 ± 0.01 [a]	0.50 ± 0.04 [b]	0.47 ± 0.02 [b]	0.41 ± 0.04 [c]
10	0.72 ± 0.01 [a]	0.71 ± 0.03 [a]	0.69 ± 0.01 [a]	0.51 ± 0.01 [c]	0.49 ± 0.03 [c]
15	0.84 ± 0.02 [a]	0.80 ± 0.03 [a]	0.74 ± 0.01 [ab]	0.65 ± 0.01 [bc]	0.57 ± 0.12 [c]
20	0.92 ± 0.01 [a]	0.89 ± 0.02 [a]	0.85 ± 0.01 [b]	0.79 ± 0.02 [c]	0.60 ± 0.03 [d]

[a–d] Means within the same row with different letters differ significantly among the treatments ($p < 0.05$). Values are given as mean ± standard deviations. Control, PO10, PO20, PO30, and PO40 were 0 wt.%, 10 wt.%, 20 wt.%, 30 wt.%, and 40 wt.% addition of PO puree, respectively.

3.10. Microbiological Analysis

During the processing and storage of sausage, microorganisms would lead to the decomposition of protein and fat, leading to rot and rancidity, which is harmful to food safety. Therefore, the total number of bacterial colonies was recognized as an important parameter to evaluate the shelf-life stability [44]. The microbiological analysis of sausages in the control group and supplement groups were tested after 0, 5, 10, 15, and 20 days of storage (Table 9). The total number of bacterial colonies in each group increased significantly after 20 days of storage ($p < 0.05$). Compared to the supplement groups and the control group, there was no significant difference in the total number of bacterial colonies of sausage, indicating that PO had no significant effect on the total number of bacterial colonies of sausage. According to the existing national standards of China (1×10^5 CFU/g) (GB2726—2016), the sausages expired after 20 days. The sausage prepared in this study was steamed sausage, which was not Chinese sausage and this kind of product normally has a high water content and short shelf life. Therefore, it is acceptable for the product to expire after 20 days.

Table 9. Microorganisms of *Pleurotus ostreatus* (PO) pork sausages (log CFU/g).

Days	Control	PO10	PO20	PO30	PO40
0	0.43 ± 0.30 [a]	0.44 ± 0.29 [a]	0.45 ± 0.22 [a]	0.44 ± 0.20 [a]	0.42 ± 0.30 [a]
5	0.86 ± 0.20 [a]	0.65 ± 0.22 [a]	0.73 ± 0.11 [a]	0.87 ± 0.19 [a]	0.76 ± 0.14 [a]
10	1.24 ± 0.61 [a]	1.43 ± 0.41 [a]	1.54 ± 0.36 [a]	1.67 ± 0.21 [a]	1.85 ± 0.62 [a]
15	3.87 ± 0.20 [a]	3.94 ± 0.11 [a]	3.97 ± 0.10 [a]	3.84 ± 0.10 [a]	3.76 ± 0.20 [a]
20	5.12 ± 0.46 [a]	5.32 ± 0.42 [a]	5.34 ± 0.41 [a]	5.36 ± 0.39 [a]	5.57 ± 0.45 [a]

[a] Means within the same row with different letters differ significantly among the treatments ($p < 0.05$). Values are given as mean ± standard error. Control, PO10, PO20, PO30, and PO40 were 0 wt.%, 10 wt.%, 20 wt.%, 30 wt.%, and 40% addition of PO puree, respectively.

4. Conclusions

A novel pork sausage incorporated with *Pleurotus ostreatus* (PO) puree was produced in this study. The PO improved the content of moisture and amino acids, lightness, springiness, and water holding capacity of the sausage while reducing the content of protein, fat, and ash, pH, redness, hardness, gumminess, and chewiness. For the sensory, the best formulation for the addition of PO was 20 wt.%, which not only enhanced the appearance and overall acceptability, but also gave it a better aroma and flavor. During 15 days of storage, the content of residual nitrite in the sausages decreased by 24.7~37.35%, and the TBARs value of the sausages was decreased by 4.76~32.14% compared with that of the control. Based on these results, the PO pork sausage in this study was successfully developed to enhance the quality and nutritional value and reduce the residual nitrite content and lipid oxidation during the storage of sausage. PO could be a competitive solution to improve flavor, safety, and overall quality of the sausages.

Author Contributions: Method, Writing—original draft, Data curation, X.W.; Writing—original draft, Writing—review and editing, Visualization, P.W.; Investigation, Q.X.; Investigation, B.J.; Writing—original draft, L.L.; Data curation, L.R.; Investigation, Data curation, X.L.; Writing—review and editing, Project administration, Supervision, L.W. All authors have read and agreed to the published version of the manuscript.

Funding: This work was supported by the Plan of Science and Technology Development of Jilin Province of China (No. 20210202066NC) and the Opening Project of Engineering Research Center of Edible and Medicinal Fungi (Ministry of Education), Jilin Agricultural University (No. JJJW2021001).

Institutional Review Board Statement: Not applicable.

Informed Consent Statement: Not applicable.

Data Availability Statement: Not applicable.

Conflicts of Interest: The authors declare no conflict of interest.

References

1. Hugo, C.J.; Hugo, A. Current trends in natural preservatives for fresh sausage products. *Trends Food Sci. Technol.* **2015**, *45*, 12–23. [CrossRef]
2. Klumbytė, N. The soviet sausage renaissance. *Am. Anthropol.* **2010**, *112*, 22–37. [CrossRef]
3. Colak, H.; Hampikyan, H.; Ulusoy, B.; Bingol, E.B. Presence of Listeria monocytogenes in Turkish style fermented sausage (sucuk). *Food Control* **2007**, *18*, 30–32. [CrossRef]
4. Honikel, K.-O. The use and control of nitrate and nitrite for the processing of meat products. *Meat Sci.* **2008**, *78*, 68–76. [CrossRef] [PubMed]
5. Perea-Sanz, L.; López-Díez, J.J.; Belloch, C.; Flores, M. Counteracting the effect of reducing nitrate/nitrite levels on dry fermented sausage aroma by Debaryomyces hansenii inoculation. *Meat Sci.* **2020**, *164*, 108103. [CrossRef] [PubMed]
6. Pini, F.; Aquilani, C.; Giovannetti, L.; Viti, C.; Pugliese, C. Characterization of the microbial community composition in Italian Cinta Senese sausages dry-fermented with natural extracts as alternatives to sodium nitrite. *Food Microbiol.* **2020**, *89*, 103417. [CrossRef] [PubMed]
7. Sebranek, J.G.; Bacus, J.N. Cured meat products without direct addition of nitrate or nitrite: What are the issues? *Meat Sci.* **2007**, *77*, 136–147. [CrossRef]
8. Weiss, J.; Gibis, M.; Schuh, V.; Salminen, H. Advances in ingredient and processing systems for meat and meat products. *Meat Sci.* **2010**, *86*, 196–213. [CrossRef] [PubMed]
9. IARC IARC Monographs on the Evaluation of Carcinogenic Risks to Humans. Available online: http://monographs.iarc.fr/ENG/Monographs/vol96/index.php (accessed on 2 April 2022).
10. Huang, L.; Zeng, X.; Sun, Z.; Wu, A.; He, J.; Dang, Y.; Pan, D. Production of a safe cured meat with low residual nitrite using nitrite substitutes. *Meat Sci.* **2020**, *162*, 108027. [CrossRef]
11. Hung, Y.; de Kok, T.M.; Verbeke, W. Consumer attitude and purchase intention towards processed meat products with natural compounds and a reduced level of nitrite. *Meat Sci.* **2016**, *121*, 119–126. [CrossRef] [PubMed]
12. Šojić, B.; Pavlić, B.; Ikonić, P.; Tomović, V.; Ikonić, B.; Zeković, Z.; Kocić-Tanackov, S.; Jokanović, M.; Škaljac, S.; Ivić, M. Coriander essential oil as natural food additive improves quality and safety of cooked pork sausages with different nitrite levels. *Meat Sci.* **2019**, *157*, 107879. [CrossRef] [PubMed]
13. Sathish Kumar, K.; Soundararajan, R.; Shanthosh, G.; Saravanakumar, P.; Ratteesh, M. Augmenting effect of infill density and annealing on mechanical properties of PETG and CFPETG composites fabricated by FDM. *Mater. Today Proc.* **2021**, *45*, 2186–2191. [CrossRef]
14. Stoica, M.; Antohi, V.M.; Alexe, P.; Ivan, A.S.; Stanciu, S.; Stoica, D.; Zlati, M.L.; Stuparu-Cretu, M. New strategies for the total/partial replacement of conventional sodium nitrite in meat products: A review. *Food Bioprocess Technol.* **2022**, *15*, 514–538. [CrossRef]
15. Tang, R.; Peng, J.; Chen, L.; Liu, D.; Wang, W.; Guo, X. Combination of Flos Sophorae and chili pepper as a nitrite alternative improves the antioxidant, microbial communities and quality traits in Chinese sausages. *Food Res. Int.* **2021**, *141*, 110131. [CrossRef] [PubMed]
16. Ozaki, M.M.; Munekata, P.E.S.; Jacinto-Valderrama, R.A.; Efraim, P.; Pateiro, M.; Lorenzo, J.M.; Pollonio, M.A.R. Beetroot and radish powders as natural nitrite source for fermented dry sausages. *Meat Sci.* **2021**, *171*, 108275. [CrossRef] [PubMed]
17. Wang, X.; Zhou, P.; Cheng, J.; Yang, H.; Zou, J.; Liu, X. The role of endogenous enzyme from straw mushroom (*Volvariella volvacea*) in improving taste and volatile flavor characteristics of Cantonese sausage. *LWT* **2022**, *154*, 112627. [CrossRef]
18. Hu, H.; Li, Y.; Zhang, L.; Tu, H.; Wang, X.; Ren, L.; Dai, S.; Wang, L. Use of tremella as fat substitute for the enhancement of physicochemical and sensory profiles of pork sausage. *Foods* **2021**, *10*, 2167. [CrossRef]
19. Stephan, A.; Ahlborn, J.; Zajul, M.; Zorn, H. Edible mushroom mycelia of Pleurotus sapidus as novel protein sources in a vegan boiled sausage analog system: Functionality and sensory tests in comparison to commercial proteins and meat sausages. *Eur. Food Res. Technol.* **2018**, *244*, 913–924. [CrossRef]
20. Pérez-Montes, A.; Rangel-Vargas, E.; Lorenzo, J.M.; Romero, L.; Santos, E.M. Edible mushrooms as a novel trend in the development of healthier meat products. *Curr. Opin. Food Sci.* **2021**, *37*, 118–124. [CrossRef]
21. Van Ba, H.; Seo, H.-W.; Cho, S.-H.; Kim, Y.-S.; Kim, J.-H.; Ham, J.-S.; Park, B.Y.; Pil Nam, S. Antioxidant and anti-foodborne bacteria activities of shiitake by-product extract in fermented sausages. *Food Control* **2016**, *70*, 201–209. [CrossRef]
22. Wang, L.; Guo, H.; Liu, X.; Jiang, G.; Li, C.; Li, X.; Li, Y. Roles of Lentinula edodes as the pork lean meat replacer in production of the sausage. *Meat Sci.* **2019**, *156*, 44–51. [CrossRef] [PubMed]
23. Marçal, S.; Sousa, A.S.; Taofiq, O.; Antunes, F.; Morais, A.M.M.B.; Freitas, A.C.; Barros, L.; Ferreira, I.C.F.R.; Pintado, M. Impact of postharvest preservation methods on nutritional value and bioactive properties of mushrooms. *Trends Food Sci. Technol.* **2021**, *110*, 418–431. [CrossRef]
24. Barbosa, J.R.; Freitas, M.M.S.; Oliveira, L.C.; Martins, L.H.S.; Almada-Vilhena, A.O.; Oliveira, R.M.; Pieczarka, J.C.; Brasil, D.D.S.B.; Carvalho Junior, R.N. Obtaining extracts rich in antioxidant polysaccharides from the edible mushroom Pleurotus ostreatus using binary system with hot water and supercritical CO_2. *Food Chem.* **2020**, *330*, 127173. [CrossRef]
25. Wang, L.; Zhang, P.; Shen, J.; Qian, Y.; Liu, M.; Ruan, Y.; Wang, X.; Zhang, S.; Ma, B. Physicochemical properties and bioactivities of original and Se-enriched polysaccharides with different molecular weights extracted from Pleurotus ostreatus. *Int. J. Biol. Macromol.* **2019**, *141*, 150–160. [CrossRef]

26. Rizzo, G.; Goggi, S.; Giampieri, F.; Baroni, L. A review of mushrooms in human nutrition and health. *Trends Food Sci. Technol.* **2021**, *117*, 60–73. [CrossRef]
27. Chirinang, P.; Intarapichet, K.-O. Amino acids and antioxidant properties of the oyster mushrooms, Pleurotus ostreatus and Pleurotus sajor-caju. *Sci. Asia* **2009**, *35*, 326. [CrossRef]
28. Wang, L.; Dong, Y.; Men, H.; Tong, J.; Zhou, J. Preparation and characterization of active films based on chitosan incorporated tea polyphenols. *Food Hydrocoll.* **2013**, *32*, 35–41. [CrossRef]
29. Horwitz, W. Official methods of analysis of AOAC International. *Trends Food Sci. Technol.* **1995**, *6*, 382.
30. dos Santos Alves, L.A.A.; Lorenzo, J.M.; Gonçalves, C.A.A.; dos Santos, B.A.; Heck, R.T.; Cichoski, A.J.; Campagnol, P.C.B. Production of healthier bologna type sausages using pork skin and green banana flour as a fat replacers. *Meat Sci.* **2016**, *121*, 73–78. [CrossRef]
31. Jridi, M.; Abdelhedi, O.; Souissi, N.; Kammoun, M.; Nasri, M.; Ayadi, M.A. Improvement of the physicochemical, textural and sensory properties of meat sausage by edible cuttlefish gelatin addition. *Food Biosci.* **2015**, *12*, 67–72. [CrossRef]
32. Serrano, A.; Cofrades, S.; Ruiz-Capillas, C.; Olmedilla-Alonso, B.; Herrero-Barbudo, C.; Jiménez-Colmenero, F. Nutritional profile of restructured beef steak with added walnuts. *Meat Sci.* **2005**, *70*, 647–654. [CrossRef]
33. Latou, E.; Mexis, S.F.; Badeka, A.V.; Kontakos, S.; Kontominas, M.G. Combined effect of chitosan and modified atmosphere packaging for shelf life extension of chicken breast fillets. *LWT—Food Sci. Technol.* **2014**, *55*, 263–268. [CrossRef]
34. Zhang, M.; Wu, J.R.; Li, X.; Yang, C.C.; Yue, X.Q. Survey of nitrite in the naturally fermented sour pickled cabbages in northeast of China. *Adv. Mater. Res.* **2012**, *610–613*, 409–412. [CrossRef]
35. Huang, R.; Huang, K.; Guan, X.; Li, S.; Cao, H.; Zhang, Y.; Lao, X.; Bao, Y.; Wang, J. Effect of defatting and extruding treatment on the physicochemical and storage properties of quinoa (Chenopodium quinoa Wild) flour. *LWT* **2021**, *147*, 111612. [CrossRef]
36. Sang, X.; Ma, X.; Hao, H.; Bi, J.; Zhang, G.; Hou, H. Evaluation of biogenic amines and microbial composition in the Chinese traditional fermented food grasshopper sub shrimp paste. *LWT* **2020**, *134*, 109979. [CrossRef]
37. Sopanrao, P.S.; Abrar, A.S.; Manoharrao, T.S.; Vaseem, B.M.M. Nutritional value of Pleurotus ostreatus (Jacq: Fr) Kumm cultivated on different lignocellulosic agro-wastes. *Innov. Rom. Food Biotechnol.* **2021**, *7*, 66–76.
38. Lee, C.H.; Chin, K.B. Effects of pork gelatin levels on the physicochemical and textural properties of model sausages at different fat levels. *LWT* **2016**, *74*, 325–330. [CrossRef]
39. da Silva, S.L.; Amaral, J.T.; Ribeiro, M.; Sebastião, E.E.; Vargas, C.; de Lima Franzen, F.; Schneider, G.; Lorenzo, J.M.; Fries, L.L.M.; Cichoski, A.J.; et al. Fat replacement by oleogel rich in oleic acid and its impact on the technological, nutritional, oxidative, and sensory properties of Bologna-type sausages. *Meat Sci.* **2019**, *149*, 141–148. [CrossRef]
40. González, A.; Nobre, C.; Simões, L.S.; Cruz, M.; Loredo, A.; Rodríguez-Jasso, R.M.; Contreras, J.; Texeira, J.; Belmares, R. Evaluation of functional and nutritional potential of a protein concentrate from Pleurotus ostreatus mushroom. *Food Chem.* **2021**, *346*, 128884. [CrossRef]
41. Monteiro, G.M.; Souza, X.R.; Costa, D.P.B.; Faria, P.B.; Vicente, J. Partial substitution of pork fat with canola oil in Toscana sausage. *Innov. Food Sci. Emerg. Technol.* **2017**, *44*, 2–8. [CrossRef]
42. Sucu, C.; Turp, G.Y. The investigation of the use of beetroot powder in Turkish fermented beef sausage (sucuk) as nitrite alternative. *Meat Sci.* **2018**, *140*, 158–166. [CrossRef] [PubMed]
43. Ferreira Ignácio Câmara, A.K.; Midori Ozaki, M.; Santos, M.; Silva Vidal, V.A.; Oliveira Ribeiro, W.; de Souza Paglarini, C.; Bernardinelli, O.D.; Sabadini, E.; Rodrigues Pollonio, M.A. Olive oil-based emulsion gels containing chia (*Salvia hispanica* L.) mucilage delivering healthy claims to low-saturated fat Bologna sausages. *Food Struct.* **2021**, *28*, 100187. [CrossRef]
44. Riel, G.; Boulaaba, A.; Popp, J.; Klein, G. Effects of parsley extract powder as an alternative for the direct addition of sodium nitrite in the production of mortadella-type sausages—Impact on microbiological, physicochemical and sensory aspects. *Meat Sci.* **2017**, *131*, 166–175. [CrossRef]
45. Choe, J.; Lee, J.; Jo, K.; Jo, C.; Song, M.; Jung, S. Application of winter mushroom powder as an alternative to phosphates in emulsion-type sausages. *Meat Sci.* **2018**, *143*, 114–118. [CrossRef] [PubMed]
46. Han, M.; Bertram, H.C. Designing healthier comminuted meat products: Effect of dietary fibers on water distribution and texture of a fat-reduced meat model system. *Meat Sci.* **2017**, *133*, 159–165. [CrossRef] [PubMed]
47. García, M.; Dominguez, R.; Galvez, M.; Casas, C.; Selgas, M. Utilization of cereal and fruit fibres in low fat dry fermented sausages. *Meat Sci.* **2002**, *60*, 227–236. [CrossRef]
48. Wang, L.; Li, C.; Ren, L.; Guo, H.; Li, Y. Production of pork sausages using pleaurotus eryngii with different treatments as replacements for pork back fat. *J. Food Sci.* **2019**, *84*, 3091–3098. [CrossRef]
49. Chau, V.; Tobias, J.W.; Bachmair, A.; Marriott, D.; Ecker, D.J.; Gonda, D.K.; Varshavsky, A. A multiubiquitin chain is confined to specific lysine in a targeted short-lived protein. *Science* **1989**, *243*, 1576–1583. [CrossRef]
50. Luan, X.; Feng, M.; Sun, J. Effect of Lactobacillus plantarum on antioxidant activity in fermented sausage. *Food Res. Int.* **2021**, *144*, 110351. [CrossRef]
51. Jo, Y.; An, K.-A.; Arshad, M.S.; Kwon, J.-H. Effects of e-beam irradiation on amino acids, fatty acids, and volatiles of smoked duck meat during storage. *Innov. Food Sci. Emerg. Technol.* **2018**, *47*, 101–109. [CrossRef]
52. Bahadoran, Z.; Mirmiran, P.; Jeddi, S.; Azizi, F.; Ghasemi, A.; Hadaegh, F. Nitrate and nitrite content of vegetables, fruits, grains, legumes, dairy products, meats and processed meats. *J. Food Compos. Anal.* **2016**, *51*, 93–105. [CrossRef]

53. Nader, M.; Hosseininezhad, B.; Berizi, E.; Mazloomi, S.M.; Hosseinzadeh, S.; Zare, M.; Derakhshan, Z.; Conti, G.O.; Ferrante, M. The residual nitrate and nitrite levels in meat products in Iran: A systematic review, meta-analysis and health risk assessment. *Environ. Res.* **2022**, *207*, 112180. [CrossRef] [PubMed]
54. JAYAKUMAR, T.; ALOYSIUSTHOMAS, P.; GERALDINE, P. Protective effect of an extract of the oyster mushroom, Pleurotus ostreatus, on antioxidants of major organs of aged rats. *Exp. Gerontol.* **2007**, *42*, 183–191. [CrossRef] [PubMed]
55. Jayakumar, T.; Thomas, P.A.; Geraldine, P. In-vitro antioxidant activities of an ethanolic extract of the oyster mushroom, Pleurotus ostreatus. *Innov. Food Sci. Emerg. Technol.* **2009**, *10*, 228–234. [CrossRef]
56. Cintya, H.; Silalahi, J.; De Lux Putra, E.; Siburian, R. The influence of fertilizer on nitrate, nitrite and vitamin C contents in vegetables. *Orient. J. Chem.* **2018**, *34*, 2614–2621. [CrossRef]
57. Lirong, W. Vitamin C for detection of nitrite in meat. *Chin. J. Health Lab. Technol.* **2008**, *18*, 2270.
58. Sebranek, J.G. Advances in the technology of nitrite use and consideration of alternatives. *Food Tech.* **1979**, *33*, 58.
59. Krichen, F.; Hamed, M., Karoud, W.; Bougatef, H.; Sila, A.; Bougatef, A. Essential oil from pistachio by-product: Potential biological properties and natural preservative effect in ground beef meat storage. *J. Food Meas. Charact.* **2020**, *14*, 3020–3030. [CrossRef]
60. Manzi, P.; Aguzzi, A.; Pizzoferrato, L. Nutritional value of mushrooms widely consumed in Italy. *Food Chem.* **2001**, *73*, 321–325. [CrossRef]
61. Sanmee, R. Nutritive value of popular wild edible mushrooms from northern Thailand. *Food Chem.* **2003**, *82*, 527–532. [CrossRef]
62. Selli, S.; Guclu, G.; Sevindik, O.; Kelebek, H. Variations in the key aroma and phenolic compounds of champignon (Agaricus bisporus) and oyster (Pleurotus ostreatus) mushrooms after two cooking treatments as elucidated by GC–MS-O and LC-DAD-ESI-MS/MS. *Food Chem.* **2021**, *354*, 129576. [CrossRef]

Article

Effect of Trehalose on the Physicochemical Properties of Freeze-Dried Powder of Royal Jelly of Northeastern Black Bee

Liangyu Li [1], Peiren Wang [2], Yanli Xu [1], Xiaoguang Wu [1,*] and Xuejun Liu [1,*]

1. College of Food Science and Engineering, Jilin Agricultural University, 2888 Xincheng Street, Changchun 130118, China; lly2115513336@163.cn (L.L.); XYL18844685763@163.cn (Y.X.)
2. Institute for Materials Discovery, University College London, London WC1E 7JE, UK; peiren.wang.20@alumni.ucl.ac.uk
* Correspondence: xiaoguangw@jlau.edu.cn (X.W.); liuxuejun@jlau.edu.cn (X.L.)

Abstract: Trehalose is known for its effect of improving the stability of freeze-dried foods. In this work, vacuum freeze-drying (VFD) technology was employed to prepare northeast black bee royal jelly into lyophilized powder and a novel method mixing trehalose into royal jelly is successfully developed to enhance the free radical scavenging ability and the nutrition stability of royal jelly lyophilized powder. The effects of different trehalose content (0, 0.1, 0.3, 0.5, 0.7 and 0.9 wt.%) on the physicochemical properties of lyophilized royal jelly powder were studied. With systematic analysis, it was found that the incorporation of suitable trehalose content in lyophilized royal jelly powder can reduce the loss of the protein, total sugar, total flavone content during the VFD process and enhance the total phenolic antioxidant capacity, solubility, angle of repose, and bulk density of the royal jelly powder. Finally, lyophilized royal jelly with 0.5 wt.% trehalose is selected as the suitable addition content which exhibits the best radical scavenging ability as well as the lowest hygroscopicity. From the perspective of sensory evaluation, all royal jelly lyophilized powders with trehalose are acceptable.

Keywords: trehalose; royal jelly; vacuum freeze drying; physicochemical characteristics

1. Introduction

Royal jelly is a thick milky-white or yellowish fluid slightly sweet and obviously acidic, which is produced and secreted by nurse honey bees from their hypopharyngeal gland [1]. Due to this kind of food being mainly used to feed to the queen bee, it is called royal jelly and also called bee milk. Royal jelly has been paid more and more attention among researchers because of its rich nutritional value, including protein, lipid, vitamins, trace elements and many other biological activities [2]. For example, 10-hydroxy-decenoic acid is one of the important active components in royal jelly [3]. Royal jelly possesses several pharmacological activities such as anti-oxidation, anti-inflammation, anti-fatigue, anti-ageing, antineoplastic, and anti-diabetes which can protect neurons and inhibit oxidative stress damage in the brain to effectively improve Alzheimer's disease and Parkinson's disease [4–6]. However, royal jelly is susceptible to the influence of light, temperature, time and other factors, resulting in the loss of components and deterioration of active substances during production, storage, and transportation, which has an impact on the quality of royal jelly. Therefore, to avoid the inactivation of active substances, fresh royal jelly is usually cryopreserved during transportation and storage, which increases the processing cost and difficulty of royal jelly seasonality. The traditional processing method has a serious impact on the quality of royal jelly and destroys its activity. Therefore, vacuum freeze drying (VFD) is widely used as the method to fabricate royal jelly lyophilized powder for convenient transportation and storage.

VFD has been widely used to fabricate products which have high quality and economic value. Freeze-dried foods can maintain almost the same color, flavor, and nutrient value compared with fresh foods [7]. Recently, the VFD process has increasingly been used to fabricate protein products [8]. However, some proteins could undergo structural changes during the VFD process. For instance, Song et al. reported that the structure of bovine serum albumin changed during the VFD process [9]. A limited number of potentials for biopharmaceuticals might undergo inactivation, denaturation, and other reactions during the fabricating process [10]. Moreover, Maillard reaction and oxidation could also happen during the VFD process, which should be avoided as much as possible during the food production process [11].

It has been reported that the stability of protein can be achieved by the formation of an amorphous glass matrix [12]. The high viscosity and stable structure of the glassy state are the important factors in preventing the protein from unfolding. To enhance the stability of the protein during the VFD process, monosaccharides and disaccharides have been widely used to prevent protein denaturation and protect the stability of the protein crystal structure [13]. Recently, different cryoprotective agents such as sucrose, lactose, mannitol, and trehalose have been used to improve the storage stability and properties of freeze-dried powders [14,15]. For instance, trehalose is a non-reducing disaccharide with a higher glass transition temperature and a lower hygroscopic ability compared with other disaccharides. Trehalose is widely used as a stabilizing agent in the food industries, which can interact intensively with the surface of macromolecules [16,17] and improve the stability of freeze-dried samples significantly [18,19]. Trehalose has been used to improve the properties of peeled shrimp protein during frozen storage and increase the total polyphenols and antioxidant activity of apple puree [20,21]. However, there are very few studies about the protective effect of trehalose on vacuum freeze drying of northeast black bee royal jelly.

In this study, the protective effect, and physicochemical properties of different trehalose content on vacuum freeze-drying of royal jelly are systemically studied. The effect of trehalose on the physicochemical property, total flavonoid content, solubility and free radical scavenging activity of royal jelly lyophilized powder are investigated. Finally, 0.5% content of trehalose is selected as the best addition content which could reduce the loss of protein and total sugar during fabrication and exhibits the best DPPH radical scavenging ability as well as the lowest hygroscopicity.

2. Materials and Methods

2.1. Materials

The frozen northeast black bee royal jelly (Jilin Hanfeng Agricultural Science and Technology Development Co., Ltd., Changchun, China) was stored at $-20\,^\circ$C and used to prepare the lyophilized powder. All the frozen royal jelly was transported by a cold chain and maintained for $-20\,^\circ$C in the whole process. Trehalose, Folin-Ciocalteu reagent, gallate acid, 2,2-diphenyl-2-picrylhydrazyl (DPPH), and 10-Hydroxy-2-decanoic acid (10-HDA) were purchased from Meilun Co., Ltd. (Dalian, China). Phenol, vitriolic acid, methanol, phosphoric acid, aluminium chloride, sodium carbonate, and potassium acetate were purchased from Beijing Beihua Co., Ltd. (Beijing, China). All chemicals and solvents were of analytical grade and used as received.

2.2. Fabrication of Lyophilized Powder

Different contents of trehalose, including 0 wt.% (Control), 0.1 wt.% (TR 1), 0.3 wt.% (TR 3), 0.5 wt.% (TR 5), 0.7 wt.% (TR 7), and 0.9 wt.% (TR 9), were added into fresh royal jelly and made into lyophilized powder via a vacuum freeze drier (Beijing Boyikang Instrument Co., Ltd., Beijing, China) (Table 1). The group with trehalose addition of 0% was the control group. First, 10 g royal jelly was defrosted at room temperature for 10 min. Then, the trehalose particles were fully dissolved in distilled water, and the visible impurities were removed from fresh royal jelly. Finally, the trehalose solution and royal jelly were poured into a centrifugal tube, mixed with a vortex oscillator for 60 s and pre-freezed at $-40\,^\circ$C for 6 h. During VFD process, the freezing time, the vacuum degree, the temperature of the heating plate, and the temperature of the cold trap were 48 h, 40 Pa, −18 °C, and −85 °C, respectively. Each formulation group was replicated three times for repeatability.

Table 1. Formulations of lyophilized powder with royal jelly add trehalose (TR).

Parameter	Control	TR 1	TR 3	TR 5	TR 7	TR 9
Fresh royal jelly (g)	10	10	10	10	10	10
Trehalose (g)	0	0.01	0.03	0.05	0.07	0.09
Water (g)	10	10	10	10	10	10
Thickness (mm)	6	6	6	6	6	6

Note: Control, TR 1, TR 3, TR 5, TR 7, and TR 9 were 0%, 0.1%, 0.3%, 0.5%, 0.7%, and 0.9% addition amount of trehalose, respectively.

2.3. Characterization of Compositional of Royal Jelly Powder

2.3.1. Protein

The protein content was measured according to the Association of Official Analytical Chemists (AOAC) (2006) method [22].

2.3.2. Total Sugars

The total sugars were evaluated according to the Phenol-Sulfuric Acid Assay method [23].

2.3.3. Fat

The fat content was determined by a fat analyser (SOX500, Hanon, Jinan, China) based on the Soxhlet extractor method.

2.3.4. 10-Hydroxy-2-Decanoic Acid (10-HDA)

The 10-HDA analysis was performed by high-performance liquid chromatography (HPLC) based on published protocols with minor modifications [24]. Prepare 100 mL of 100 μg mL^{-1} 10-HDA and standard substance 0, 5, 10, 20, 30, and 40 Cg mL^{-1} reserve solution in methanol: water (50:50, V/V).

Approximate 50 mg lyophilized powder was dissolved in 25 mL methanol: water (50:50, V/V) solvent and then treated with 35 kHz ultrasound in an ultrasonic cleaner (Barker, Shanghai, China) for 30 min. After ultrasonic treatment, the sample solution was filtered (0.45 μm filter) and transferred to 2 mL autosampler vials prior for injection.

HPLC separation was performed on a 100 mm × 4.6 mm × 3.5 μm C18 column (Meilun Co., Ltd., Dalian, China) at 30 °C with a mobile phase flow rate of 1 mL min^{-1} (using isometric conditions). The mobile phase consists of methanol: water: phosphoric acid (50:50:0.3, V/V/V). The maximum absorbance of 10-HDA is 210 nm and the injection volume was 10 μL.

2.3.5. Moisture

The moisture content was determined by a rapid moisture meter (HB43-S, Mettler Toledo, Greifensee, Switzerland).

2.3.6. pH Value

The pH value was measured by a pH meter (FE-28, Mettler Toledo, Greifensee, Switzerland) according to the method reported by Balkanska et al. [25].

2.3.7. Ash

The ash content was measured according to the AOAC (942.05) method [26].

2.3.8. Water Activity

The water activity (a_w) of lyophilized powder was determined by a pre-calibrated water activity meter (Rotronic, Bassersdorf, Switzerland) at 26 ± 0.5 °C.

2.4. Characterization of Royal Jelly Lyophilized Powder

2.4.1. Angle of Repose

Herein, the angle of repose was measured via a fixed funnel method which was similar to the method reported by Alanazi et al. [27]. Firstly, a funnel was fixed on the iron rack vertically above a piece of blank paper. Then, the lyophilized powder was poured into the funnel and slipped on the blank paper until the peak of the powder pile touched the outlet of the funnel. Finally, the angle of repose can be calculated by:

$$\theta = \tan^{-1}(H/R) \tag{1}$$

where θ is the angle of repose, H is the distance between the blank paper and the funnel outlet, and R is the radius at the bottom of the powder cone. The measurement was replicated three times to calculate the average angle.

2.4.2. Bulk Density and Tapped Density

The measurements of bulk density and tapped density were carried out with the method reported by Erdem et al. with slight modifications [28]. First, the powder was poured into a clean and dry graduated cylinder (5 mL) which was with a mass of m. Tap the cylinder once slightly to remove all the adhering powder on the wall, and then record the volume of the powder v_1 and weigh the mass of cylinder and powder m_1. Then, tap the cylinder until the powder reached a constant volume, record the volume of the powder v_2 and weigh the mass of cylinder and powder m_2. Finally, the bulk density D_b and the tapped density D_t can be calculated by:

$$D_b = (m_1 - m)/v_1 \tag{2}$$

$$D_t = (m_2 - m)/v_2 \tag{3}$$

2.5. Solubility

The solubility of lyophilized powder was carried out with the method reported by Erdem et al. with slight modifications [28]. In total, 2.0 g royal jelly lyophilized powder and 50 mL distilled water were added into a glass beaker and mixed for 5 min at 800 rpm with a stirrer. The obtained suspensions stood at room temperature for 10 min and centrifuged at $500 \times g$ for 5 min. Finally, the supernatant was dried at 102 °C in an incubator oven until obtained a constant weight. The solubility can be calculated by:

$$\text{Solubility} = (m_s/m_t) \times 100\% \tag{4}$$

where m_s is the mass of soluble solid, and m_t is the mass of the total solid before dissolution.

2.6. Hygroscopicity

Hygroscopicity was determined by the method reported by Tonon et al. [29]. One gram of each powder sample was exposed at 75% RH (saturated sodium chloride solution) for 192 h. The hygroscopic ability was presented as a percentage of the water absorption of the dried powder:

$$\text{Hygroscopicity} = \frac{m_{\text{wet}} - m_{\text{dry}}}{m_{\text{dry}}} \times 100\% \tag{5}$$

where m_{wet} is the mass of powder after water absorption, and m_{dry} is the mass of the dry powder.

2.7. Measurement of Total Flavonoid Content (TFC)

The TFC of royal jelly lyophilized powder was determined by the aluminium chloride colorimetric method reported by Liu et al. with slight modifications [30]. An amount of 0.5 g lyophilized powder sample was mixed with 15 mL of 95% ethanol and ultra-sonicated for 1 h. Then, the ethanol extract of lyophilized royal jelly powder was obtained by filtration. An amount of 1 mL of ethanol extract, 1 mL of $AlCl_3$ solution (1 M), 1 mL potassium acetate (1 M), and 15 mL ethanol (95 wt.%) were mixed and added distilled water to 50 mL. The solution was kept at room temperature for 40 min and absorbance was recorded at 420 nm wavelength against blank via a UV-visible spectrophotometer (TU-1901L, Beijing Universal Instrument Co., Ltd., Beijing, China). Various concentrations of rutin were prepared in alcohol as a reference flavonoid. Finally, the same process and measurement were repeated and the calibration curve was plotted. The TFC of royal jelly lyophilized powder was expressed as rutin equivalent per gram of royal jelly ($mgRE \cdot g^{-1}$).

2.8. Measurement of Total Phenolic Contents (TPC)

The total phenol in ethanolic extract of royal jelly lyophilized powder was measured via a colorimetric method. The total phenolic contents were measured by the Folin-Ciocalteau method reported by Singleton et al. with slight modifications [31]. In total, 10 mg lyophilized powder sample was mixed with 100 mL of ethanol to prepare the ethanol extract of powder (100 $mg \cdot L^{-1}$). Moreover, 100 μL of the ethanol extract and 100 μL of 7.5 wt.% Na_2CO_3 were fully mixed. Then, 50 μL of Folin-Ciocalteu reagent was added and the obtained mixture and kept at room temperature in the dark for 30 min. The absorbance of all the samples was measured at 760 nm via a UV-visible spectrophotometer (TU-1901L, Beijing Universal Instrument Co., Ltd., Beijing, China). Under the same conditions, gallate acid was used as a reference phenolic. Finally, the same process and measurement were repeated and the calibration curve was plotted. TPC of the extract was evaluated based on gallic acid equivalent ($mgGAE \cdot g^{-1}$).

2.9. The Free Radical Scavenging Activity Assays

Antioxidant activities were measured by DPPH radical-scavenging assay based on the method reported by Brand-Williams et al. with slight modification [32]. One gram of lyophilized powder was added to 20 mL of ethanol. The mixture was shocked in the dark for 1 h and filtrated. 3.6 mL of DPPH (0.1 mmol L^{-1} in ethanol) and 0.66 mL extract were mixed. The mixture was incubated in the dark for 30 min at room temperature and then the scavenging capacity was measured at 517 nm.

2.10. Color Measurements

Color values (L^*, lightness; a^*, redness; and b^*, yellowness) of lyophilized powder were measured using a colorimeter (Xinlian Creation Electronic Co. Ltd., Shanghai, China). A standard plate CX 2064 was used as standard ($L^* = 94.52$, $a^* = -0.86$, $b^* = 0.68$). The ΔE^* was calculated according to the following formula:

$$\begin{cases} \Delta E^* = \left(\Delta a^2 + \Delta b^2 + \Delta L^2\right)^{0.5} \\ \Delta a = a* - a*_{sample} \\ \Delta b = b* - b*_{sample} \\ \Delta L = L* - L*_{sample} \end{cases} \quad (6)$$

where ΔL^*, Δa^* and Δb^* are the differences between the L^*, a^*, and b^* values of the treatment and those of standard samples, respectively.

2.11. Particle Size

The particle size distribution of lyophilized powder was determined by a dry powder laser particle sizer (Mastersizer 2000, Malvern Panalytical GmbH, Kassel, Germany) at 25 °C. Before the measurement, all the samples were ground through a 100-mesh sieve.

2.12. Scanning Electron Microscopy (SEM)

The microstructure of lyophilized powder without and with trehalose were analyzed by a scanning electron microscope (SEM) (ZEISS EVO 18, Zeiss, Oberkochen, Germany).

The SEM images were captured using a voltage of 20 kV and a 1500× magnification or a 6000× magnification.

2.13. Differential Scanning Calorimetry (DSC)

According to the method reported by Butreddy et al., the glass transition temperature (Tg) of lyophilized powder could be analyzed by differential scanning calorimetry [33]. DSC is used to measure stability, shelf life, denaturation, and other irreversible changes during transportation. About 2 mg of lyophilized powder was hermetically sealed in an alumina crucible with the temperature ranges of 30–250 °C at a heating rate of 10 °C min^{-1}. Calibration was performed with indium before performing the analysis.

2.14. X-ray Diffraction (XRD)

The crystal type and crystallinity of lyophilized powder were measured by X-ray diffractometer (Ultima IV, Rigaku, Tokyo, Japan) with Cu-Kα radiation operating (λ = 1.5406 Å) at 45 kV and 40 mA.

2.15. Fourier Transform Infrared Spectroscopy (FTIR)

ATR-FTIR (IRAffinity-1, Shimadzu, Tokyo, Japan) was used to indicate the chemical structure of lyophilized royal jelly powder. The FTIR spectra of the lyophilized powder were recorded within the range of 400–4000 cm^{-1} at ambient temperature (20 °C) with 32 times scanning.

2.16. Statistical Analysis

A randomized complete block design was applied and all the experiment was replicated three times. The difference between factors and levels were submitted to the analysis of variance (ANOVA). The determine significant differences ($p < 0.05$) among the means used Tukey tests. The analysis was taken by SPSS software version 19. All the data were presented as mean ± standard deviation.

3. Results

3.1. Chemical Composition and Water Activity of Royal Jelly Powder

Table 2 showed the chemical composition of lyophilized powder in the control group and TR groups. It shows that TR samples had more proteins and total sugar than the control sample ($p < 0.05$). The reason why the protein contents increased from 33.39% to 35.60% in the lyophilized powder with trehalose was that the trehalose could prevent the loss of active substances such as proteins from drying-related stresses during the VFD process [34]. With the increase of trehalose content of the samples, the total sugar content in the lyophilized powder increased from 33.80% to 36.80%. There were no significant differences in fat, ash, pH value, and 10-HDA between the control group and TR groups ($p > 0.05$). The moisture content and a_w of lyophilized powder were 3.32%–4.03%, 0.203–0.235, respectively. The moisture content of all the samples was found to be lower than 5%, which was important for stability. The a_w is an important parameter to predict the stability of lyophilized powder. The a_w of freeze-dried production is generally between 0.28–0.11 and all a_w values of the lyophilized powders in this research were within this range which could help to control non-enzymatic browning and inhibit microbial growth [18]. The a_w decreased with the increase of trehalose content, which was due to trehalose having an ability to enhance the hydrogen bond strength. Commonly, the water molecule had the disposition to move to the molecules which were easy to form hydrogen bonds. Due to the existence of trehalose, there was a weaker interaction between water molecules and royal jelly lyophilized powder leading to lower a_w [35].

Table 2. Proximate composition (%), water activity (a_v) and acidity of trehalose-royal jelly lyophilized powder (TR).

Parameter	Control	TR 1	TR 3	TR 5	TR 7	TR 9
Protein (%)	33.39 ± 0.10[f]	33.72 ± 0.10[d]	33.19 ± 0.10[e]	34.67 ± 0.10[c]	35.08 ± 0.10[b]	35.60 ± 0.10[a]
Fat (%)	15.63 ± 0.11[a]	15.62 ± 0.10[a]	15.63 ± 0.11[a]	15.62 ± 0.10[a]	15.64 ± 0.11[a]	15.63 ± 0.12[a]
Total sugar (%)	33.80 ± 0.14[f]	34.13 ± 0.15[e]	34.80 ± 0.17[d]	35.47 ± 0.17[c]	36.13 ± 0.17[b]	36.80 ± 0.15[a]
Ash (%)	3.7 ± 0.11[a]	3.7 ± 0.10[a]	3.7 ± 0.11[a]	3.7 ± 0.12[a]	3.7 ± 0.10[a]	3.7 ± 0.11[a]
pH	4.62 ± 0.09[a]	4.53 ± 0.07[a]	4.54 ± 0.09[a]	4.58 ± 0.08[a]	4.60 ± 0.10[a]	4.60 ± 0.09[a]
Moisture (%)	3.32 ± 0.06[f]	3.47 ± 0.04[e]	3.56 ± 0.05[d]	3.71 ± 0.07[c]	3.88 ± 0.07[b]	4.03 ± 0.06[a]
a_w	0.235 ± 0.004[a]	0.229 ± 0.003[b]	0.221 ± 0.002[c]	0.213 ± 0.003[d]	0.210 ± 0.003[d]	0.203 ± 0.004[e]
10-HDA (%)	5.16 ± 0.01[a]	5.16 ± 0.01[a]	5.17 ± 0.01[a]	5.17 ± 0.01[a]	5.16 ± 0.01[a]	5.16 ± 0.01[a]

Notes: Values are given as mean ± standard error. The letter from a to f shows the different value in significant differences from high value to low value ($p < 0.05$).

3.2. Evaluation of Royal Jelly Freeze-Dried Powder

The angle of repose, bulk density and tapped density were used to determine powder fluidity. The flow behavior of the powder is useful to predict its quality characteristics during processing, packaging, and storage [36]. Table 3 shows the parameters of royal jelly lyophilized powder.

Table 3. Bulk density, tapped density and angle of repose of control and trehalose-royal jelly lyophilized powder (TR).

Parameter	Control	TR 1	TR 3	TR 5	TR 7	TR 9
Bulk density (g mL^{-1})	0.353 ± 0.010[e]	0.384 ± 0.007[d]	0.394 ± 0.017[c]	0.432 ± 0.006[b]	0.458 ± 0.009[a]	0.458 ± 0.009[a]
Tapped density (g mL^{-1})	0.592 ± 0.006[f]	0.612 ± 0.006[e]	0.625 ± 0.006[d]	0.668 ± 0.008[c]	0.704 ± 0.006[b]	0.802 ± 0.007[a]
Angle of respose (°)	62.24 ± 0.09[f]	62.73 ± 0.11[e]	63.77 ± 0.09[d]	64.54 ± 0.11[c]	65.06 ± 0.10[b]	65.65 ± 0.08[a]

Notes: Values are given as mean ± standard error. The letter from a to f shows the different value in significant differences from high value to low value ($p < 0.05$).

3.2.1. Angle of Repose

In general, the angles of repose <30°, 30°–45°, 45°–55°, and >55° indicated good flowability, some cohesiveness, true cohesiveness, and high cohesiveness (very limited flowability), respectively [37]. As shown in Table 3, the control group had the smallest angle of repose but also >50° among all the groups, indicating that the lyophilized powder had a high cohesion. The higher angle of repose of the TR groups might be due to the smaller particle size and higher cohesion of the lyophilized powder in TR groups. With the increase of trehalose content, the angle of repose and the fluidity of the powder gradually increased.

3.2.2. Bulk Density and Tapped Density

Table 3 shows that the bulk density and tapped density of royal jelly lyophilized powder increased with the increase of the trehalose content. The increased density could be explained by the moisture content and particle shape of the lyophilized powder. First, the moisture content of the powder affects the bulk density and tapped density directly. If the powder contains lower moisture content, the total solid density will increase. Second, the particle shape and size also significantly affect the bulk density and tapped density. For instance, the particles with irregular shapes have low bulk density [18]. According to SEM images, the structure of lyophilized powder with trehalose had a broken glass shape, and the sample in the control group was more complete. Therefore, the royal jelly lyophilized powder with trehalose had a higher bulk density and tapped density than the control sample.

3.3. Solubility

In the food industry, solubility can be regarded as the rate of dissolution to describe powder reconstitution properties [38]. According to the research reported by Jayasundera et al., the low a_w value and relatively high moisture content (due to the high water holding capacity of the protein) of powder also contributed to higher solubility [39]. Haque et al. reported that high a_w values exacerbate protein denaturation which harmed the solubility [40]. Quek et al. proved that there was a positive correlation between the solubility of spray-dried watermelon powder and water content [41]. As shown in Figure 1, the trehalose could improve the solubility of lyophilized powder and the solubility of powder increased with the increase of trehalose addition level. The control group and TR 9 group were ~70.80% and ~78%, respectively.

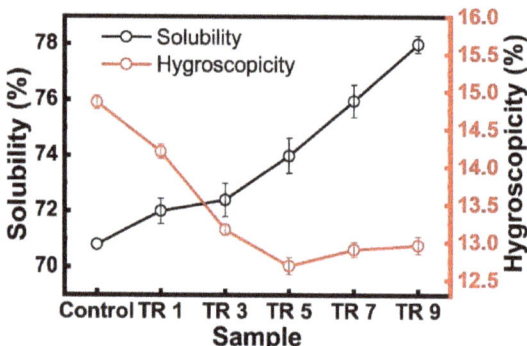

Figure 1. Solubility and hygroscopicity of control sample and trehalose-royal jelly lyophilized powder (TR).

3.4. Hygroscopicity

Hygroscopicity is the ability of a material to absorb moisture from the environment [42]. Hygroscopic powders will absorb water from the air, increasing their cohesion and decreasing their flowability which adversely affects the quality of the powder. Powders with hygroscopic properties less than 20% are generally considered as not very hygroscopic [43]. It is reported that the powders are considered hygroscopic if the hygroscopicity value of powder is in the range of 15%–20% (determined at 75% of RH) [44]. According to Figure 1, all samples could be considered as not hygroscopic. The hygroscopicity of the powder dropped to the bottom with 0.5 wt.% content of trehalose (TR 5) and then increased slightly.

3.5. Total Flavonoids and Total Phenols Contents (TFC and TPC)

TFC and TPC of royal jelly have a variety of pharmacological properties, including antibacterial, anticancer, anti-inflammatory, immunomodulatory and antioxidant activities [45]. The TFC and TPC content of royal jelly lyophilized powder are shown in Figure 2. The content of flavonoids and phenols in TR groups were higher than those in the control group ($p < 0.05$) and reached the peak in the TR 5 group of which the TPC and TFC were 2.08 mgGAE·g^{-1} and 11.2 mgRE·g^{-1}, respectively. This phenomenon could be due to the lower content of flavonoids and phenols degradation caused by trehalose during the VFD process. Trehalose can interact with bioactive compounds in royal jelly and form complexes [46]. According to the research describing the effect of adding sugars during storage, trehalose has no direct internal hydrogen bonds compared with most other disaccharides. All the four internal hydrogen bonds are connected indirectly via two water molecules which can form part of the natural dihydrate structure. This arrangement makes the molecule unusual flexibility around the disaccharide bond, which may allow trehalose to adhere more closely to the irregular surface of the macromolecule than other disaccharides and protect the bioactive substance [47].

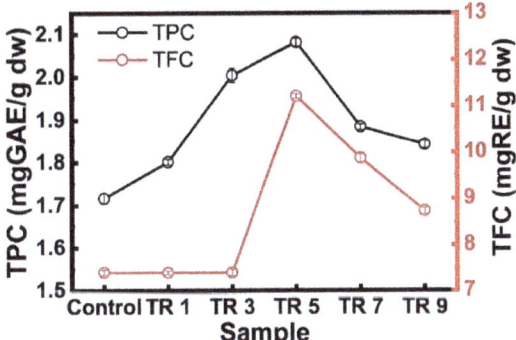

Figure 2. TPC and TFC of the control sample and trehalose royal jelly lyophilized powder (TR).

3.6. The Free Radical Scavenging Activity of Lyophilized Powder

Figure 3 shows the radical-scavenging effect of samples in control and TR groups upon hydroxyl radicals. All lyophilized powder inhibited the formation of hydroxyl radicals in varying degrees. The DPPH radical-scavenging effect of the control group was 8% and TR samples showed significantly higher free radical scavenging activity than the control group ($p < 0.05$). TR 5 showed the highest radical-scavenging effect compared to other TR samples. However, trehalose had no free radical scavenging activity. It was speculated that the strongest antioxidant capacity of the TR 5 group might be due to the highest TPC and TFC among the TR groups. According to the studies of Guo et al. and Nurcholis et al., TPC of royal jelly showed certain antioxidants and the TFC also had a significant positive correlation with royal jelly antioxidant activity from ethanol [2,48].

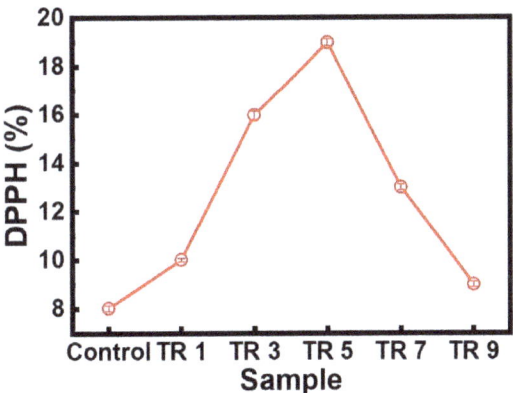

Figure 3. The free radical scavenging activity of the control sample and trehalose-royal jelly lyophilized powder (TR).

Therefore, lyophilized powder with 0.5% trehalose was then selected for further analysis because of the good particle properties, low moisture, and high antioxidant activity and solubility.

3.7. Color

A higher L^* value means the brighter color of the lyophilized powder. The negative a^* value indicates that the color of freeze-dried powder was close to the green. The higher value of b^* shows lyophilized powder is close to yellow. Table 4 shows that the L^* values of royal jelly lyophilized powder with different trehalose content were significantly different

($p < 0.05$). The L^* value of TR groups was slightly decreased and the b^* value of the TR groups was increased compared with the control sample. The ΔE^* represented the color difference between the TR groups and the control group. With the addition of trehalose, ΔE^* of lyophilized powder increased. The addition of trehalose increased the content of phenolic substances in the lyophilized powder, which resulted in the color deepening of lyophilized powder [49]. However, the color of the TR 5 group was closest to the control group. It was speculated that this difference might be due to non-enzymatic browning. Phenolic substances could promote the non-enzymatic browning of royal jelly to a certain extent and form brown substances through their oxidation and condensation which could affect the color of royal jelly lyophilized powder [50].

Table 4. Color comparison of royal jelly lyophilized powder (TR) and control group.

Sample	L^*	a^*	b^*	ΔE^*
Control	89.80 ± 0.04[a]	−0.53 ± 0.01[a]	20.54 ± 0.04[e]	-
TR 1	88.80 ± 0.05[f]	−0.52 ± 0.01[a]	21.47 ± 0.06[c]	1.36 ± 0.05[c]
TR 3	89.06 ± 0.03[e]	−0.55 ± 0.02[a]	22.52 ± 0.08[a]	2.11 ± 0.07[a]
TR 5	89.56 ± 0.05[c]	−0.53 ± 0.03[a]	20.82 ± 0.02[d]	0.37 ± 0.01[d]
TR 7	89.71 ± 0.03[b]	−0.54 ± 0.05[a]	22.10 ± 0.07[b]	1.56 ± 0.09[b]
TR 9	89.28 ± 0.02[d]	−0.52 ± 0.02[a]	21.96 ± 0.05[b]	1.51 ± 0.04[b]

Notes: Values are given as mean ± standard error. The letter from a to f shows the different value in significant differences from high value to low value ($p < 0.05$). Control, TR 1, TR 3, TR 5, TR 7, and TR 9 were 0%, 0.1%, 0.3%, 0.5%, 0.7%, and 0.9% addition amount of trehalose, respectively.

3.8. Particle Size

The particle size distribution affects many properties such as bulk behavior and the homogeneity of powder [51]. As shown in Section 3.6, the lyophilized powder with 0.5% trehalose was selected for further analysis. Figure 4 shows that TR 5 appeared approximately as Log-normal distribution. There was a small peak around 10 μm which might be the fractions generation during handling or the sieving process. The median diameter (d50) of lyophilized powder with trehalose (124 μm) was slightly smaller than that of the control group (134 μm). The size of particles could depend on the milling type, operating conditions, and milling time of fabrication [52]. Almost 80% of royal jelly lyophilized powder fell within the range of the opening sizes of the sieve screens.

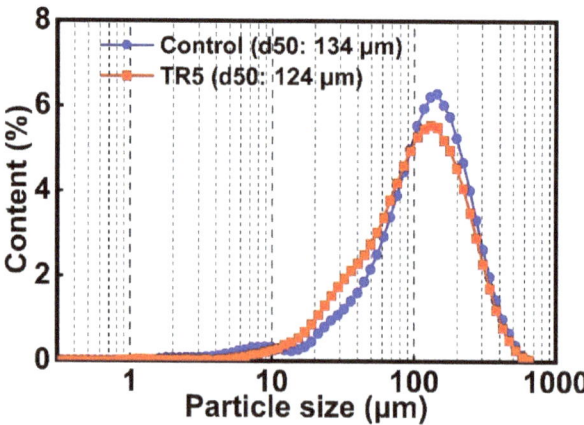

Figure 4. The particle size of the control sample and royal jelly lyophilized powder with 0.5% addition of trehalose (TR).

3.9. SEM

SEM was employed to observe the microstructure of the lyophilized powder (Figure 5). The figures indicated significant differences between the powder with and without the trehalose. The structure of powders with trehalose was broken glass shaped and had miscellaneous irregular sizes and shapes. However, the surfaces were smoother than that of samples in the control group. The visible holes and fractures could be observed in the outer surfaces of powder which was consistent with the characteristics of added sugars and reflected the presence of trehalose.

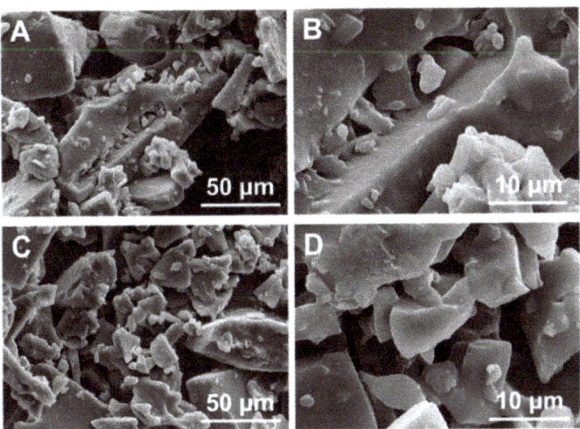

Figure 5. The SEM images of the pure royal jelly lyophilized powder (**A,B**) and trehalose-royal jelly lyophilized powder (TR 5) (**C,D**).

3.10. DSC

Figure 6 is the DSC result of the lyophilized royal jelly without trehalose and with 0.5% trehalose. The thermal stability of the samples could be evaluated by the peak temperature and enthalpy change parameters in the DSC curve. The better thermal stability of the sample could be represented by the higher peak temperature and larger enthalpy change. In Figure 6, the first endothermic peak at ~56 °C may be related to the water released. The second endothermic peak at 145 and 162 °C were related to the melting of crystals in the control group and TR 5 group, respectively. It was proven that the lyophilized royal jelly powder with trehalose had better stability than the control sample. In addition, the glass transition was observed in the TR 5 group at around 65 °C because of the addition of the trehalose.

3.11. XRD

The XRD result of lyophilized royal jelly powder with 0.5% and without trehalose were plotted in Figure 7. It could be observed that the TR 5 sample and control sample showed similar particularly sharp diffraction peaks at 22.32° and 23.74°. It was proven that the addition of trehalose did not change the crystal structure of the lyophilized royal jelly powder obviously and all the samples had a complete crystal state. The vibration intensity of the diffraction peak reflected the degree of crystallization of the substance. Compared with the control group, the strength of the diffraction peak of the lyophilized royal jelly powder with trehalose was significantly enhanced. The crystallinity of the lyophilized royal jelly powder with trehalose was changed, which could lead to the enhancement of the vibration intensity of the characteristic peak.

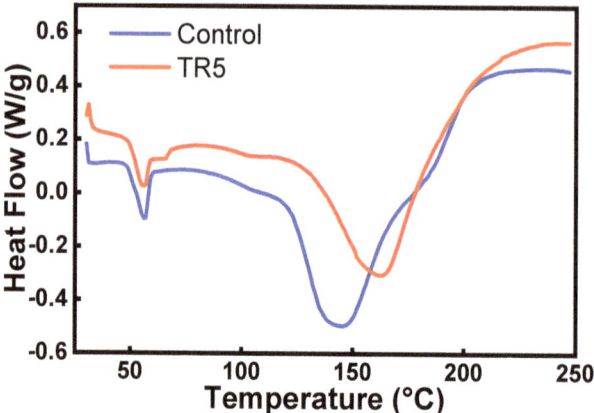

Figure 6. The DSC result of the control sample and royal jelly lyophilized powder with 0.5% addition of trehalose (TR 5).

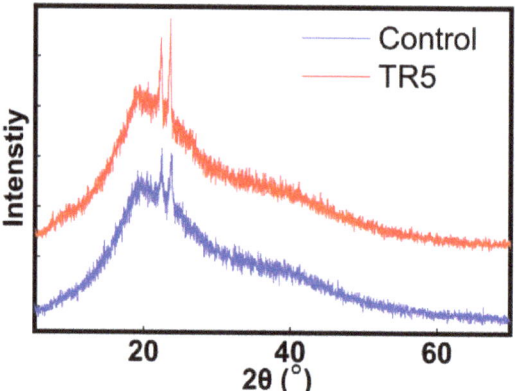

Figure 7. X-ray diffraction result of the control sample and royal jelly lyophilized powder with 0.5% addition of trehalose (TR 5).

3.12. FTIR

The FTIR spectrum obtained from 4000 to 400 cm^{-1} was employed to represent the lyophilized powder of royal jelly. As shown in Figure 8, the band at 3426 cm^{-1} was identified as the vibration of the –OH of carbohydrates, water, and organic acids [53]. The peak of all samples at 2936 cm^{-1} was attributed to C–H stretching in carboxylic acid and NH$_3$ stretching in free amino acids. The most important information used to distinguish the sample was in the 1800–750 cm^{-1} region. The absorption peaks at 1643, 1527, 1338 and 1238 cm^{-1} represented the of the protamine group [54]. High specific for amino acids and proteins were the peaks near 1338 cm^{-1} and the peak at 1045 cm^{-1} corresponded to the C–O stretch of the carbohydrates [55]. The FTIR spectral graph showed that the chemical component of the lyophilized royal jelly with trehalose had no change.

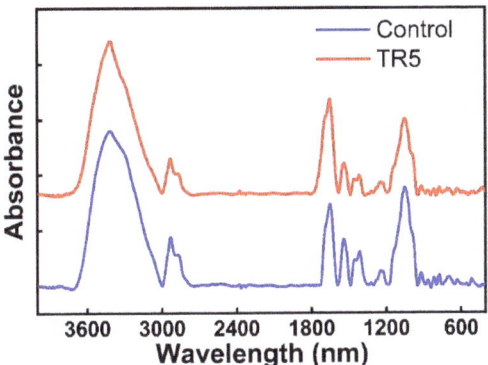

Figure 8. X-ray diffraction patterns of the control sample and royal jelly lyophilized powder with 0.5% addition of trehalose (TR 5).

4. Conclusions

Herein, a novel method combining trehalose and royal jelly is successfully developed to enhance the free radical scavenging ability and the nutrition stability of royal jelly lyophilized powder after the VFD process. With systematic analysis, 0.5 wt.% of trehalose is selected as the best addition content which can reduce the loss of TFC and TPC during fabrication and exhibit the best DPPH radical scavenging ability as well as the lowest hygroscopicity. TFC and TPC contents of the powder increase obviously from 1.71 to 2.08 mgGAE·g^{-1} and 7.4 to 11.2 mgRE·g^{-1}, respectively. Moreover, the bulk density and tapped density increase from 0.353 to 0.432 g·mL^{-1} and from 0.592 to 0.668 g·mL^{-1}, respectively, which enhance conducive to the storage, processing, and transportation of lyophilized powder. The addition of trehalose can also improve the solubility of lyophilized powder and reduce the hygroscopic property and water activity of powder, which are the key factors to the flowability and of stability the lyophilized powder. According to the FTIR and XRD results, the chemical composition and crystal structure of the lyophilized royal jelly powder have no obvious change. The DSC result shows that trehalose can improve the stability of royal jelly lyophilized powder.

Author Contributions: Investigation, Methodology, Formal analysis, Writing—original draft, L.L.; Formal analysis, Visualization, Writing—review and editing, P.W.; Data curation, Methodology, Investigation, Y.X.; Project administration, Supervision, Conceptualization X.W.; Project administration, Supervision, Data curation, Funding acquisition, X.L. All authors have read and agreed to the published version of the manuscript.

Funding: This work was supported by the Science and Technology Key Projects of Jilin Provincial Department of Science and Technology in 2019 (20200402069NC), "The key technology of vacuum freeze drying of royal jelly of northeast black bee and the development of convenient and nutritious food industrialization".

Institutional Review Board Statement: Not applicable.

Informed Consent Statement: Not applicable.

Data Availability Statement: Data are contained within the current manuscript.

Conflicts of Interest: The authors declare that they have no known competing financial interests or personal relationships that could have appeared to influence the work reported in this paper.

References

1. Ramanathan, A.; Nair, A.J.; Sugunan, V.S. A review on royal jelly proteins and peptides. *J. Funct. Foods* **2018**, *44*, 255–264. [CrossRef]
2. Guo, J.; Wang, Z.; Chen, Y. Active components and biological functions of royal jelly. *J. Funct. Foods* **2021**, *82*, 104514. [CrossRef]

3. Duong, H.A.; Vu, M.T.; Nguyen, T.D. Determination of 10-hydroxy-2-decenoic acid and free amino acids in royal jelly supplements with purpose-made capillary electrophoresis coupled with contactless conductivity detection. *J. Food. Compos. Anal.* **2020**, *87*, 103422. [CrossRef]
4. Sari, E.; Mahira, K.F.; Patel, D.N.; Chua, L.S.; Pratami, D.K.; Sahlan, M. Metabolome analysis and chemical profiling of Indonesian royal jellies as the raw material for cosmetic and bio-supplement products. *Heliyon* **2021**, *5*, e06912. [CrossRef]
5. Park, M.J.; Kim, B.Y.; Deng, Y.; Park, H.G.; Choi, Y.S.; Lee, K.S.; Jin, B.R. Antioxidant capacity of major royal jelly proteins of honeybee (*Apis mellifera*) royal jelly. *J. Asia Pac. Entomol.* **2020**, *2*, 445–448. [CrossRef]
6. Mokaya, H.O.; Njeru, L.K.; Lattorff, H.M.G. African honeybee royal jelly: Phytochemical contents, free radical scavenging activity, and physicochemical properties. *Food Biosci.* **2020**, *37*, 100733. [CrossRef]
7. Zhang, L.; Qiao, Y.; Wang, C.; Liao, L.; Shi, D.; An, K.; Shi, L. Influence of high hydrostatic pressure pretreatment on properties of vacuum-freeze dried strawberry slices. *Food Chem.* **2020**, *331*, 127203. [CrossRef]
8. Peters, B.H.; Staels, L.; Rantanen, J.; Molnár, F.; De Beer, T.; Lehto, V.P.; Ketolainen, J. Effects of cooling rate in microscale and pilot scale freeze-drying–variations in excipient polymorphs and protein secondary structure. *Eur. J. Pharm. Sci.* **2016**, *95*, 72–81. [CrossRef]
9. Song, J.G.; Lee, S.H.; Han, H.K. Biophysical evaluation of aminoclay as an effective protectant for protein stabilization during freeze-drying and storage. *Int. J. Nanomed.* **2016**, *11*, 6609. [CrossRef]
10. Tu, Z.; Zhong, B.; Wang, H. Identification of glycated sites in ovalbumin under freeze-drying processing by liquid chromatography high-resolution mass spectrometry. *Food Chem.* **2017**, *226*, 1–7. [CrossRef]
11. Cicerone, M.T.; Pikal, M.J.; Qian, K.K. Stabilization of proteins in solid form. *Adv. Drug Deliv. Rev.* **2015**, *93*, 14–24. [CrossRef] [PubMed]
12. Franks, F.; Hatley, R.H.M.; Friedman, H.L. The thermodynamics of protein stability: Cold destabilization as a general phenomenon. *Biophys. Chem.* **1988**, *3*, 307–315. [CrossRef]
13. Allison, S.D.; Dong, A.; Carpenter, J.F. Counteracting effects of thiocyanate and sucrose on chymotrypsinogen secondary structure and aggregation during freezing, drying, and rehydration. *Biophys. J.* **1996**, *4*, 2022–2032. [CrossRef]
14. Barbiroli, A.; Marengo, M.; Fessas, D.; Ragg, E.; Renzetti, S.; Bonomi, F.; Iametti, S. Stabilization of beta-lactoglobulin by polyols and sugars against temperature-induced denaturation involves diverse and specific structural regions of the protein. *Food Chem.* **2017**, *234*, 155–162. [CrossRef] [PubMed]
15. Wu, H.Y.; Sun, C.B.; Liu, N. Effects of different cryoprotectants on microemulsion freeze-drying. *Innov. Food Sci. Emerg. Technol.* **2019**, *54*, 28–33. [CrossRef]
16. Ohtake, S.; Wang, Y.J. Trehalose: Current use and future applications. *J. Pharm. Sci.* **2011**, *6*, 2020–2053. [CrossRef] [PubMed]
17. Walayat, N.; Xiong, H.; Xiong, Z. Role of cryoprotectants in surimi and factors affecting surimi gel properties: A review. *Food Rev Int.* **2020**, 1–20. [CrossRef]
18. Muhoza, B.; Xia, S.; Wang, X. The protection effect of trehalose on the multinuclear microcapsules based on gelatin and high methyl pectin coacervate during freeze-drying. *Food Hydrocoll.* **2020**, *105*, 105807. [CrossRef]
19. Sakhaee, N.; Sakhaee, S.; Takallou, A.; Mobaraki, A.; Maddah, M.; Moshrefi, R. Hydrodynamic volume of trehalose and its water uptake mechanism. *Biophys. Chem.* **2019**, *249*, 106145. [CrossRef]
20. Zhang, B.; Hao, G.; Cao, H.; Tang, H.; Deng, S. The cryoprotectant effect of xylooligosaccharides on denaturation of peeled shrimp (*Litopenaeus vannamei*) protein during frozen storage. *Food Hydrocoll.* **2018**, *77*, 228–237. [CrossRef]
21. Loncaric, A.; Dugalic, K.; Mihaljevic, I.; Jakobek, L.; Pilizota, V. Effects of sugar addition on total polyphenol content and antioxidant activity of frozen and freeze-dried apple purée. *J. Agric. Food Chem.* **2014**, *7*, 1674–1682. [CrossRef] [PubMed]
22. AOAC. *International Official Methods of Analysis of AOAC International*, 18th ed.; Association of Analytical Chemists: Arlington, VA, USA, 2006.
23. Nielsen, S.S. *Phenol-Sulfuric Acid Method for Total Carbohydrates*; Springer: Boston, MA, USA, 2010; pp. 47–53.
24. Garcia-Amoedo, L.H.; Almeida-Muradian, L.B. Determination of trans-10-hydroxy-2-decenoic acid (10-HDA) in royal jelly from São Paulo State, Brazil. *LWT* **2003**, *23*, 62–65. [CrossRef]
25. Balkanska, R. Correlations of physicochemical parameters, antioxidant activity and total polyphenol content of fresh royal jelly samples. *Int. J. Curr. Microbiol. Appl. Sci.* **2018**, *4*, 3744–3750. [CrossRef]
26. Thiex, N.; Novotny, L.; Crawford, A. Determination of ash in animal feed: AOAC official method 942.05 revisited. *J. AOAC Int.* **2012**, *5*, 1392–1397. [CrossRef]
27. Alanazi, F.K. Utilization of date syrup as a tablet binder, comparative study. *Saudi Pharm. J.* **2010**, *2*, 81–89. [CrossRef]
28. Erdem, B.G.; Kaya, S. Production and application of freeze dried biocomposite coating powders from sunflower oil and soy protein or whey protein isolates. *Food Chem.* **2021**, *339*, 127976. [CrossRef]
29. Tonon, R.V.; Brabet, C.; Hubinger, M.D. Influence of process conditions on the physicochemical properties of açai (*Euterpe oleraceae Mart.*) powder produced by spray drying. *J. Food Eng.* **2008**, *3*, 411–418. [CrossRef]
30. Liu, J.R.; Yang, Y.C.; Shi, L.S.; Peng, C.C. Antioxidant properties of royal jelly associated with larval age and time of harvest. *J. Agric. Food Chem.* **2008**, *23*, 11447–11452. [CrossRef]
31. Singleton, V.L.; Orthofer, R.; Lamuela-Raventós, R.M. Analysis of total phenols and other oxidation substrates and antioxidants by means of folin-ciocalteu reagent. *Method Enzymol.* **1999**, *299*, 152–178.

32. Brand-Williams, W.; Cuvelier, M.E.; Berset, C. Use of a free radical method to evaluate antioxidant activity. *LWT* **1995**, *1*, 25–30. [CrossRef]
33. Butreddy, A.; Janga, K.Y.; Ajjarapu, S.; Sarabu, S.; Dudhipala, N. Instability of therapeutic proteins—An overview of stresses, stabilization mechanisms and analytical techniques involved in Lyophilized proteins. *Int. J. Biol. Macromol.* **2021**, *167*, 309–325. [CrossRef] [PubMed]
34. Dissanayake, M.; Kasapis, S.; George, P.; Adhikari, B.; Palmer, M.; Meurer, B. Hydrostatic pressure effects on the structural properties of condensed whey protein/lactose systems. *Food Hydrocoll.* **2013**, *2*, 632–640. [CrossRef]
35. Ding, S.; Yang, J. The effects of sugar alcohols on rheological properties, functionalities, and texture in baked products–A review. *Trends Food Sci. Technol.* **2021**, *111*, 670–679. [CrossRef]
36. Allen, T. *Powder Sampling and Particle Size Determination*; Elsevier: Houston, TX, USA, 2003; pp. 295–358.
37. Geldart, D.; Abdullah, E.C.; Hassanpour, A.; Nwoke, L.C.; Wouters, I.J.C.P. Characterization of powder flowability using measurement of angle of repose. *China Particuol.* **2006**, *4*, 104–107. [CrossRef]
38. Fang, Y.; Selomulya, C.; Chen, X.D. On measurement of food powder reconstitution properties. *Dry Technol.* **2007**, *1*, 3–14. (In Chinese) [CrossRef]
39. Jayasundera, M.; Adhikari, B.; Howes, T.; Aldred, P. Surface protein coverage and its implications on spray-drying of model sugar-rich foods: Solubility, powder production and characterisation. *Food Chem.* **2011**, *4*, 1003–1016. [CrossRef]
40. Haque, E.; Bhandari, B.R.; Gidley, M.J.; Deeth, H.C.; Møller, S.M.; Whittaker, A.K. Protein conformational modifications and kinetics of water−protein interactions in milk protein concentrate powder upon aging: Effect on solubility. *J. Agric. Food Chem.* **2010**, *13*, 7748–7755. [CrossRef]
41. Quek, S.Y.; Chok, N.K.; Swedlund, P. The physicochemical properties of spray-dried watermelon powders. *Chem. Eng. Process.* **2007**, *5*, 386–392. [CrossRef]
42. Tontul, I.; Topuz, A. Spray-drying of fruit and vegetable juices: Effect of drying conditions on the product yield and physical properties. *Trends Food Sci. Technol.* **2017**, *63*, 91–102. [CrossRef]
43. Nurhadi, B.; Andoyo, R.; Indiarto, R. Study the properties of honey powder produced from spray drying and vacuum drying method. *Food Res. Int.* **2012**, *3*, 907.
44. Ng, M.L.; Sulaiman, R. Development of beetroot (*Beta vulgaris*) powder using foam mat drying. *LWT* **2018**, *88*, 80–86. [CrossRef]
45. Nagai, T.; Inoue, R. Preparation and the functional properties of water extract and alkaline extract of royal jelly. *Food Chem.* **2004**, *2*, 181–186. [CrossRef]
46. Betoret, E.; Mannozzi, C.; Dellarosa, N.; Laghi, L.; Rocculi, P.; Dalla Rosa, M. Metabolomic studies after high pressure homogenization processed low pulp mandarin juice with trehalose addition. Functional and technological properties. *J. Food Eng.* **2017**, *200*, 22–28. [CrossRef]
47. Kopjar, M.; Tiban, N.N.; Pilizota, V.; Babic, J. Stability of anthocyanins, phenols and free radical scavenging activity through sugar addition during frozen storage of blackberries. *J. Food Process. Preserv.* **2009**, *33*, 1–11. [CrossRef]
48. Nurcholis, W.; Putri, D.N.S.B.; Husnawati, H.; Aisyah, S.I.; Priosoeryanto, B.P. Total flavonoid content and antioxidant activity of ethanol and ethyl acetate extracts from accessions of Amomum compactum fruits. *Ann. Agric. Sci.* **2021**, *1*, 58–62. [CrossRef]
49. Kek, S.P.; Chin, N.L.; Yusof, Y.A.; Tan, S.W.; Chua, L.S. Total phenolic contents and colour intensity of Malaysian honeys from the *Apis* spp. and *Trigona* spp. bees. *Agric. Agric. Sci. Procedia* **2014**, *2*, 150–155. [CrossRef]
50. Li, X.J.; Bai, W.D.; Zhao, W.H. Study on Non-enzymatic Browning Simulation System of Semi-sweet Guangdong Yellow Rice Wine in the Process of Decocting. *Food Ind.* **2018**, *39*, 91–95.
51. Shenoy, P.; Viau, M.; Tammel, K.; Innings, F.; Fitzpatrick, J.; Ahrné, L. Effect of powder densities, particle size and shape on mixture quality of binary food powder mixtures. *Powder Technol.* **2015**, *272*, 165–172. [CrossRef]
52. Nogueira, G.F.; Fakhouri, F.M.; de Oliveira, R.A. Incorporation of spray dried and freeze dried blackberry particles in edible films: Morphology, stability to pH, sterilization and biodegradation. *Food Packag. Shelf Life* **2019**, *20*, 100313. [CrossRef]
53. Ganaie, T.A.; Masoodi, F.A.; Rather, S.A.; Gani, A. Exploiting maltodextrin and whey protein isolate macromolecules as carriers for the development of freeze dried honey powder. *Carbohydr. Polym.* **2021**, *2*, 100040. [CrossRef]
54. Yang, L.; Guo, J.; Yu, Y.; An, Q.; Wang, L.; Li, S.; Qi, S. Hydrogen bonds of sodium alginate/Antarctic krill protein composite material. *Carbohydr. Polym.* **2016**, *142*, 275–281. [CrossRef] [PubMed]
55. Zhang, L.; Wei, Y.; Liao, W.; Tong, Z.; Wang, Y.; Liu, J.; Gao, Y. Impact of trehalose on physicochemical stability of β-carotene high loaded microcapsules fabricated by wet-milling coupled with spray drying. *Food Hydrocoll.* **2021**, *121*, 106977. [CrossRef]

Article

The Effects of a Gum Arabic-Based Edible Coating on Guava Fruit Characteristics during Storage

Sherif Fathy El-Gioushy [1,*], Mohamed F. M. Abdelkader [2], Mohamed H. Mahmoud [3], Hanan M. Abou El Ghit [4], Mohammad Fikry [5], Asmaa M. E. Bahloul [6], Amany R. Morsy [7], Lo'ay A. A. [8,*], Adel M. R. A. Abdelaziz [9], Haifa A. S. Alhaithloul [10], Dalia M. Hikal [11], Mohamed A. Abdein [12,*], Khairy H. A. Hassan [13] and Mohamed S. Gawish [14]

1. Horticulture Department, Faculty of Agriculture (Moshtohor), Benha University, Benha 13736, Egypt
2. Department of Plant Production, College of Food and Agriculture, King Saud University, Riyadh 12372, Saudi Arabia; mohabdelkader@ksu.edu.sa
3. Department of Biochemistry, College of Science, King Saud University, Riyadh 11451, Saudi Arabia; mmahmoud2@ksu.edu.sa
4. Botany and Microbiology Department, Faculty of Science, Helwan University, Cairo 11795, Egypt; hanan8760@yahoo.com
5. Department of Agricultural and Biosystems Engineering, Faculty of Agriculture, Benha University, Moshtohor 13736, Egypt; moh.eltahlawy@fagr.bu.edu.eg
6. Department of Agricultural Economics, Faculty of Agriculture, Banha University, Banha 13736, Egypt; asmaa.bahlol@fagr.bu.edu.eg
7. Plant Protection Department, Faculty of Agriculture, Benha University, Benha 13736, Egypt; amani.alzoheri@fagr.bu.edu.eg
8. Pomology Department, Faculty of Agriculture, Mansoura University, Mansoura 35516, Egypt
9. Central Lab. of Organic Agriculture, Agriculture Research Center, Giza 12411, Egypt; dr.adel.organic@gmail.com
10. Biology Department, Collage of Science, Jouf University, Sakaka 72388, Saudi Arabia; haifasakit@ju.edu.sa
11. Nutrition and Food Science, Home Economics Department, Faculty of Specific Education, Mansura University, Mansoura 35516, Egypt; dr.daliahikal@mans.edu.eg
12. Biology Department, Faculty of Arts and Science, Northern Border University, Rafha 91911, Saudi Arabia
13. Tropical Fruits Research Department, Horticulture Research Institute, Agricultural Research Centre, Giza 12619, Egypt; kh11191@ymail.com
14. Pomology Department, Faculty of Agriculture, Damietta University, Damietta 34511, Egypt; msagawishaa@gmail.com
* Correspondence: sherif.elgioushy@fagr.bu.edu.eg (S.F.E.-G.); Loay_Arafat@mans.edu.eg (L.A.A.); abdeingene@yahoo.com (M.A.A.)

Abstract: Guava is a nutritious fruit that has perishable behavior during storage. We aimed to determine the influences of some edible coatings (namely, cactus pear stem (10%), moringa (10%), and henna leaf (3%) extracts incorporated with gum Arabic (10%)), on the guava fruits' properties when stored under ambient and refrigeration temperatures for 7, 14, and 21 days. The results revealed that the coating with gum Arabic (10%) only, or combined with the natural plant extracts, exhibited a significant reduction in weight loss, decay, and rot ratio. Meanwhile, there were notable increases in marketability. Moreover, among all tested treatments, the application of gum Arabic (10%) + moringa extract (10%) was the superior treatment for most studied parameters, and exhibited for the highest values for maintaining firmness, total soluble solids, total sugars, and total antioxidant activity. Overall, it was suggested that coating guava with 10% gum Arabic combined with other plant extracts could maintain the postharvest storage quality of the cold-storage guava.

Keywords: guava; edible coating; gum Arabic; natural plant extracts; storability

1. Introduction

Guava (*Psidium guajava* L.) is a popular fruit with a climacteric nature feature. It has a relatively short shelf life (3–4 days) at tropical ambient temperature (28 ± 2 °C), due to its physiology, disorder, postharvest infection, and aging [1,2]. The stability of guava could be influenced by numerous factors, such as storage temperatures, relative humidity, packaging materials, and coating nature [3,4]. Producers of guava fruits store them in traditional packs, such as paper and/or plastic materials. Although these packaging materials present some merits, they could cause severe environmental difficulties because they are non-recyclable and nonedible resources [4–6].

Physiologically, guava is a climacteric fruit with high inhalation and transpiration degrees that are analogous to different products such as bananas [5] and mushrooms [7]. This creates the need to develop novel technologies for extending its shelf life [8], providing better storage conditions, and enhancing its visual features.

The edible coatings are applied directly on the surface of the fruit, consisting of thin membranes, invisible to the naked eye. They can carry natural additives and are important in extending the shelf life of foods, as they enhance the protective action of the fruit epidermis in preventing water loss, color changes, mechanical lesions and even microbial deterioration, and generally give surface glossy appearance [3,9]. Such technology has exposed a great possibility with low cost and proper features for related usage in food plants.

Previously, edible coatings were fabricated from biomaterials—namely polysaccharides, proteins and lipids—and their results were utilized to formulate edible coatings [10]. Formerly, coating materials efficiently prolonged a fruit's shelf-life, and increased the consumers' health and environment. Numerous studies have reported the efficiency of edible coatings in expanding the shelf-life of different kinds of fruits by decreasing their weight loss [10,11], respiration [12,13], oxidative reaction rates [14], and physiological disorders [15]. Edible coatings are an excellent alternative to chemical preservation [16].

Among the natural polymers used to formulate edible coatings, gum Arabic (GA) and cactus pear extract have been deemed the most promising materials, mostly because their high accessibility, low price, and good performance. However, their high hydrophilicity permits H_2O to easily form. GA is one of the ecological biopolymers acquired from the branches and stems of Acacia trees (*Acacia* spp.). It is comprised of rhamnose, galactose, arabinose, and glucuronic acid with Ca, Mg, and K ions [17,18]. In addition, GA is commercially and securely utilized as a food additive, due to its film shaping, emulsification, and encapsulation attributes [18,19].

Numerous studies have been conducted on the application of GA for nullity purposes, such as its ability to postpone the physicochemical alterations of bananas during cold storage [20], and minimize the fungal infection of anthracnose on banana and papaya fruits [20]. GA coatings have efficiently kept the antioxidative polyphenols of tomatoes [21] and papayas [22], and lowered browning, vitamin C, and the polyphenols of cold-stored mangos [23]. Moreover, GA-coating significantly decreases weight loss, chilling damage, membrane leakage, and decay prevalence, with slight increases in total soluble solids, pH, and sugar [24,25].

M. oleifera is widely in demand for its nutritional and medicinal properties, due to its content of vitamins B and C, amino acids, crude protein, and its low anti-nutritional and antimicrobial agents with film organic and shelf-life-boosting properties [26–28]. Moreover, its extracts have been utilized to formulate the coating that effectively preserves the postharvest quality of citrus fruit [29]. Moringa extract maintains postharvest quality, keeping and elongating the shelf life of avocado fruit by lowering respiration, ethylene production rates and higher firmness during storage [28,30,31]. It also significantly inhibits (30–33%) radial mycelial growth of the pathogen with a combination of GA edible coatings with moringa leaf extract. There is evidence for the incorporation of moringa leaf extract with edible coatings improving antimicrobial activity [32,33].

The mucilage films from the cactus pear plant prolong shelf life and maintain guava quality attributes. Further research is wanted to realize whether mucilage is possible for use as an edible film under cold storage [28]. In guava and cactus pear, the film prolonged the fruit's skin color and retained superior total soluble solids concentration (TSSC), firmness

(F), and dry matter concentration (DMC), lowered the fruit's weight loss and prolonged its skin color. Firmness, TSSC, and DMC of fruit were comparable among treatments [34].

Henna (*Lawsonia inermis*) leaf extracts are a natural plant product with a projecting function against pathogens. Their expansion and propagation inhibit toxic activity [35].

There were no experiments undertaken to estimate the impact of combining the GA with each of the leaf extracts of Moringa, cactus pear, and Henna plants, in prior edible-coating investigations. Therefore, this study aimed to investigate the effects of edible coatings based on GA with cactus pear, moringa, and Henna leaf extract on the storability and shelf life of guava.

2. Materials and Methods

2.1. Fruit Materials

The guava trees were grown in a commercial orchard in El-Qalubia Governorate, Egypt (latitude, 30°17" N and longitude, 31°20" E). The trees were about 12 years of age and planted at a space of 5 × 5 M apart in loamy clay soil under an immersion irrigation system and subjected to all ideal agriculture traditions. The maturity stage (yellowish-green), was the second week of August based on Mercado-Silva et al. [36], the trees and guavas were similar in size and free of obvious signs of infection.

2.2. Postharvest Treatments and Storage

The current study was performed during two seasons, 2020 and 2021, at the laboratory of the Department of Horticultural Crops Technology, National Research center, Giza, Egypt. Fine-feature guava fruits of the "Maamoura" cultivar, that were deep green color, uniform size, firm, and free from blemishes and mechanical damage, were harvested.

2.3. Preparation of Plant Extracts

Cactus pear stems were skinned and chopped. Samples were steamed with H_2O in the ratio of 1:10 in an autoclave at 160 °C for 1 h. The boiled pulp was filtered and cooled. The slurry was centrifuged for 10 min and the supernatant gained was utilized as a coating substance (cactus pear mucilage). The filtrate (pulping liquor) was also utilized as a coating solution. Polyethylene glycol of molecular weight 2000 as plasticizer was mixed to the coating solution 5% w/v. The pH value of this solution was modified at 7, using small drops of ammonia solution.

Moringa leaf extract was made by soaking 100 g of air-dried moringa leaves in 1 L of dH_2O for 24 h. Then, it was diluted after being filtered with H_2O. Concentrations of 20% were prepared by dispersing 20 mL of filtered solution and 3 mL glycerol in 100 mL dH_2O in a beaker.

GA powder of food-grade was acquired from Sigma Co., Cairo, Egypt. GA solution at 10% (w/v) was arranged by dissolving 500 g of GA powder in 5 L of dH_2O. The solution of GA was agitated with low heat at 40 °C for 60 min, using a hot plate with a magnetic stirrer (Model: 502P-2 USA), then filtered using a muslin cloth to eliminate impurities and any nameless materials. After cooling the solution to 20 °C, glycerol monostearate at 1% was mixed as a plasticizer to increase the intensity and elasticity of the coating solution. The pH of the solution was altered to 5.6 with 1 N NaOH using a digital pH-meter (Model: AD1000, Bucharest, Romania). Then, 3% Henna was formulated; we soaked 100 g of air-dried henna leaves in 1 L of dH_2O for 24 h. Then, it was diluted after being filtered with dH_2O. A concentration of 3% was prepared by dispersing 3 mL of filtered solution and 3 mL glycerol in 100 mL dH_2O in a beaker.

2.4. Treatment of Guava Fruit

A total of 600 clean complete fruits were chosen and haphazardly allocated into 5 treatments with 3 replicates (40 fruits/replicate) and expressed as control: 10% of GA; 10% of GA + 10% of moringa leaves extract; 10% of GA + 10% of cactus pear stems extract; and 10% of GA + 3% of Henna leaves extract. All guava fruits were dipped in the different extract solutions for 2.5 min. Then, these fruits were dropped in 10% GA solutions for another 2.5 min. After dropping treatments, the fruits were allowed to dry for 60 min at RT by an electric fan. After that, the fruits in each treatment were wrapped in foam sheets enclosed with punched polyethylene layers with a thickness of 0.04 mm and then boxed in cardboard boxes with measurements of 35 × 25 × 10 cm. All untried boxes were kept at 7 ± 1 °C and 90 ± 5% RH for 24 days. the physical measurements were analyzed at harvest time and then every 7 days they had breaks from the cold-storage time.

2.5. Determination of Fruit Physical Properties

Fruit weight loss (%) was estimated by the next formula:

$$\frac{\text{Fruit weight before storage} - \text{fruit weight after each period of cold storage}}{\text{Fruit weight before storage}} \times 100 \quad (1)$$

Fruit decay (%) was documented every 8 days of cold storage by calculating the number of rotten fruits owing to fungus, or any microorganism's infection, and expressed as a percentage of the original number of kept fruits using the next equation:

$$\frac{\text{Number of decayed fruit at specific storage period}}{\text{Intial number of stored fruits}} \times 100 \quad (2)$$

Marketable fruit (%) was calculated by the next formula:

$$\frac{\text{Number of sound fruit at specific storage period}}{\text{Intial number of stored fruits}} \times 100 \quad (3)$$

Fruit firmness was determined in three guava fruits per replicate at two equatorials at differing sites, to measure the penetration force using a hand-held fruit firmness tester (FT-327, Valencia, Italy) fortified with an 8 mm cylindrical stainless steel plunger tip (Watkins and Harman, 1981). Two senses were shown on each fruit flesh after peeling. The firmness value was calculated in terms of kilogram-force (kg_f) and data was assessed as Newton (N).

2.6. Fruit Chemical Characteristics

2.6.1. Fruit Pigments

Chlorophyll and carotenoid contents in the pulp of guavas (three replicates) were spectrophotometrically measured using the assay of Wellburn (1994). The absorbance of the extract was determined at a spectrum of 663 nm for chlorophyll a, 646 nm for chlorophyll b and 470 for carotene using a spectrophotometer (UV1901PC spectrophotometer). Pigment contents were calculated by the next equations:

$$\text{Chlorophyll a } (\mu g\ mL^{-1}) = 12.21\ E663 - 2.81\ E646 \quad (4)$$

$$\text{Chlorophyll b } (\mu g\ mL^{-1}) = 20.13\ E646 - 5.03\ E663 \quad (5)$$

$$\text{Total chlorophyll } (\mu g\ mL^{-1}) = \text{Chlorophyll a} + \text{chlorophyll b} \quad (6)$$

$$\text{Total carotenoide } (\mu g\ mL^{-1}) = [(1000\ E470) - (3.27 \times \text{Chlorophyll a} + 104 \times \text{Chlorophyll b})]/198 \quad (7)$$

where E = Optical density at the specified spectrum length and findings were calculated as mg/100 g of fresh weight (FW).

2.6.2. Ascorbic Acid, TSS, Titratable Acidity (TA), and TSS/TA Ratio, TPC and TAA of the Pulp

After each cold-storage period, 15 guava fruits from each treatment (3 replicates) were separated by pounding the pulp; then, the juice was drained via a muslin cloth and utilized for quantifying interior fruit quality [37], as indicated.

For ascorbic acid analysis, sections of fruit juice were utilized, the oxalic acid solution was mixed to each sample, titrated with 2,6-dichlorophenol-indophenol dye solution, calculated as a mg of ascorbic acid, and expressed as mg/100 mL of the juice.

Fruit juice TSS was quantified using a hand refractometer, 0–32 scale (ATAGO N-1E, Japan) and conveyed in Brix after making the temperature correction at 20°.

For juice TA, an aliquot of fruit juice was taken and titrated against 0.1 N NaOH with phenolphthalein as a marker to the endpoint, and stated as (%) of citric acid.

Fruit TSS/TA ratio was expressed from the values noted for fruit juice TSS and TA revealed.

Total phenolic content (TPC) and total antioxidant activity (TAA) were quantified in the MeOH extract (80%) of dried guava pulp. TPC was assessed using the Folin–Ciocalteu reagent, as explained by Kähkönen et al. [38]. TAA was assessed based on the scavenging activity of the stable 1,1-diphenyl-2-picrylhydrazyl (DPPH) radical and conveyed as IC_{50} (mg/mL), which signified the quantity of the plant sample extract essential to decrease the early concentration of DPPH radicals by 50% [39].

2.6.3. Total Pectin

Quantities of pectic substances in guava were calorimetrically quantified by the carbazole sulfuric acid, using the assay of Yu et al. [40]. The results were conveyed as g anhydro galacturonic acid (A. G. A) per 100 gm on a dry weight basis.

2.7. Shelf-Life Period

After 24 days of cold storage at 7 ± 1 °C, the guava (15 fruit per replicate) were obtained and arranged at ambient conditions (22–24 °C and $65 \pm 5\%$ RH), till 30% of fruits became bad, or rotting happened. Then, the number of days was documented as deemed for the shelf-life period of the guava.

2.8. Inoculation of Fruits

Inoculation of guava was performed by squirting fruits with spore suspension (10^6 spores/mL) of *R. stolonifer* as causative agents of soft rot disease, then air-dried at RT. All were treated with each dose, and the control treatment were exemplified at 3 replicates, with 15 fruits for each replicate. All fruits were kept at 20 ± 2 °C for 15 days and infection (%) was noted. All the above-noted parameters were quantified in 20 fruits, and the average values were selected at the early time (Table 1).

Table 1. Characteristics of guava fruit prior to storage.

Parameters	Values
Weight loss (%)	0.00
Decay (%)	0.00
Marketable (%)	100.00
Firmness (N)	67.91
TSS (%)	7.01
Acidity (%)	1.11
TSS/Acid ratio	6.32
Ascorbic acid (mg/100 mL juice)	105.15
Total chlorophyll (mg/100 g FW)	16.15
Total Carotenoids (mg/100 g FW)	5.39
Total sugars (%)	1.65
Total Pectin (%)	0.69
Total Phenolic (IC_{50} mg/mL)	39.11
Antioxidants (IC_{50} mg/mL)	210.17
Rhizopus Rot infection (%)	0.00

2.9. Statistical Analysis

This experiment was assembled in a wholly randomized design, with 3 replicates consisting of two factors: postharvest treatments and storage period. This experiment was analyzed as a factorial experiment. Data calculated as (%) were altered to the arcsine of the square root before statistical analysis, and non-transformed means are displayed as they were. The effects of postharvest treatments and cold-storage period on several properties were statistically analyzed by ANOVA, using the MSTAT-C statistical package. Comparisons among means were performed by Duncan's multiple range test (DMRT) at probability ≤ 0.05.

3. Results

3.1. Weight Loss (%), Decay (%), Rot and Marketable (%)

With the advance of the cold-storage period, fruit weight loss and fruit decay percentages were significantly increased, whereas marketable fruit (%) significantly decreased (Figure 1). Weight loss increased during storage and recorded the highest values in control fruit after 21 days. Weight loss (%) was significantly lowered in samples that were coated with GA (10%), GA (10%) + moringa (10%), GA (10%) + cactus (10%), and GA (10%) + henna (3%). Moreover, the lowest values of weight loss were coupled with the treatment GA (10%) + moringa extract (10%), during cold-storage and both experimental seasons. Additionally, concerning decay percentage, all edible-coating treatments recorded significant reductions compared with the uncoated samples. Among all treatments, fruits that were coated with GA (10%) + moringa (10%) obtained the lowest values of decay percentage compared with all treatments. Furthermore, rot (%) increased as storage days progressed, and the control treatment scored the highest values during both seasons. Alternatively, fruit samples coated with GA (10%) + (10%) moringa, and GA (10%) + henna (10%) obtained the lowest averages in rot percentages during the 2019 and 2020 experimental seasons. On the contrary, the marketability fruit percentage significantly decreased in all treatments, as well as with increasing storage days. In this respect, the natural edible coatings kept the marketability of guava higher than uncoated fruits, and fruits samples coated with GA (10%) and GA (10%) + moringa (10%) also kept the fruit at a higher percent, although there were the significant differences among treatments.

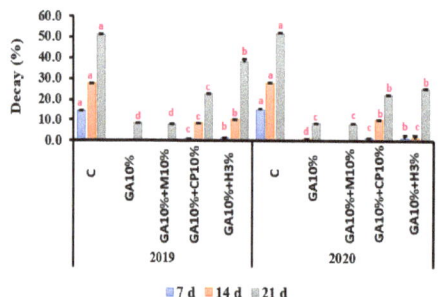 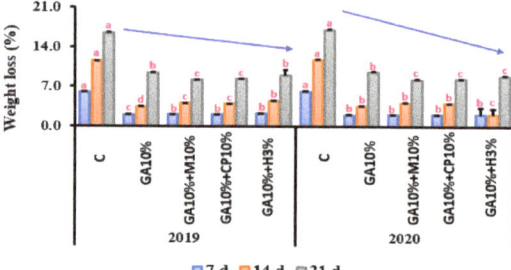

Figure 1. Cont.

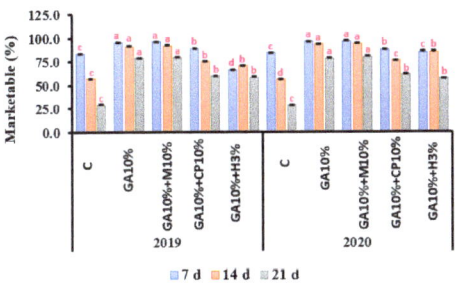
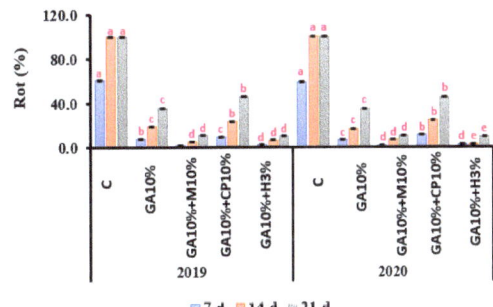

Figure 1. Effect of edible coatings of moringa extract, cactus pear, and henna, and GA, on weight loss, decay, rot, and marketable percentages of guava fruits during cold storage for 21 days throughout two seasons (2019–2020). The superscript letters, present the significantly between effect of treatments using Duncan's multiple range tests.

3.2. Firmness (N), TSS (%), Acidity (%), and TSS/Acid (%)

Data in Figure 2 suggest there were significant differences among the treatments. Respecting firmness, the fruit samples handled with GA (10%) and GA (10%) + moringa (10%) appeared to have higher values of firmness during storage periods, as well as during both experimental seasons, compared with the other coated and uncoated treatments which recorded the lowest firmness values. Moreover, there were notable decreases in firmness when storage days were increased.

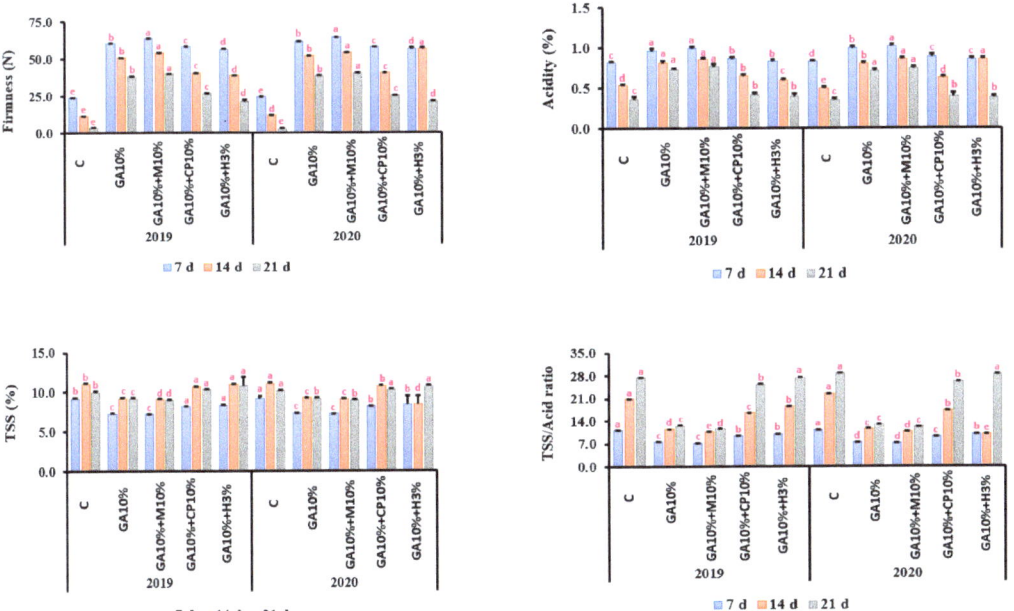

Figure 2. Effect of edible coatings moringa extract, cactus pear, henna and GA, on firmness (N), TSS (%), acidity (%), and TSS/acid ratio of guava during cold storage for 21 days throughout two seasons (2019–2020). The superscript letters, present the significantly between effect of treatments using Duncan's multiple range tests.

As for both TSS and TSS/acid ratio, the studied treatments showed the differences among them in this respect, but there was no specific direction in these traits during both seasons. Concerning acidity, the data indicated that there were significant differences among treatments during storability days as well as during both seasons. From these data, the highest values of acidity were obtained with GA (10%) and GA (10%) + moringa (10%) in the 2019 and 2020 seasons.

3.3. Total Sugars and Pectin

Respecting the effects of used edible coatings on total sugars and total pectin (%), Figure 3 shows the differences among the studied treatments. Control treatments (uncoated samples) were coupled with the highest total sugar percentage followed by GA (10%) + henna (3%), and the other treatments showed intermediate values. As for total pectin (%), Figure 3 shows the opposite effect than total sugar content. In this regard, fruit samples coated with GA (10%) and GA (10%) + moringa (10%) scored the higher averages of pectin content than the others. There were gradual increases in pectin content in contents of storage days from 7 to 21 days in both treatments and both seasons.

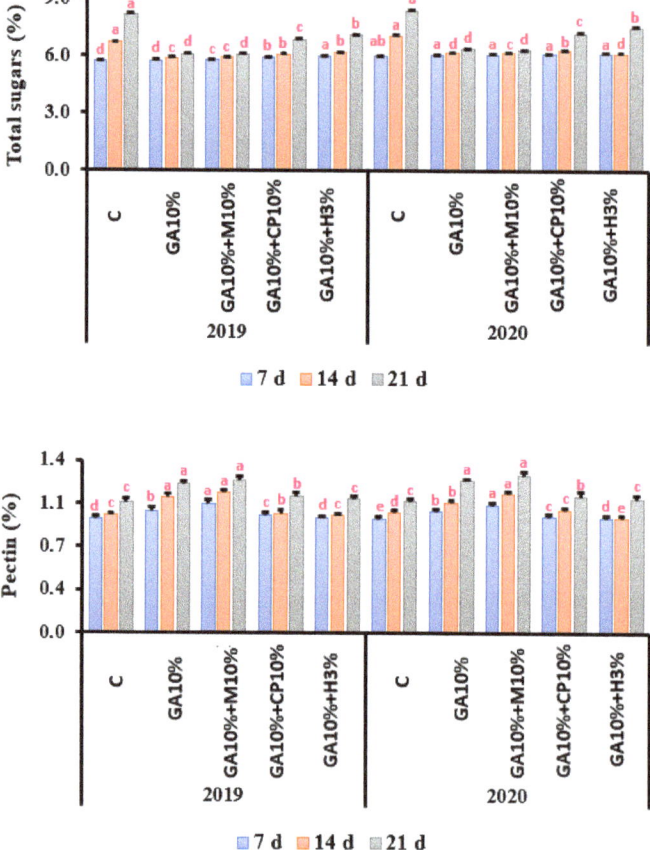

Figure 3. Effect of edible coatings—moringa extract, cactus pear, henna, and GA—on total sugars and pectin (%) of guava during cold storage for 21 days throughout two seasons (2019–2020). The superscript letters, present the significantly between effect of treatments using Duncan's multiple range tests.

3.4. Effects on Total Carotenoid and Total Chlorophyll

Figure 4 shows the effects of edible coatings based on GA on the total carotenoid and total chlorophyll contents of the treated guava fruits. Regarding total carotin, we noticed that for the exception of the control treatment (uncoated fruits) the carotin values were in the lower content on the 7th day, which increased on the 14th day, and then decreased on day 21. The uncoated fruits' carotin content gradually decreased during the storage periods. Concerning the chlorophyll content, all edible-coating treatments were kept in higher averages than uncoated fruits. GA (10%) and GA (10%) + moringa (10%) recorded the higher values compared with the others.

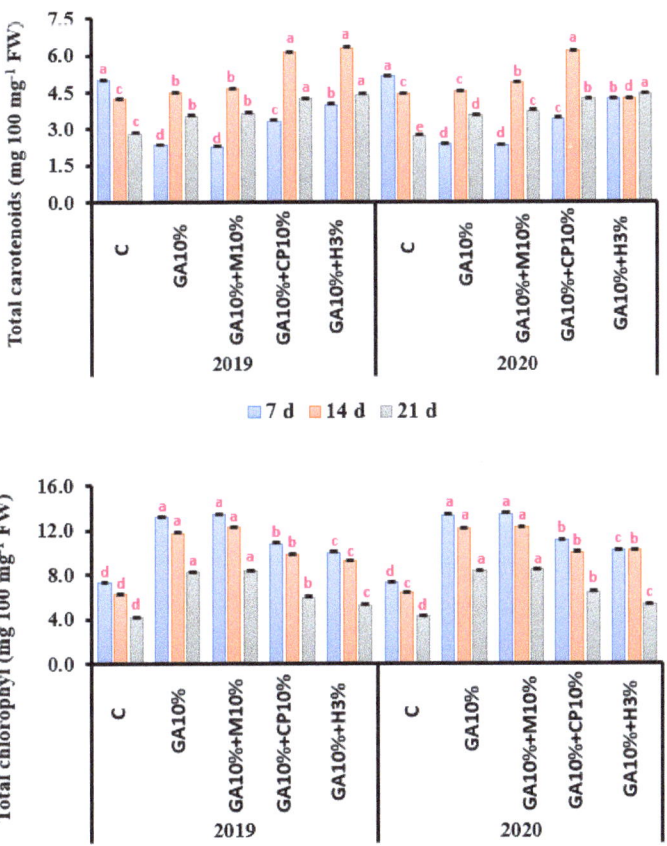

Figure 4. Effect of edible coatings—moringa extract, cactus pear, henna, and GA—on total carotenoid and total chlorophyll (mg 100 mg^{-1} FW) of guava during cold storage for 21 days throughout two seasons (2019–2020). The superscript letters, present the significantly between effect of treatments using Duncan's multiple range tests.

3.5. Effects on VC Juice Content, Total Phenols, and Antioxidants Activity

With respect to VC content (Figure 5), all edible coatings treatments were kept it in higher averages compared with the uncoated fruits, and GA (10%) and GA (10%) + moringa (10%) recorded the higher values compared with the others. Regarding the total phenol content, there were slight significant differences among all treatments (Figure 5). From these data, control treatment of uncoated samples recorded higher values compared with the other treatments, in both 2019 and 2020 seasons. Meanwhile, the treatments GA (10%) and GA (10%) + moringa (10%) reduced total phenol contents, compared with all treated and untreated samples during the two experimental seasons. Moreover, the total phenol contents were in the lower content at the 7th day, which increased at the 14th day, and then decreased on day 21. For all treatments during 2019 and 2020 seasons, this carotenoid trend was exhibited. As for antioxidant activity, the control exhibited bigger averages than all treatments. The opposite trend also appeared for total carotenoids during storage days; there was higher content at the 7th day, which decreased on the 14th day, and then increased at day 21 in both treatments during the two studied seasons.

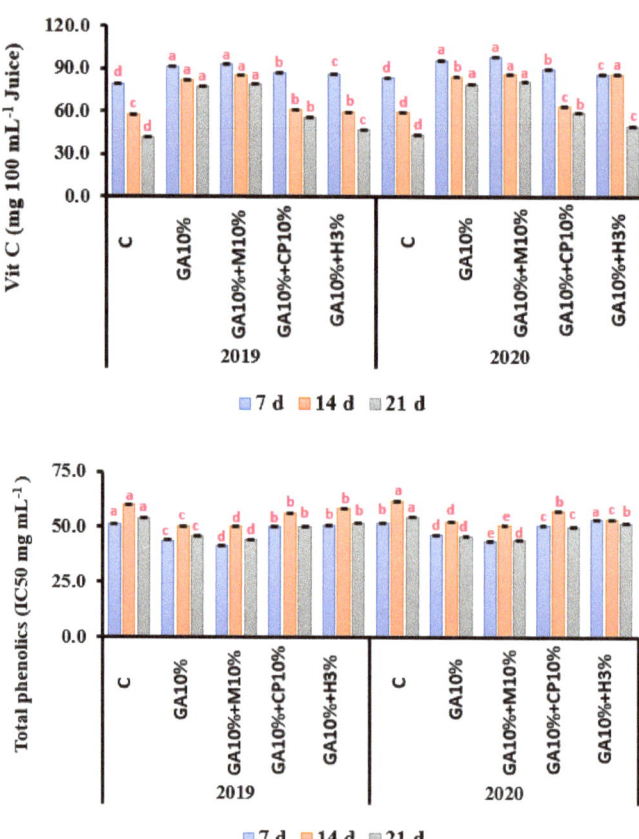

Figure 5. *Cont.*

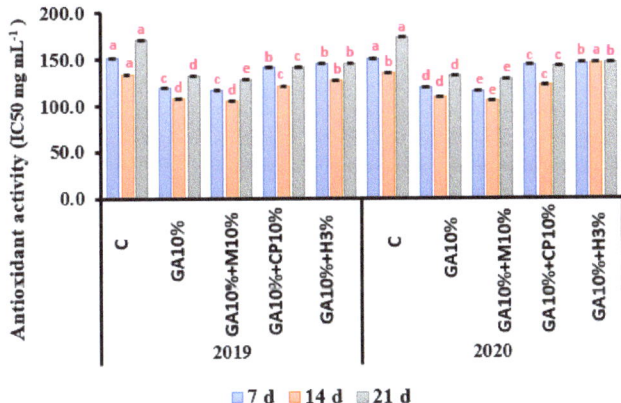

Figure 5. Effect of edible coatings—moringa extract, cactus pear, henna, and GA—on VC juice content, total phenols, carotenoid, and antioxidants activity (IC_{50} mg 100 mg^{-1} FW) of guava during cold storage for 21 days throughout two seasons (2019–2020). The superscript letters, present the significantly between effect of treatments using Duncan's multiple range tests.

3.6. Effects on the Shelf Life

Data revealed that, all treatments enhanced the shelf-life periods of guava fruit coated with the natural edible materials over the control, and the superior treatments in this respect were GA (10%) and GA (10%) + moringa leaf extract (10%) during two studied seasons, 2019 and 2020, respectively (Figure 6).

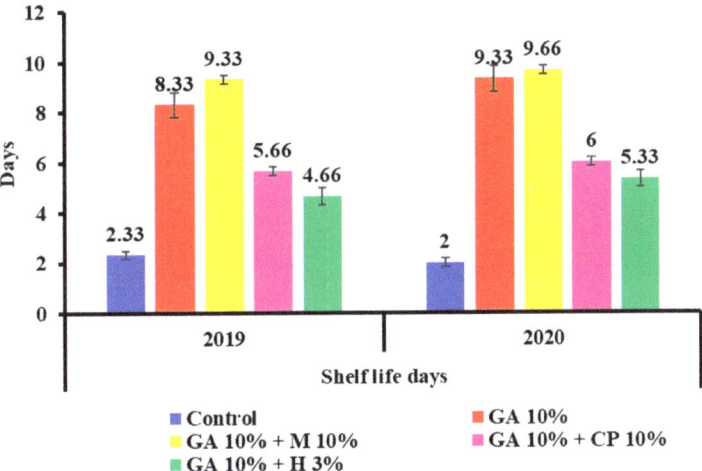

Figure 6. Effect of edible coatings—moringa extract, cactus pear, henna and GA—on shelf life of guava during cold storage.

3.7. Relationship among Fruit Characteristics

Correlation analysis was applied for investigating the interdependence of the guava characteristics (Figure 7). The results indicated that weight loss, decay, TSS, TSS/acid, total sugar, antioxidant, and rot had a negative relationship to firmness. This strong relationship revealed that greater guava had lesser antioxidant activity, which agrees with the resulted relationship for blackberries [41,42]. Strong positive correlations were found between firmness and acidity, V.C, and total chlorophyl. The marketable percentage was positively linked to firmness, acidity, V.C. and total chlorophyl; thus, these indicators could be used to forecast other results. Meanwhile, marketable percentage was negatively correlated with weight loss, decay, TSS, TSS/acid, total sugar, antioxidant, and rot, verifying that some physicochemical changes could cause lower acceptance by the consumers, and consequently create lesser marketability. PCA was utilized to discover the connection among the variables. Thus, PCA was utilized to discover the relationships among parameters in different treatments. PCA analysis revealed that the first principal component (PC_1) and the second principal component (PC_2) were 82.5 and 17.5%, respectively, with the accumulative variance contribution rate of 99% (>75–85%). PC_1 was positively associated with the variables: VC, pectin, acidity, sugars, antioxidants, weight loss, firm, and decay area. PC_2 was positively associated with the variables of TSS, phenolics, carotene, and chlorophyl. Furthermore, locations of the combined GA with natural extracts and the functional and freshness properties were close, showing the impact of GA-natural extracts on the quality of guava during the cold storage.

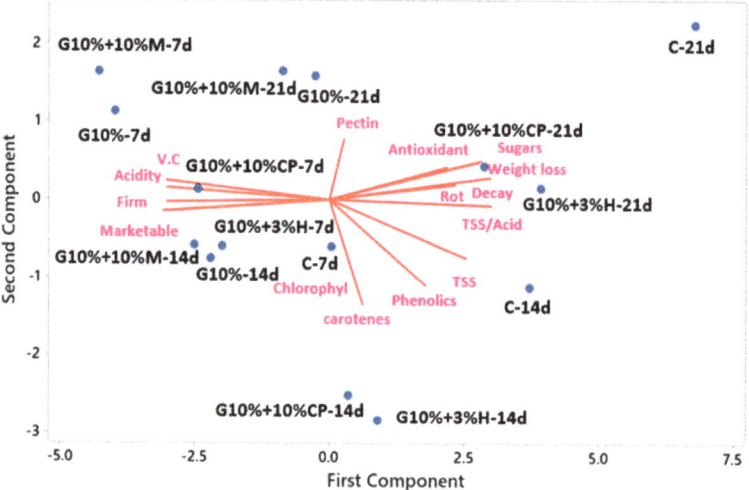

Figure 7. PCA analysis of the different treatments with different parameters.

4. Discussion

Edible coatings have barrier features that decrease a fruits' surface permeability to oxygen and carbon dioxide, resulting in a change in internal gas composition that reduces oxidative metabolism and increases the fruit's shelf life [43]. We proposed that coating fruits with GA would result in significant differences, and increase the fruit shelf-life [20,44].

Water exchange between the interior and exterior atmospheres is believed to be the primary cause of fruit weight loss and decay percentage, resulting in a low marketable percentage during cold storage. With the advancement of cold storage, the growth in transpiration ratio, ethylene making, and the cellular interruption of fruits resulted in a rise in the physiological failure of weight and decay prevalence, and a reduction in saleable guavas [45–47].

Consequently, the use of GA can decrease the gases exchange among orange fruits and the environment by accumulating carbon dioxide in fruits with low O_2-availability for respiration and, as a result, inhibit respiratory enzymes. Furthermore, GA coating can close openings in the peel. Furthermore, the coating can impede the fungi growth in a wide range of horticultural products [48]. This application improved membrane integrity, postponed fruit senescence, and reduced transpiration and respiration [49]. However, coating alters the atmosphere and inhibits the gaseous exchange of the fruit, which prevents ascorbic acid oxidation by limiting the entry of oxygen into the fruit's interior [50]. The lower TSS values in treated samples compared with the controls could be attributed to the conversion of organic acids to sugars via gluconeogenesis and the solubilization of cell wall ingredients by galactosidases and glucosidases found in guava fruit [51].

These findings are matched with those obtained by [25]; they indicate that guava fruits coated with 10% GA showed a significant reduction in weight loss (%) and a delay in the change in firmness, titratable acidity, soluble solid concentration color, maintaining the sensory quality during storage at room temperature, as compared with the uncoated control fruit of banana and papaya fruits, as well as Khaliq, Mohamed, Ali, Ding and Ghazali [23] in 'Choke Anan' mangoes, who reported that treated fruits with edible GA coatings had significantly higher firmness than uncoated fruits during cold storage. Moreover, in avocado fruit (Maluma), the application of GA at 10 or 15% + Moringa leaf extracts maintained higher firmness than other treatments [33]. A similar effect was found in mass loss, with fruit covered with the previously stated coatings showing minimal change. It may be argued that the prevention of moisture loss was the primary reason why coated fruit remained firmer than uncoated fruit. This theory is supported by [29], who found that moisture loss is not only associated with mass loss, but also fruit softening.

Cactus pear extracts and henna leaf extracts significantly decreased weight loss, compared with the control, during cold storage in both seasons. The antimicrobial and antifungal properties of henna leaf extract, as well as its antioxidant activity, contribute to its promising effect in delaying fruit weight loss and decay percentage. In addition, prickly pear has antimicrobial and antioxidant properties.

Lawsone makes up about 0.5 to 1.5 percent of the ingredients in henna. The main constituent responsible for the plant's dyeing properties is lawsone (2-dihydroxynaphthoquinone). Henna, on the other hand, contains mannitol, tannic acid, mucilage, and gallic acid. These substances are present in henna as a mixture. The antimicrobial activity could be attributed to many free hydroxyls that can combine with carbohydrates and proteins in the bacterial cell wall. They may become entangled with enzyme sites, proformance them inactive [52]. When compared with alcoholic and oily extracts, water extracts had no antibacterial activity. This could be due to a lack of solvent properties, which are key in antibacterial effectiveness.

Phenolic compounds are antioxidants that act as protective mechanisms in fruit. TPC content has a defense mechanism against plant pathogen invasion and plays an important role in plant resistance [2,53]. Furthermore, trapping the lipid alkoxyl radicals, antioxidants and phenols could significantly reduce reactive oxygen species (ROS) and prevent lipid peroxidation in plant tissue [2,54,55]. Maintaining TPC and increasing TAA could be attributed to postharvest treatments' ability to scavenge excess ROS and, as a result, reduce oxidative damage to the fruits [23,56,57].

Hence, the use of GA can reduce gas exchange among orange fruits and the environment by accumulating carbon dioxide in the fruits, resulting in low availability of O_2 for respiration and, as a result, the reticence of respiratory enzymes. Furthermore, GA-coating can plug openings in the peel. Furthermore, a coating can prevent the growth of fungi in a variety of horticultural products [48].

5. Conclusions

Overall, our findings indicated that both postharvest applications "cactus pear stem (10%), moringa (10%), and henna leaf (3%) extracts incorporated with gum Arabic (10%)" and their combinations had a positive impact on the quality characteristics of 'Maamoura' guavas during cold storage. The combined treatments of 10% GA + 10% moringa leaf extract were the most effective coating for fresh guava after long periods of cold storage. These applications significantly reduced weight loss, decay and *Rhizopus* rot infection (%), while also increasing marketable percentage, and delaying fruit softening during cold storage. Furthermore, these applications delayed color development by significantly retaining total chlorophyll content, maintaining fruit content in vitamin C and acidity, and slowing the accumulation of fruit contents in TSS and TSS/acid ratios compared with untreated fruits during the cold-storage period. Finally, compared with the control, these applications significantly increased the shelf-life period at ambient conditions after the end of the cold-storage period.

Author Contributions: Conceptualization, M.F.M.A., M.S.G., A.M.E.B. and L.A.A.; methodology, M.F., A.R.M., M.H.M., H.M.A.E.G. and A.M.R.A.A.; software, H.M.A.E.G. and K.H.A.H.; validation, S.F.E.-G., M.S.G., A.M.E.B. and M.H.M. formal analysis, S.F.E.-G.; investigation, M.H.M. and H.M.A.E.G.; data curation, K.H.A.H.; writing—original draft preparation, S.F.E.-G., M.S.G., A.M.E.B., M.F.M.A. and K.H.A.H.; writing—review and editing, S.F.E.-G., M.S.G., A.M.E.B., M.F.M.A., M.F., A.R.M., A.M.R.A.A. and H.A.S.A.; funding acquisition, M.F.M.A., M.A.A., D.M.H. and L.A.A. All authors have read and agreed to the published version of the manuscript.

Funding: The authors extend their appreciation to King Saud University for supporting this work. Researchers Supporting Project under project number (RSP-2021/406), King Saud University, Riyadh, Saudi Arabia.

Institutional Review Board Statement: Not applicable.

Informed Consent Statement: Not applicable.

Data Availability Statement: Relevant data applicable to this research are within the paper.

Conflicts of Interest: The authors declare no conflict of interest.

References

1. Valencia-Chamorro, S.A.; Palou, L.; Del Río, M.A.; Pérez-Gago, M.B. Antimicrobial edible films and coatings for fresh and minimally processed fruits and vegetables: A review. *Crit. Rev. Food Sci. Nutr.* **2011**, *51*, 872–900. [CrossRef] [PubMed]
2. Lo'ay, A.A.; Doaa, M.H. The potential of vine rootstocks impacts on 'Flame Seedless' bunches behavior under cold storage and antioxidant enzyme activity performance. *Sci. Hortic.* **2020**, *260*, 108844. [CrossRef]
3. Lo'ay, A.A.; El-Khateeb, A.Y. Impact of chitosan/PVA with salicylic acid, cell wall degrading enzyme activities and berries shattering of 'Thompson seedless' grape vines during shelf life. *Sci. Hortic.* **2018**, *238*, 281–287. [CrossRef]
4. Kocira, A.; Kozłowicz, K.; Panasiewicz, K.; Staniak, M.; Szpunar-Krok, E.; Hortyńska, P. Polysaccharides as Edible Films and Coatings: Characteristics and Influence on Fruit and Vegetable Quality—A Review. *Agronomy* **2021**, *11*, 813. [CrossRef]
5. Murmu, S.B.; Mishra, H.N. Optimization of the arabic gum based edible coating formulations with sodium caseinate and tulsi extract for guava. *LWT* **2017**, *80*, 271–279. [CrossRef]
6. Lo'ay, A.A.; Dawood, H.D. Minimize browning incidence of banana by postharvest active chitosan/PVA Combines with oxalic acid treatment to during shelf-life. *Sci. Hortic.* **2017**, *226*, 208–215. [CrossRef]
7. Mahajan, P.V.; Oliveira, F.; Macedo, I. Effect of temperature and humidity on the transpiration rate of the whole mushrooms. *J. Food Eng.* **2008**, *84*, 281–288. [CrossRef]
8. Forato, L.A.; de Britto, D.; de Rizzo, J.S.; Gastaldi, T.A.; Assis, O.B. Effect of cashew gum-carboxymethylcellulose edible coatings in extending the shelf-life of fresh and cut guavas. *Food Packag. Shelf Life* **2015**, *5*, 68–74. [CrossRef]
9. Lo'ay, A.A.; Dawood, H.D. Active chitosan/PVA with ascorbic acid and berry quality of 'Superior seedless' grapes. *Sci. Hortic.* **2017**, *224*, 286–292. [CrossRef]
10. Yan, J.; Luo, Z.; Ban, Z.; Lu, H.; Li, D.; Yang, D.; Aghdam, M.S.; Li, L. The effect of the layer-by-layer (LBL) edible coating on strawberry quality and metabolites during storage. *Postharvest Biol. Technol.* **2019**, *147*, 29–38. [CrossRef]
11. Klangmuang, P.; Sothornvit, R. Active coating from hydroxypropyl methylcellulose-based nanocomposite incorporated with Thai essential oils on mango (cv. Namdokmai Sithong). *Food Biosci.* **2018**, *23*, 9–15. [CrossRef]
12. Perdones, A.; Sánchez-González, L.; Chiralt, A.; Vargas, M. Effect of chitosan–lemon essential oil coatings on storage-keeping quality of strawberry. *Postharvest Biol. Technol.* **2012**, *70*, 32–41. [CrossRef]
13. Saberi, B.; Golding, J.B.; Marques, J.R.; Pristijono, P.; Chockchaisawasdee, S.; Scarlett, C.J.; Stathopoulos, C.E. Application of biocomposite edible coatings based on pea starch and guar gum on quality, storability and shelf life of 'Valencia' oranges. *Postharvest Biol. Technol.* **2018**, *137*, 9–20. [CrossRef]
14. Thakur, R.; Pristijono, P.; Golding, J.; Stathopoulos, C.E.; Scarlett, C.; Bowyer, M.; Singh, S.; Vuong, Q. Development and application of rice starch based edible coating to improve the postharvest storage potential and quality of plum fruit (*Prunus salicina*). *Sci. Hortic.* **2018**, *237*, 59–66. [CrossRef]
15. Chen, H.; Sun, Z.; Yang, H. Effect of carnauba wax-based coating containing glycerol monolaurate on the quality maintenance and shelf-life of Indian jujube (*Zizyphus mauritiana* Lamk.) fruit during storage. *Sci. Hortic.* **2019**, *244*, 157–164. [CrossRef]
16. Nair, M.S.; Saxena, A.; Kaur, C. Effect of chitosan and alginate based coatings enriched with pomegranate peel extract to extend the postharvest quality of guava (*Psidium guajava* L.). *Food Chem.* **2018**, *240*, 245–252. [CrossRef]
17. Prakash, A.; Joseph, M.; Mangino, M. The effects of added proteins on the functionality of gum arabic in soft drink emulsion systems. *Food Hydrocoll.* **1990**, *4*, 177–184. [CrossRef]

18. Motlagh, S.; Ravines, P.; Karamallah, K.; Ma, Q. The analysis of Acacia gums using electrophoresis. *Food Hydrocoll.* **2006**, *20*, 848–854. [CrossRef]
19. Anderson, D.; Eastwood, M. The safety of gum arabic as a food additive and its energy value as an ingredient: A brief review. *J. Hum. Nutr. Diet.* **1989**, *2*, 137–144. [CrossRef]
20. Maqbool, M.; Ali, A.; Alderson, P.G.; Zahid, N.; Siddiqui, Y. Effect of a novel edible composite coating based on gum arabic and chitosan on biochemical and physiological responses of banana fruits during cold storage. *J. Agric. Food Chem.* **2011**, *59*, 5474–5482. [CrossRef]
21. Ali, A.; Maqbool, M.; Alderson, P.G.; Zahid, N. Effect of gum arabic as an edible coating on antioxidant capacity of tomato (*Solanum lycopersicum* L.) fruit during storage. *Postharvest Biol. Technol.* **2013**, *76*, 119–124. [CrossRef]
22. Addai, Z.R.; Abdullah, A.; Mutalib, S.; Musa, K. Effect of gum Arabic on quality and antioxidant properties of papaya fruit during cold storage. *Int. J. ChemTech Res.* **2013**, *5*, 2854–2862.
23. Khaliq, G.; Mohamed, M.T.M.; Ali, A.; Ding, P.; Ghazali, H.M. Effect of gum arabic coating combined with calcium chloride on physico-chemical and qualitative properties of mango (*Mangifera indica* L.) fruit during low temperature storage. *Sci. Hortic.* **2015**, *190*, 187–194. [CrossRef]
24. Ullah, A.; Abbasi, N.; Shafique, M.; Qureshi, A. Influence of edible coatings on biochemical fruit quality and storage life of bell pepper cv. "Yolo Wonder". *J. Food Qual.* **2017**, *2017*. [CrossRef]
25. Anjum, M.A.; Akram, H.; Zaidi, M.; Ali, S. Effect of gum arabic and *Aloe vera* gel based edible coatings in combination with plant extracts on postharvest quality and storability of 'Gola' guava fruits. *Sci. Hrotic.* **2020**, *271*, 109506. [CrossRef]
26. Moyo, B.; Masika, P.J.; Muchenje, V. Antimicrobial activities of *Moringa oleifera* Lam leaf extracts. *Afr. J. Biotechnol.* **2012**, *11*, 2797–2802. [CrossRef]
27. Tesfay, S.; Modi, A.; Mohammed, F. The effect of temperature in moringa seed phytochemical compounds and carbohydrate mobilization. *S. Afr. J. Bot.* **2016**, *102*, 190–196. [CrossRef]
28. Tesfay, S.Z.; Magwaza, L.S. Evaluating the efficacy of moringa leaf extract, chitosan and carboxymethyl cellulose as edible coatings for enhancing quality and extending postharvest life of avocado (*Persea americana* Mill.) fruit. *Food Packag. Shelf Life* **2017**, *11*, 40–48. [CrossRef]
29. Adetunji, C.; Fawole, O.; Arowora, K.; Nwaubani, S.; Ajayi, E.; Oloke, J.; Majolagbe, O.; Ogundele, B.; Aina, J.; Adetunji, J. Quality and safety of Citrus Sinensis coated with Hydroxypropylmethylcellulose edible coatings containing Moringa oleifera extract stored at ambient temperature. *Glob. J. Sci. Front. Res* **2012**, *12*, 6.
30. Ncama, K.; Magwaza, L.S.; Mditshwa, A.; Tesfay, S.Z. Plant-based edible coatings for managing postharvest quality of fresh horticultural produce: A review. *Food Packag. Shelf Life* **2018**, *16*, 157–167. [CrossRef]
31. Yousef, A.; El-Moniem, E.; Saleh, M. The effect of some natural products on storability and fruit properties of Fuerte avocado. *Int. J. ChemTech Res.* **2015**, *8*, 1454–1462.
32. Salehi, F. Edible Coating of Fruits and Vegetables Using Natural Gums: A Review. *Int. J. Fruit Sci.* **2020**, *20*, S570–S589. [CrossRef]
33. Kubheka, S.F.; Tesfay, S.Z.; Mditshwa, A.; Magwaza, L.S. Evaluating the Efficacy of Edible Coatings Incorporated with Moringa Leaf Extract on Postharvest of 'Maluma' Avocado Fruit Quality and Its Biofungicidal Effect. *HortScience* **2020**, *55*, 410–415. [CrossRef]
34. Zegbe, J.A.; Mena-Covarrubias, J.; Domínguez-Canales, V.S.I. Cactus mucilage as a coating film to enhance shelf life of unprocessed guavas (*Psidium guajava* L.). *Acta Hortic.* **2015**, *1067*, 423–427. [CrossRef]
35. Habbal, O.; Hasson, S.S.; El-Hag, A.H.; Al-Mahrooqi, Z.; Al-Hashmi, N.; Al-Bimani, Z.; Al-Balushi, M.S.; Al-Jabri, A.A. Antibacterial activity of *Lawsonia inermis* Linn (Henna) against *Pseudomonas aeruginosa*. *Asian Pac. J. Trop Biomed.* **2011**, *1*, 173–176. [CrossRef]
36. Mercado-Silva, E.; Benito-Bautista, P.; de los Angeles García-Velasco, M. Fruit development, harvest index and ripening changes of guavas produced in central Mexico. *Postharvest Biol. Technol.* **1998**, *13*, 143–150. [CrossRef]
37. A.O.A.C. *Official Methods of Analysis*; Association of Official Analytical Chemists Washington: Washington, DC, USA, 2005.
38. Kähkönen, M.P.; Hopia, A.I.; Vuorela, H.J.; Rauha, J.-P.; Pihlaja, K.; Kujala, T.S.; Heinonen, M. Antioxidant activity of plant extracts containing phenolic compounds. *J. Agric. Food Chem.* **1999**, *47*, 3954–3962. [CrossRef]
39. Abe, N.; Murata, T.; HIRoTA, A. Novel DPPH radical scavengers, bisorbicillinol and demethyltrichodimerol, from a fungus. *Biosci. Biotechnol. Biochem.* **1998**, *62*, 661–666. [CrossRef]
40. Yu, L.; Reitmeier, C.; Love, M. Strawberry texture and pectin content as affected by electron beam irradiation. *J. Food Sci.* **1996**, *61*, 844–846. [CrossRef]
41. Wu, R.; Frei, B.; Kennedy, J.A.; Zhao, Y. Effects of refrigerated storage and processing technologies on the bioactive compounds and antioxidant capacities of 'Marion' and 'Evergreen' blackberries. *LWT-Food Sci. Technol.* **2010**, *43*, 1253–1264. [CrossRef]
42. Milosevic, T.; Mratinic, E.; Milosevic, N.; Glisic, I.; Mladenovic, J. Segregation of blackberry cultivars based on the fruit physico-chemical attributes. *J. Agric. Sci.* **2012**, *18*, 100–109.
43. Dhall, R. Advances in edible coatings for fresh fruits and vegetables: A review. *Crit. Rev. Food Sci. Nutr.* **2013**, *53*, 435–450. [CrossRef]
44. Ali, A.; Maqbool, M.; Ramachandran, S.; Alderson, P.G. Gum arabic as a novel edible coating for enhancing shelf-life and improving postharvest quality of tomato (*Solanum lycopersicum* L.) fruit. *Postharvest Biol. Technol.* **2010**, *58*, 42–47. [CrossRef]

45. Barman, K.; Asrey, R.; Pal, R. Putrescine and carnauba wax pretreatments alleviate chilling injury, enhance shelf life and preserve pomegranate fruit quality during cold storage. *Sci. Hortic.* **2011**, *130*, 795–800. [CrossRef]
46. El-Khalek, A.; Fathy, A. Effectiveness of gum arabic, potassium salts and their incorporation in the control of postharvest diseases and maintaining quality of 'Washington' navel oranges during long term cold storage. *Egypt. J. Hortic.* **2018**, *45*, 185–203. [CrossRef]
47. Razzaq, K.; Khan, A.S.; Malik, A.U.; Shahid, M.; Ullah, S. Role of putrescine in regulating fruit softening and antioxidative enzyme systems in 'Samar Bahisht Chaunsa' mango. *Postharvest Biol. Technol.* **2014**, *96*, 23–32. [CrossRef]
48. Tripathi, P.; Dubey, N. Exploitation of natural products as an alternative strategy to control postharvest fungal rotting of fruit and vegetables. *Postharvest Biol. Technol.* **2004**, *32*, 235–245. [CrossRef]
49. Lester, G.E.; Grusak, M.A. Field application of chelated calcium: Postharvest effects on cantaloupe and honeydew fruit quality. *HortTechnology* **2004**, *14*, 29–38. [CrossRef]
50. Baldwin, E.; Burns, J.; Kazokas, W.; Brecht, J.; Hagenmaier, R.; Bender, R.; Pesis, E. Effect of two edible coatings with different permeability characteristics on mango (*Mangifera indica* L.) ripening during storage. *Postharvest Biol. Technol.* **1999**, *17*, 215–226. [CrossRef]
51. Echeverria, E.; Ismail, M. Changes in sugars and acids of citrus fruits during storage. In Proceedings of the Annual Meeting of the Florida State Horticulture Society, Florida, FA, USA; 1988; pp. 50–52.
52. El-Banna, M.F.; Lo'ay, A.A. Evaluation berries shattering phenomena of 'Flame seedless' vines grafted on different rootstocks during shelf life. *Sci. Hortic.* **2019**, *246*, 51–56. [CrossRef]
53. Beckman, C.H. Phenolic-storing cells: Keys to programmed cell death and periderm formation in wilt disease resistance and in general defence responses in plants? *Physiol. Mol. Plant Pathol.* **2000**, *57*, 101–110. [CrossRef]
54. Blokhina, O.; Virolainen, E.; Fagerstedt, K.V. Antioxidants, oxidative damage and oxygen deprivation stress: A review. *Ann. Bot.* **2003**, *91*, 179–194. [CrossRef] [PubMed]
55. Lo'ay, A.A.; Ameer, N.M. Performance of calcium nanoparticles blending with ascorbic acid and alleviation internal browning of 'Hindi Be-Sennara' mango fruit at a low temperature. *Sci. Hortic.* **2019**, *254*, 199–207. [CrossRef]
56. Devi, R. Role of Gibberellic Acid and Calcium Chloride in Ripening Related Biochemical Changes in Guava (*Psidium guajava* L.) Fruit. Ph.D. Thesis, CCSHAU, Hisar, India, 2016.
57. Lo'ay, A.A.; El-Ezz, S.F.A. Performance of 'Flame seedless' grapevines grown on different rootstocks in response to soil salinity stress. *Sci. Hortic.* **2021**, *275*, 109704. [CrossRef]

Article

The Combined Effect of Hot Water Treatment and Chitosan Coating on Mango (*Mangifera indica* L. cv. Kent) Fruits to Control Postharvest Deterioration and Increase Fruit Quality

Hoda A. Khalil [1], Mohamed F. M. Abdelkader [2], A. A. Lo'ay [3,*], Diaa O. El-Ansary [4], Fatma K. M. Shaaban [5], Samah O. Osman [5], Ibrahim E. Shenawy [6], Hosam-Eldin Hussein Osman [7], Safaa A. Limam [8], Mohamed A. Abdein [9,*] and Zinab A. Abdelgawad [10]

Citation: Khalil, H.A.; Abdelkader, M.F.M.; Lo'ay, A.A.; El-Ansary, D.O.; Shaaban, F.K.M.; Osman, S.O.; Shenawy, I.E.; Osman, H.-E.H.; Limam, S.A.; Abdein, M.A.; et al. The Combined Effect of Hot Water Treatment and Chitosan Coating on Mango (*Mangifera indica* L. cv. Kent) Fruits to Control Postharvest Deterioration and Increase Fruit Quality. *Coatings* 2022, 12, 83. https://doi.org/10.3390/coatings12010083

Academic Editors: Lili Ren and Stefano Farris

Received: 6 December 2021
Accepted: 31 December 2021
Published: 12 January 2022

Publisher's Note: MDPI stays neutral with regard to jurisdictional claims in published maps and institutional affiliations.

Copyright: © 2022 by the authors. Licensee MDPI, Basel, Switzerland. This article is an open access article distributed under the terms and conditions of the Creative Commons Attribution (CC BY) license (https://creativecommons.org/licenses/by/4.0/).

1. Department of Pomology, Faculty of Agriculture (EL-Shatby), University of Alexandria, Alexandria P.O. Box 21545, Egypt; hoda.khalil@alexu.edu.eg
2. Department of Plant Production, College of Food and Agriculture, King Saud University, Riyadh 12372, Saudi Arabia; mohabdelkader@ksu.edu.sa
3. Pomology Department, Faculty of Agriculture, Mansoura University, Mansoura, El-Mansoura P.O. Box 35516, Egypt
4. Precision Agriculture Laboratory, Department of Pomology, Faculty of Agriculture (El-Shatby), University of Alexandria, Alexandria P.O. Box 21545, Egypt; diaa.elansary@alexu.edu.eg
5. Horticulture Research Institute, ARC, Giza 12619, Egypt; dr.fatmakorany@yahoo.com (F.K.M.S.); ayatosman012@gmail.com (S.O.O.)
6. Pomology Department, Faculty of Agriculture, Cairo University, Cairo P.O. Box 12613, Egypt; Dr.shenawy@hotmail.com
7. Anatomy Department, College of Medicine, Taif University, P.O. Box 11099, Taif 21944, Saudi Arabia; h.hussein@tu.edu.sa
8. Food Science and Technology Department, Faculty of Agriculture, Assiut University, Assiut 71515, Egypt; Limamsafaa@gmail.com
9. Biology Department, Faculty of Arts and Science, Northern Border University, Rafha 91911, Saudi Arabia
10. Botany Department, Women's College, Ain Shams University, Cairo 11566, Egypt; Zinababdelgawad@women.asu.edu.eg
* Correspondence: Loay_Arafat@mans.edu.eg (A.A.L.); abdeingene@yahoo.com (M.A.A.)

Abstract: The synergistic effect of dipping in 55 °C for 5 min of hot water (HW) and 1% chitosan coating during the storage of mango at 13 ± 0.5 °C and 85%–90% relative humidity for 28 days was investigated. The combined treatment significantly suppressed the fruit decay percentage compared with both the single treatment and the control. In addition, the specific activities of key plant defense-related enzymes, including peroxidase (POD) and catalase (CAT), markedly increased. The increase occurred in the pulp of the fruits treated with the combined treatment compared to those treated with HW or chitosan alone. While the control fruits showed the lowest values, the combination of pre-storage HW treatment and chitosan coating maintained higher values of flesh hue angle (h°), vitamin C content, membrane stability index (MSI) percentage, as well as lower weight loss compared with the untreated mango fruits. The combined treatment and chitosan treatment alone delayed fruit ripening by keeping fruit firmness, lessening the continuous increase of total soluble solids (TSS), and slowing the decrease in titratable acidity (TA). The results showed that the combined application of HW treatment and chitosan coating can be used as an effective strategy to suppress postharvest decay and improve the quality of mango fruits.

Keywords: hot water; chitosan; mango; decay; storage; membrane stability index

1. Introduction

Mango fruits (*Mangifera indica* L.) are recognized as one of the most desirable fruits due to their appealing color, delectable flavor, and superior nutritional value. However, mangoes, a climacteric fruit, ripen shortly after harvest and are susceptible to anthracnose caused by *Colletotrichum* species, which results in significant postharvest losses and restrictions on mango fruit storage, handling, and transportation [1]. Mango fruits are currently stored in a controlled (or modified) atmosphere to control postharvest decay and delay the ripening process [2]. In addition, fungicides are widely used to minimize postharvest decay and prolong the shelf life of mango fruits. However, fungicides are being limited as pathogens gain resistance to them, and consumers are concerned about the risks associated with fungicide residue [3]. Therefore, alternative and safe techniques are needed to slow the ripening of mango fruit and reduce postharvest decay.

Hot water (HW) treatment is the oldest and simplest form of heat treatment for controlling postharvest decay that uses a combination of appropriate temperatures (typically over 40 °C) and exposure durations to avoid fruit quality loss [4,5]. In a variety of fruits, HW treatment can effectively inhibit many important postharvest pathogens [6]. Treatment with HW at 55 °C for 35 min reduced the incidence of anthracnose in mango cultivars 'Tu Shien' [7], 'Kent', 'Keitt', and 'Tommy Atkins' [8]. Likewise, Dessalegn et al. [6], working on mango cv. 'Amba Kurfa', found that HW treatment at 51 °C for 3 min decreased the amount of anthracnose disease. Additionally, HW treatment has been identified as an elicitor for the activation of the defensive response in harvested fruits [9].

Although HW treatment helps prevent postharvest decay in mango fruits, there have been reports of detrimental impacts on the quality of the fruit, including accelerated fruit ripening, fruit skin browning, and mango fruit softening [10]. Based on the results presented above, it would be better to develop new treatments that may mitigate the negative effects of HW treatment, prevent postharvest decay, and delay the ripening process of mango fruits.

Several biopolymers, including chitosan, pectin, alginate, starch, carrageenan, zein, soy protein, and gelatin, have been applied in the development of coating formulations for fruit shelf life extension. Edible coating is simple, biodegradable, and ecologically friendly, it is a good alternative for synthetic materials, and it may be consumed by humans [11]. Edible coatings have been highlighted as a potential technology to prevent postharvest infection and the associated fungal degradation of fruits [12–14]. The application of fruit coatings has demonstrated technological advantages such as better appearance, antibacterial and antioxidant properties, and improved taste [15]. Some coatings have already been tested on tropical fruits, including avocados and mangoes, with different degrees of effectiveness. For instance, Daisy et al. [16] found that gum Arabic (15%) preserved ascorbic acid and carotenoids in 'Apple' mango kept at room temperature for 15 days. Likewise, Moalemiyan et al. [17] reported that coating mango with pectin, sorbitol, monoglyceride, and beeswax combinations resulted in an increased shelf life, especially at decreased color development, weight loss, softness, and acid production compared with the control. In addition, a study by Bambalele et al. [15] reported that moringa leaf extract (1%) and carboxymethyl cellulose (1%) maintained the ascorbic acid and membrane integrity and delayed fruit softening in 'Keitt' mango after storage at 10 °C for 21 days.

Chitosan is one of the polysaccharide-based coatings. It is a high molecular weight cationic polysaccharide commonly formed by the alkaline deacetylation of chitin found in the crustacean exoskeleton, fungal cell walls, and other biological components [18]. It is composed of poly-1,4-β-D-glucopyranos amine and 2-amino-2-deoxy-(1->4)-β-D-glucopyranan. Chitosan has great potential as a film or a biodegradable edible coating for food packaging [19], with good biocompatibility, nontoxicity [18], and film-forming characteristics [20]. Chitosan has been used on a variety of fruits such as mango as a semipermeable coating to prolong storage life and decrease postharvest decay [21,22]. The application of chitosan in mango [23] has been demonstrated to enhance fruit quality, keep firmness, decrease ethylene production and mold contamination, delay the ripening process and senescence, and decrease color changes.

The combination of edible coating and HW treatment has been examined to maintain fruit quality and minimize unanticipated damage [24,25]. Keeping fruit quality and controlling postharvest decay in fruit cannot be entirely controlled by HW treatment or chitosan treatment alone. For several reasons, combining HW and chitosan may have a synergistic impact on fruit: (1) The fruit surface may be partly disinfected by the HW treatment; (2) Pathogen resistance may be induced by chitosan; (3) Combining HW with chitosan can improve the effectiveness of postharvest disease control. In the previous studies, researchers used a combination of these two treatments on sweet cherry [26], papaya [27], and dragon fruit [28], and the results indicated that the combination treatment reduced postharvest disease and preserved fresh fruit quality better than HW or chitosan.

Although HW and edible coatings have been intensively investigated in recent years, the combined effect of these treatments has received less attention, especially on mango fruits. To our knowledge, there are no published data about the use of hot water treatment and chitosan coatings for maintaining fruit quality and prolonging the shelf life of "Kent" mangoes. Thus, our research aimed to see if HW treatment, followed by chitosan coating, may help keep mango fruit fresh, maintain quality indices, and extend the shelf life of "Kent" mangoes while also reducing postharvest decay. In addition, HW treatment-induced resistance in mango fruit was also investigated to better understand the defense mechanism of HW treatment against pathogens during storage.

2. Materials and Methods

2.1. Plant Materials and Treatments

The present study was carried out during the 2020 and 2021 seasons on mature green stage mango fruits (*Mangifera indica* L. cv. "Kent"). The fruits were harvested from a private orchard at Alexandria–Cairo desert road and were immediately transported to the Horticulture Lab in the Faculty of Agriculture, Alexandria University, Egypt. The coating solution preparation was done as 1 g of chitosan (40 kDa in thickness, from crab shells) which was gently dissolved in 100 mL of 2 percent acetic acid solution (v/v) on a magnetic stirrer (350 rpm INTLLAB, New York, NY, USA) to get 1 percent chitosan solution (w/v), and Tween 20 (0.2 g) was added in the chitosan solution. The fruits were selected for uniformity of size, ripeness, and being free of defects, and were divided into four groups (50 fruit for each). The first group of fruits was washed with distilled water (control). The second group was dipped in hot water (HW) at 55 °C for 5 min. The third group was dipped in 1% chitosan solution for 1 min. The fourth group was dipped in hot water (HW) at 55 °C for 5 min, and then dipped in 1% chitosan solution for 1 min. After being air dried, all the fruits were stored at 13 °C, 85%–90% RH for 28 days. The initial physio-chemical properties were determined in ten mango fruits and the changes were followed up in 7 day intervals during the storage period.

2.2. Measurements of Fruit Physical and Chemical Features

2.2.1. Decayed Fruit Percentage

Decay due to browning skin, shriveling, and diseases were recorded and estimated based on the initial number of fruits in each sample and reported as a percentage.

2.2.2. Fruit Firmness

Fruit firmness was measured using an Effegi pressure tester (Effegi, 48011 Alfonsine, Alfonsine, Italy) connected to a flat probe (8 mm diameter). At the equator of the fruit, the skins were removed in four pieces and four independent measurements were recorded for each fruit.

2.2.3. Weight Loss Percentage

Weight loss was assessed by weighing ten labeled treated fruits at 7-day intervals during the storage. The percentage of weight loss was assessed by the following equation:

$$\text{Weight loss \%} = \frac{\text{The initial weight} - \text{fruit weight at examination date}}{\text{The initial weight}} \times 100$$

2.2.4. Fruit Content of Total Soluble Solids (TSS) and Titratable Acidity (TA)

For each treatment, four fruit pulp samples were squeezed out, and the resulting juice was used to measure the TSS percentage using a hand refractometer (Atago, Japan), and the titratable acidity in grams of citric acid per 100 mL of fruit juice [29].

2.2.5. Vitamin C

Vitamin C was determined by oxidizing ascorbic acid with 2,6-dichlorophenol endophenoldye and the results were reported in mg/100 g on a fresh weight basis [30].

2.2.6. Fruit Color Index

Flesh color was measured using a Minolta Chroma Meter CR-200 (Minolta Co. Ltd., Osaka, Japan). Flesh color measurements were expressed as hue angle chromaticity values (h°). Four readings were obtained at different points on each mango fruit for each data observation [31,32].

2.2.7. Defense-Related Enzymes Activities

The peroxidase activity of *M. indica* kernels was determined using a colorimetric test based on the initial increase in absorbance at 420 nm in the presence of a constant volume of hydrogen peroxide and a crude extract of pyrogallol. POD activity was expressed as U/g FW, where one unit of peroxidase activity was defined as the amount of extract that caused a 0.001 per minute change in absorbance at 420 nm [33].

Catalase (CAT) activity was measured using the method of Beers and Sizer [34], with certain modifications. The reaction mixture consisted of 0.2 crude extracted from 50 mM sodium phosphate buffer (pH 7.0) and 150 mL of 20 mM of H_2O_2. The action of CAT on hydrogen peroxide caused a decrease in absorbance at 240 nm. The change in absorbance per minute was defined as one unit of CAT.

2.2.8. Membrane Stability Index Percentage (MSI)

Ion leakage from fruit peels was assessed in peel discs using the method described by Sairam et al. [35], with modification, and was represented as a percentage of membrane stability index (MSI). For each replicate/treatment, 3 g of wash disks were randomly selected and placed in 30 mL of deionized water at room temperature on a shaker for 4 h. Before boiling (C_1), a digital conductivity meter (Orion 150A +, Thermo Electron Corporation, Colorado, CO, USA) was used to check the conductivity. The same disk was placed in a boiling water bath (100 °C) for 30 min to release all electrolytes, cooled to 22 ± 2 °C in running water, and boiled to measure conductivity (C_2). MSI was calculated as a percentage using the following formula: $[1 - (C_1/C_2)] \times 100$. Fruit softening or the appearance of chilling injury signs indicated the termination of the trial.

2.3. Statistical Analysis

The two-way analysis of variance (ANOVA) method was used to evaluate the data for the effects of the treatments on the investigated parameters. According to Snedecor and Cochran [36], the treatment means were separated and compared using the least significant differences (L.S.D.) at the 0.05 level of significance. SPSS 18.0 software was used for all statistical analyses (SPSS Inc., Chicago, IL, USA). In the figures, data were presented as means of standard errors (SE).

3. Results

3.1. Effect of Chitosan and Hot Water (HW) Treatments on Fruit Quality
3.1.1. Decay

As for the storage period, regardless of the treatments, the obtained results showed that the incidence of fruit decay appeared after 14 days of cold storage, followed by a consistent increase with increasing the storage period up to 28 days (Tables 1 and 2, Figure 1). Data also indicated that all treatments significantly decreased the decay incidence of mango fruits during the two seasons of the study. The chitosan treatment did not significantly differ from the combined treatment in both seasons. The incidence of decay was higher in the control mango fruits than in the other treatments. The control fruit had an 8.9% decay incidence after 14 days of storage, but the HW treatment, chitosan, and the combination treatments had no decay incidence (Table 1). In the first season, at the end of the storage date, the control treatment had a decay incidence of 30.20%, followed by the HW-treated samples (19.35%), while the chitosan and combined treatments were only 3.6 and 2.4%, respectively (Table 1). A significant interaction effect between the treatments and the storage period on the percentage of fruit decay was obtained in both seasons. Data revealed that the fruit decay percentage was least with all treatments, especially when those treatments were accompanied by the shortest storage period in comparison with the same treatments accompanied by the longest storage period.

Table 1. Effect of hot water treatment and chitosan coating on postharvest decay (%) of mango fruits during storage at 13 ± 0.5 °C and 85%–90% relative humidity (RH) for 28 days during 2020 season.

Treatments	Storage Period (Days)					Mean
	0	7	14	21	28	
	Season 2020					
Control	0.00 h	0.00 h	8.96 d	17.5 c	30.20 a	11.33 a
Hot water (HW)	0.00 h	0.00 h	0.00 h	6.13 e	19.35 b	5.09 b
1% chitosan	0.00 h	0.00 h	0.00 h	0.00 h	3.68 f	0.73 c
HW + 1% chitosan	0.00 h	0.00 h	0.00 h	0.00 h	2.43 g	0.48 c
Mean	0.00 d	0.00 d	2.24 c	5.90 b	13.91 a	–

Means followed by the same letters within treatments, storage period, and their interactions in 2020 season are not significantly different at level $p \leq 0.05$.

Table 2. Effect of hot water treatment and chitosan coating on postharvest decay (%) of mango fruits during storage at 13 ± 0.5 °C and 85%–90% relative humidity (RH) for 28 days during 2021 season.

Treatments	Storage Period (Days)					Mean
	0	7	14	21	28	
	Season 2021					
Control	0.00 h	0.00 h	7.58 d	15.66 c	28.32 a	10.31 a
Hot water (HW)	0.00 h	0.00 h	0.00 h	6.30 e	21.50 b	5.56 b
1% chitosan	0.00 h	0.00 h	0.00 h	0.00 h	4.13 g	0.82 c
HW + 1% chitosan	0.00 h	0.00 h	0.00 h	0.00 h	5.12 f	1.02 c
Mean	0.00 d	0.00 d	1.89 c	5.49 b	14.76 a	–

Means followed by the same letters within treatments, storage period, and their interactions in 2021 season are not significantly different at level $p \leq 0.05$.

Figure 1. Visual appearance of mango fruit after 7 days (**A**) and 21 days (**B**) of storage at 13 ± 0.5 °C and 85%–90% relative humidity.

3.1.2. Weight Loss

The percentage of mango fruit weight loss increased dramatically when the storage duration was extended to 28 days (Figure 2A and Table 3). All treatments considerably decreased weight loss of 'Kent' mango fruits during cold storage when compared to untreated control fruits (Figure 2A and Table 3). Fruit treated with chitosan or the combined treatment (Figure 2A) lost less weight (4.13% and 5.16%, respectively) than the control or the HW-treated fruit (6.80% and 6.16%, respectively). As for the interaction effect, data showed that all the treatments at the same storage period decreased weight loss.

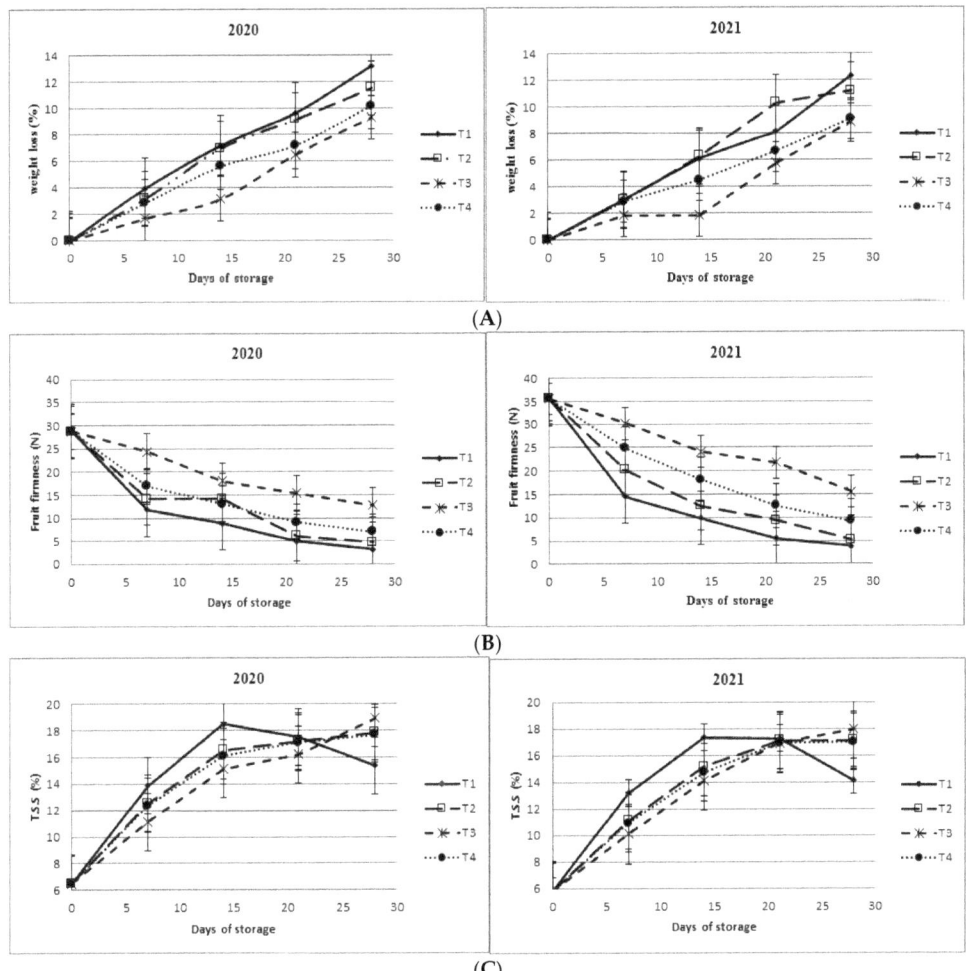

Figure 2. Effect of hot water treatment and chitosan coating on weight loss (**A**), fruit firmness (**B**), and total soluble solids (T.S.S) (**C**) of mango fruits during storage at 13 ± 0.5 °C and 85%–90% relative humidity (RH) for 28 days during 2020 and 2021 seasons. Values are means ± SE from three replicates. Statistical analysis was performed using LSD test.

Table 3. Results of the analysis of variance with mean square testing the effects of treatments (T), storage period (S), and their interactions on fruit firmness (N), weight loss (%), TSS (%), acidity (%), vitamin C (mg/100 g FW), color index (h°), peroxidase activity (units/mg FW), catalase activity (units/mg FW), and MSI (%) during 2020 and 2021 seasons.

Season 2020	Fruit Firmness (N)	Weight Loss (%)	TSS (%)	Acidity (%)	Ascorbic Acid (mg/100 g FW)	Color Index (h°)	Peroxidase Activity (Units/mg FW)	Catalase Activity (Units/mg FW)	MSI (%)
Treatments (T)	38.00 ***	38.20 ***	3.17 *	969.00 ***	21 ***	121.40 ***	199.80 ***	3.39 *	199.80 ***
Storage period (S)	414.30 ***	416.30 ***	505 ***	8362.10 ***	541.50 ***	1755.10 ***	771.80 ***	73.11 ***	771.80 ***
T X S	3.74 ***	3.70 ***	7.77 ***	118.80 ***	2.32 *	13.70 ***	58.80 ***	7.80 ***	58.80 ***

Season 2021	Fruit Firmness (N)	Weight Loss (%)	TSS (%)	Acidity (%)	Ascorbic Acid (mg/100 g FW)	Color Index (h°)	Peroxidase Activity (O.D)	Catalase Activity	MSI (%)
Treatments (T)	158.40 ***	4.25 *	2.77 ns	389.20 ***	191.30 ***	332.80 ***	52.70 ***	25.26 ***	52.70 ***
Storage period (S)	600.10 ***	90.87 ***	701.28 ***	2349.80 ***	3115.30 ***	3395.80 ***	67.30 ***	37.51 ***	67.30 ***
T X S	12.30 ***	2.89 **	13.19 ***	61.90 ***	18.80 ***	32.58 ***	13.80 ***	2.90 **	13.80 ***

ns, *, **, *** nonsignificant, or significant at p = 0.05, 0.01, and 0.001, respectively.

3.1.3. Firmness

Data presented in Table 3 and Figure 2B reflected the reduction in fruit firmness with the progress of the storage period for all treatments. However, after 7 days of storage, the control fruits and the fruits treated with HW softened faster than the other treatments. At the end of storage, the control fruit and the fruit treated with HW showed low values of 11.48 and 12.53 N, respectively, whereas the combined-treated fruits and the chitosan-treated fruits showed a high fruit firmness value of 14. 96 and 19.84 N, respectively. Data also showed that there was a significant interaction between the treatments and the storage period for mango fruit firmness in both seasons. The ripening of mango fruits leads to a loss in firmness with the progress of the storage period.

3.1.4. Total Soluble Solids (TSS) and Titratable Acidity (TA)

Total soluble solids and titratable acidity, as well as a comparison of the means over the storage period for both experimental seasons, are shown in Figures 2C and 3A and Table 3. TSS increase in all treatments over time, while TA decreased over the fruit's storage period, which is a climacteric fruit feature during the ripening process. At the end of the storage period, all postharvest treatments significantly increased TSS and decreased TA compared to the control treatment. In both seasons, fruits treated with chitosan had the lowest TSS values (13.57 and 12.97, respectively) and the highest TA values (0.79 and 0.77, respectively).

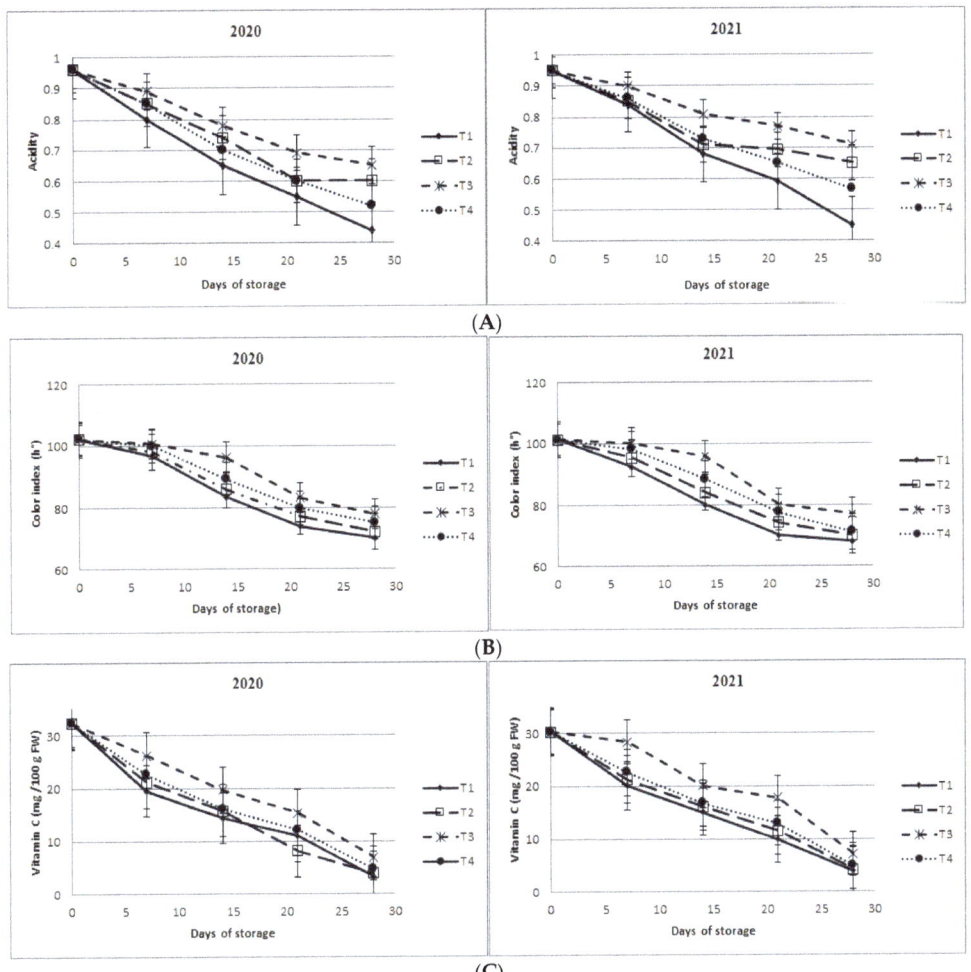

Figure 3. Effect of hot water treatment and chitosan coating on acidity (**A**), color index (**B**), and vitamin C (**C**) of mango fruits during storage at 13 ± 0.5 °C and 85%–90% relative humidity (RH) for 28 days during 2020 and 2021 seasons. Values are means ± SE from three replicates. Statistical analysis was performed using LSD test.

3.1.5. Changes in Flesh Color (h°)

Data shown in Figures 1 and 3B and Table 3 presented the changes in flesh hue angle according to chitosan, HW treatment, and the cold storage period. Chitosan-treated fruits had the highest significant values of flesh hue angle, followed by the combined treatment and HW, whereas the control treatment showed a lower flesh hue angle content than the other treatments.

3.1.6. Changes in Vitamin C Content, Peroxidase (POD) Activity, Catalase (CAT) Activity, and Membrane Stability Index Percentage (MSI%)

Peroxidase and catalase activities increased during the storage period in all treatments. However, after one week of the storage period, POD and CAT activities were lower in all treatments, including the control (Figures 3C and 4A–C, and Table 3). At the end of the storage period, all the treatments showed higher POD and CAT activities than the control.

The chitosan and combined treatment presented the highest POD and CAT activities. Vitamin C content was lower than initial in all treatments and decreased during storage. At the end of the storage period (28 days), all treatments obtained a higher vitamin C content than the control. However, the chitosan and combined treatments showed higher vitamin C content than the other treatments. In all treatments, the membrane stability index (MSI) recorded lower values than initially and declined throughout the storage period (Figure 4C and Table 3). All treatments applied at the end of the storage period maintained a higher MSI than the control. The chitosan treatment and the combination treatment had a greater MSI in both seasons than the other treatments, including the control.

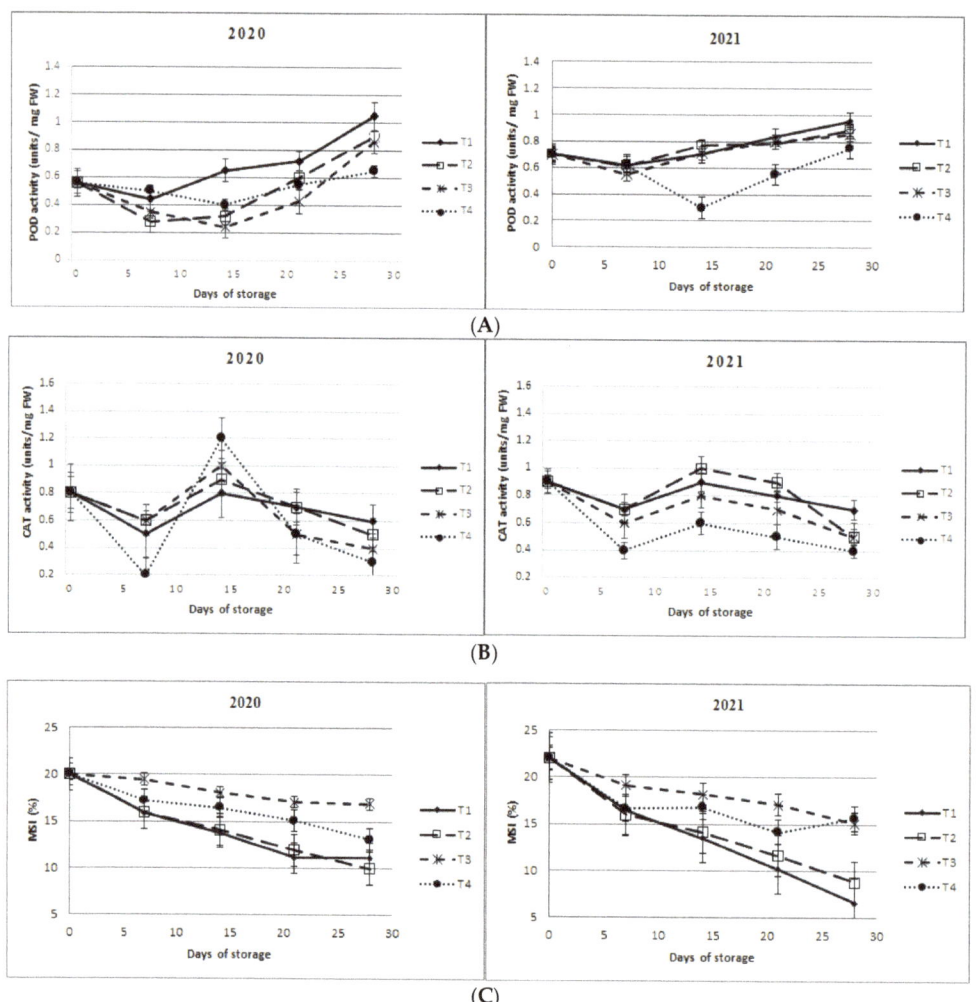

Figure 4. Effect of hot water treatment and chitosan coating on peroxidase (POD) activity (**A**), catalase (CAT) activity (**B**), and membrane stability index (MSI) percentage (**C**) of mango fruits during storage at 13 ± 0.5 °C and 85%–90% relative humidity (RH) for 28 days during 2020 and 2021 seasons. Values are means ± SE from three replicates. Statistical analysis was performed using LSD test.

4. Discussion

As a climacteric fruit, mango fruit exhibits relatively high levels of bioactivity, including high respiration rates and postharvest ethylene production, thus reducing its shelf life. Due to consumer concerns about the use of synthetic chemicals, various natural coatings are currently being investigated throughout the shelf life for their effectiveness in slowing ripening and maintaining fruit quality [37–39]. Our investigation demonstrated that chitosan coating, alone or in combination with HW treatment, could efficiently delay ripening, increase postharvest quality, and control the decay of mango fruits. These results revealed that chitosan coating could be an alternate and effective technique for prolonging the postharvest life of mango fruits. In the current study, chitosan coating, alone or combined with HW treatment, significantly decreased the decay percentage after 3 weeks of cold storage compared to the control and the other treatments (Tables 1 and 2). The fruit coating with a concentration of 0.2% chitosan significantly inhibited the decay incidence of mango fruits caused by disease and maintained the fruit quality [5]. Chitosan postharvest application is known to influence the host–pathogen as antibacterial and antifungal activity [22,40]. It can destroy the plasma membrane of the spore of several pathogens, inhibit mycelial growth, and induce damage to the fungal cytoplasm [41]. Moreover, chitosan may also induce a defense mechanism in host tissues [42]. Several studies have shown that the use of HW is beneficial for tropical fruits [43–46]. The main cause of fruit weight loss is water loss induced by respiration and transpiration processes [47]. Chitosan has been reported to decrease the respiration rate in mango fruit [47–49]. The lower weight loss observed in chitosan-treated fruits may be related to the higher vapor barrier of chitosan. On the contrary, one study reported that 'Ataulfo' mango coated with chitosan film had greater weight loss [50]. Chitosan coating is reported to minimize transpiration losses by forming a semipermeable layer on the fruit's surface and acting as a selective permeability to water vapor [51]. In addition, it is reported to decrease the transpiration rate and retard senescence by modifying the internal atmosphere of the fruits [52]. The application of HW treatment has been reported to reduce or increase the weight loss of fruits. In this study, the HW treatment decreased fruit weight loss. Consistent with our result, Fawaz [5,53] mentioned that the weight loss of 'Alphones' mango fruits was reduced by a 45 °C HW treatment, suggesting that a mild heat treatment would dissolve the cuticle wax and decrease water loss [54]. The ripening of mango fruit is marked by a softening of the texture and a change in the color of the surface. The results of our study showed that the chitosan coating alone or in combination with the HW treatment effectively slowed the ripening of mango fruit, as evidenced by the retention of firmness and the delayed color change. In addition, TA and vitamin C reduction, TSS increase, and the weight loss of mango fruit were significantly suppressed by the chitosan coating. Thus, the application of the chitosan coating alone or in combination with the HW treatment effectively maintained the postharvest quality of the mango fruits according to the result obtained on the mango fruits treated by chitosan-based coating after harvest [49,55]. In the previous studies, chitosan maintained the firmness of the mango fruit [47,56,57]. Amin et al. [58] reported that mango fruit firmness was found to decrease linearly with storage time. The rate of firmness loss, however, was consistently decreased with the addition of a up to 2% chitosan–Aloe vera coating. The results of the HW treatment are consistent with [9,43]. The slowing of the softening might be attributed to the prevention of the formation of cell wall hydrolysis enzymes, which maintain membrane stability and reduce firmness [59]. Consistent with our result, chitosan has been shown to reduce fruit TSS during storage in mango [47,56,60]. Chitosan-treated fruits had the highest significant values of flesh hue angle, followed by the combined treatment and the HW treatment. These findings are consistent with those of Zhu et al. and Djioua et al. [25,47] on mango. They reported that the application of a 2% chitosan coating and a HW treatment at 50 °C for 30 min were effective treatments to maintain firmness and delay color change during fruit storage. Chitosan formed a semipermeable layer over the fruit peel and altered the atmosphere by elevating CO_2 and lowering O_2, which inhibited ethylene production and delayed ripening [61]. As the ripening process was delayed, it consequently reduced the color changes by decreasing carotenoid biosynthesis and preserving the chlorophyll content. The hue values decreased with the advancing of the storage period and the values were significant for all treatments, with the change of flesh color from creamy to orange. Previously, the use of 1% chitosan was shown to improve the ascorbic acid content of mango [57,62]. Similar results have been reported in a previous study using the HW treatment [5]. The ascorbic acid content was found to be increased 1.14-fold in HW-treated fruits in contrast with the control fruit. The results showed that the chitosan treatment and the combination treatment had a greater MSI in both seasons than the other treatments, including the control (Figure 3C and Table 3). Ripening is a potentially oxidative process in which the transition from the maturing stage to the ripening stage is accompanied by a dynamic move to an oxidative state [63]. Likewise, excessive ROS generation can lead to the oxidation of the cell membrane lipids and proteins involved in mango ripening, which results in a gradual loss of membrane stability because of the changes in the biophysical and biochemical characteristics of the cell membranes. The expression of genes encoding enzymes involved in the fruit antioxidant system, including POD, PPO, and catalase, increases during ripening [64,65], and endogenous defense against the accumulation of harmful ROS has also been reported [49,66,67]. This might demonstrate that these treatments, especially chitosan and combined treatments, improved the antioxidant network of the fruit [67,68], allowing for the more effective regulation of metabolic free radical levels, hence preserving peel cell membrane integrity and maintaining better flesh firmness. Although many of the previous publications have indicated that both hot water and chitosan could be applied for different fruit protections, the novelty of our study is that, under the synergistic effect of the coating of the chitosan solution with a low concentration (1%) and the hot water treatment, the postharvest decay was suppressed, the quality of the mango fruits was improved, and the shelf life was extended by increasing the temperature of the hot water and shortening the treatment time of the hot water. Moreover, different fruits have different profiles regarding their storage capacity, with different treatment applications.

5. Conclusions

Edible coatings combined with HW treatment were used to induce fruit decay resistance and improve fruit quality parameters. The present study evaluated the effect of a chitosan edible coating combined with a HW treatment on mango fruits during the storage time for 18 days at 13 ± 0.5 °C and 85%–90% relative humidity for 28 days. The results revealed that a combination of HW and chitosan treatments dramatically decreased the decay incidence percentage and improved the quality in mango fruits while also elevating the specific activity of POD and CAT defense-related enzymes. The combination of the prestorage HW treatment and chitosan coating maintained higher values of flesh hue angle (h°), vitamin C content, membrane stability index (MSI) percentage, as well as lower weight loss compared with the untreated mango fruits. Fruits treated with HW ripened the fastest, comparable to the control fruits and the other treatments. However, a combination of the HW and chitosan treatments slowed down the fruits' ripening. As a result, combining HW and the chitosan coating improved the effects of each treatment alone. This application might be a promising technology and a novel strategy for controlling fruit decay, thus maintaining mango fruit quality during the storage period.

Author Contributions: Conceptualization, M.F.M.A., S.O.O., I.E.S., H.-E.H.O. and Z.A.A.; Data curation, H.A.K., D.O.E.-A., F.K.M.S., S.O.O., I.E.S., S.A.L., M.A.A. and Z.A.A.; Formal analysis, S.A.L., M.A.A. and Z.A.A.; Funding acquisition, I.E.S. and H.-E.H.O.; Investigation, H.A.K.; Methodology, H.A.K., M.F.M.A., D.O.E.-A., F.K.M.S., S.O.O. and Z.A.A.; Project administration, D.O.E.-A. and F.K.M.S.; Resources, M.F.M.A., S.A.L. and Z.A.A.; Software, A.A.L., F.K.M.S., S.O.O., M.A.A. and Z.A.A.; Supervision, A.A.L., H.-E.H.O. and M.A.A.; Visualization, H.-E.H.O. and M.A.A.; Writing—original draft, M.F.M.A. and D.O.E.-A. All authors have read and agreed to the published version of the manuscript.

Funding: The authors extend their appreciation to Taif University for supporting this work, the Researchers Supporting Project under Project No. (TURSP-2020/116), Taif University, Taif, Saudi Arabia.

Institutional Review Board Statement: Not applicable.

Informed Consent Statement: Not applicable.

Data Availability Statement: Relevant data applicable to this research are within the paper.

Acknowledgments: The authors extend their appreciation to Taif University for supporting current work by Taif University Researchers Supporting Project number (TURSP-2020/116), Taif University, Taif, Saudi Arabia.

Conflicts of Interest: The authors declare no conflict of interest.

References

1. Zeng, K.; Cao, J.; Jiang, W. Enhancing disease resistance in harvested mango (*Mangifera indica* L. cv.'Matisu') fruit by salicylic acid. *J. Sci. Food Agric.* **2006**, *86*, 694–698. [CrossRef]
2. Noomhorm, A.; Tiasuwan, N. Controlled atmosphere storage of mango fruit, *Mangifera indica* L. cv. Rad. *J. Food Process Preserv.* **1995**, *19*, 271–281. [CrossRef]
3. Onyeani, C.; Osunlaja, S. Comparative effect of Nigerian indigenous plants in the control of anthracnose disease of mango fruits. *Int. J. Sci. Technol. Res.* **2012**, *1*, 80–85.
4. Usall, J.; Ippolito, A.; Sisquella, M.; Neri, F. Physical treatments to control postharvest diseases of fresh fruits and vegetables. *Postharvest Biol. Technol.* **2016**, *122*, 30–40. [CrossRef]
5. Hasan, M.U.; Malik, A.U.; Khan, A.S.; Anwar, R.; Muhammad, L.; Amjad, A.; Shah, M.S.; Amin, M. Impact of postharvest hot water treatment on two commercial mango cultivars of Pakistan under simulated air freight conditions for China. *Pak. J. Agric. Res. Sci.* **2020**, *57*, 1381–1391.
6. Dessalegn, Y.; Ayalew, A.; Woldetsadik, K. Integrating plant defense inducing chemical, inorganic salt and hot water treatments for the management of postharvest mango anthracnose. *Postharvest Biol. Technol.* **2013**, *85*, 83–88. [CrossRef]
7. Shiesh, C.; Lin, H. Effect of vapor heat and hot water treatments on disease incidence and quality of Taiwan native strain mango fruits. *Int. J. Agric. Biol.* **2010**, *12*, 673–678.
8. Mansour, F.; Abd-El-Aziz, S.; Helal, G. Effect of fruit heat treatment in three mango varieties on incidence of postharvest fungal disease. *J. Plant Pathol.* **2006**, *88*, 141–148. Available online: https://www.jstor.org/stable/41998304 (accessed on 1 July 2006).
9. Huan, C.; Han, S.; Jiang, L.; An, X.; Yu, M.; Xu, Y.; Ma, R.; Yu, Z. Postharvest hot air and hot water treatments affect the antioxidant system in peach fruit during refrigerated storage. *Postharvest Biol. Technol.* **2017**, *126*, 1–14. [CrossRef]
10. Kim, Y.; Brecht, J.K.; Talcott, S.T. Antioxidant phytochemical and fruit quality changes in mango (*Mangifera indica* L.) following hot water immersion and controlled atmosphere storage. *Food Chem.* **2007**, *105*, 1327–1334. [CrossRef]
11. Kumar, N.; Neeraj. Polysaccharide-based component and their relevance in edible film/coating: A review. *Nutr. Food Sci.* **2019**, *49*, 793–823. [CrossRef]
12. El-Mohamedy, R.; El-Gamal, N.G.; Bakeer, A. Application of chitosan and essential oils as alternatives fungicides to control green and blue moulds of citrus fruits. *Int. J. Curr. Microbiol. Appl. Sci.* **2015**, *4*, 629–643.
13. Sivakumar, D.; Bautista-Baños, S. A review on the use of essential oils for postharvest decay control and maintenance of fruit quality during storage. *Crop Prot.* **2014**, *64*, 27–37. [CrossRef]
14. Bautista-Baños, S.; Sivakumar, D.; Bello-Pérez, A.; Villanueva-Arce, R.; Hernández-López, M. A review of the management alternatives for controlling fungi on papaya fruit during the postharvest supply chain. *Crop Prot.* **2013**, *49*, 8–20. [CrossRef]
15. Bambalele, N.L.; Mditshwa, A.; Magwaza, L.S.; Tesfay, S.Z. The Effect of Gaseous Ozone and Moringa Leaf–Carboxymethyl Cellulose Edible Coating on Antioxidant Activity and Biochemical Properties of 'Keitt' Mango Fruit. *Coatings* **2021**, *11*, 1406. [CrossRef]
16. Daisy, L.L.; Nduko, J.M.; Joseph, W.M.; Richard, S.M. Effect of edible gum Arabic coating on the shelf life and quality of mangoes (*Mangifera indica*) during storage. *J. Food Sci. Technol.* **2020**, *57*, 79–85. [CrossRef]
17. Moalemiyan, M.; Ramaswamy, H.S.; Maftoonazad, N. Pectin-based edible coating for shelf-life extension of ataulfo mango. *J. Food Process Eng.* **2012**, *35*, 572–600. [CrossRef]
18. Prabaharan, M.; Mano, J. Chitosan derivatives bearing cyclodextrin cavitiesas novel adsorbent matrices. *Carbohydr. Polym.* **2006**, *63*, 153–166. [CrossRef]
19. Arvanitoyannis, I.S. Totally and partially biodegradable polymer blends based on natural and synthetic macromolecules: Preparation, physical properties, and potential as food packaging materials. *J. Macromol. Sci.* **1999**, *39*, 205–271. [CrossRef]
20. Arvanitoyannis, I.S.; Nakayama, A.; Aiba, S.-I. Chitosan and gelatin based edible films: State diagrams, mechanical and permeation properties. *Carbohydr. Polym.* **1998**, *37*, 371–382. [CrossRef]
21. Kumar, N.; Petkoska, A.T.; AL-Hilifi, S.A.; Fawole, O.A. Effect of Chitosan–Pullulan Composite Edible Coating Functionalized with Pomegranate Peel Extract on the Shelf Life of Mango (*Mangifera indica*). *Coatings* **2021**, *11*, 764. [CrossRef]

22. Basumatary, I.B.; Mukherjee, A.; Katiyar, V.; Kumar, S.; Dutta, J. Chitosan-Based Antimicrobial Coating for Improving Postharvest Shelf Life of Pineapple. *Coatings* **2021**, *11*, 1366. [CrossRef]
23. Chien, P.-J.; Sheu, F.; Yang, F.-H. Effects of edible chitosan coating on quality and shelf life of sliced mango fruit. *J. Food Eng.* **2007**, *78*, 225–229. [CrossRef]
24. Ban, Z.; Wei, W.; Yang, X.; Feng, J.; Guan, J.; Li, L. Combination of heat treatment and chitosan coating to improve postharvest quality of wolfberry (*Lycium barbarum*). *Int. J. Food Sci.* **2015**, *50*, 1019–1025. [CrossRef]
25. Djioua, T.; Charles, F.; Freire, M., Jr.; Filgueiras, H.; Ducamp-Collin, M.N.; Sallanon, H. Combined effects of postharvest heat treatment and chitosan coating on quality of fresh-cut mangoes (*Mangifera indica* L.). *Int. J. Food Sci.* **2010**, *45*, 849–855. [CrossRef]
26. Chailoo, M.J.; Asghari, M.R. Hot water and chitosan treatment for the control of postharvest decay in sweet cherry (*Prunus avium* L.) cv. Napoleon (Napolyon). *J. Stored Prod. Postharvest Res.* **2011**, *2*, 135–138.
27. Vilaplana, R.; Chicaiza, G.; Vaca, C.; Valencia-Chamorro, S. Combination of hot water treatment and chitosan coating to control anthracnose in papaya (*Carica papaya* L.) during the postharvest period. *Crop Prot.* **2020**, *128*, 105007. [CrossRef]
28. Nguyen, H.T.; Boonyaritthongchai, P.; Buanong, M.; Supapvanich, S.; Wongs-Aree, C. Postharvest Hot Water Treatment Followed by Chitosan-and κ-Carrageenan-Based Composite Coating Induces the Disease Resistance and Preserves the Quality in Dragon Fruit (*Hylocereus undatus*). *Int. J. Fruit Sci.* **2020**, *20*, S2030–S2044. [CrossRef]
29. Chen, P.; PM, C.; Mellenthin, W.M. Effects of harvest date on ripening capacity and postharvest life of "D'Anjou" pears. *J. Am. Soc. Hortic. Sci.* **1981**, *106*, 38–42.
30. Ranganna, S. *Manual of Analysis of Fruit and Vegetable Products*; Tata McGraw-Hill Education: New York, NY, USA, 1977; p. 634.
31. McGuire, R.G. Reporting of objective color measurements. *HortScience* **1992**, *27*, 1254–1255. [CrossRef]
32. Voss, D.H. Relating colorimeter measurement of plant color to the Royal Horticultural Society Colour Chart. *HortScience* **1992**, *27*, 1256–1260. [CrossRef]
33. Ebiloma, U.; Arogba, S.; Aminu, O. Some activities of peroxydase from mango (*Mangifera indica* L. Var. Mapulehu) kernel. *Int. J. Biol. Chem.* **2011**, *5*, 200–206. [CrossRef]
34. Beers, R.F.; Sizer, I.W. A spectrophotometric method for measuring the breakdown of hydrogen peroxide by catalase. *J. Biol. Chem.* **1952**, *195*, 133–140. [CrossRef]
35. Sairam, R.K.; Rao, K.V.; Srivastava, G. Differential response of wheat genotypes to long term salinity stress in relation to oxidative stress, antioxidant activity and osmolyte concentration. *Plant Sci.* **2002**, *163*, 1037–1046. [CrossRef]
36. Snedecor, G.; Cochran, W. *Statistical Methods*, 6th ed.; Iowa State University Press: Ames, IA, USA, 1990; p. 507.
37. Lo'ay, A.A.; Dawood, H. Minimize browning incidence of banana by postharvest active chitosan/PVA Combines with oxalic acid treatment to during shelf-life. *Sci. Hortic.* **2017**, *226*, 208–215. [CrossRef]
38. Lo'ay, A.A.; Dawood, H. Active chitosan/PVA with ascorbic acid and berry quality of 'Superior seedless' grapes. *Sci. Hortic.* **2017**, *224*, 286–292. [CrossRef]
39. Lo'ay, A.A.; Taher, M.A. Influence of edible coatings chitosan/PVP blending with salicylic acid on biochemical fruit skin browning incidence and shelf life of guava fruits cv. 'Banati'. *Scientia Horticulturae* **2018**, *235*, 424–436. [CrossRef]
40. Lo'ay, A.A.; El-Khateeb, A. Impact of chitosan/PVA with salicylic acid, cell wall degrading enzyme activities and berries shattering of 'Thompson seedless' grape vines during shelf life. *Sci. Hortic.* **2018**, *238*, 281–287. [CrossRef]
41. El Hadrami, A.; Adam, L.R.; El Hadrami, I.; Daayf, F. Chitosan in plant protection. *Mar. Drugs* **2010**, *8*, 968–987. [CrossRef]
42. Zeng, K.; Deng, Y.; Ming, J.; Deng, L. Induction of disease resistance and ROS metabolism in navel oranges by chitosan. *Sci. Hortic.* **2010**, *126*, 223–228. [CrossRef]
43. Vilaplana, R.; Hurtado, G.; Valencia-Chamorro, S. Hot water dips elicit disease resistance against anthracnose caused by Colletotrichum musae in organic bananas (*Musa acuminata*). *LWT-Food Sci. Technol.* **2018**, *95*, 247–254. [CrossRef]
44. Wijeratnam, R.W.; Hewajulige, I.; Abeyratne, N. Postharvest hot water treatment for the control of Thielaviopsis black rot of pineapple. *Postharvest Biol. Technol.* **2005**, *36*, 323–327. [CrossRef]
45. Lo'ay, A.A.; Mostafa, N.A.; Al-Qahtani, S.M.; Al-Harbi, N.A.; Hassan, S.; Abdein, M.A. Influence of the Position of Mango Fruit on the Tree (*Mangifera indica* L. CV. 'Zibda') on Chilling Sensitivity and Antioxidant Enzyme Activity. *Horticulturae* **2021**, *7*, 515. [CrossRef]
46. Lo'ay, A.A.; Taher, M.A. Effectiveness salicylic acid blending in chitosan/PVP biopolymer coating on antioxidant enzyme activities under low storage temperature stress of 'Banati' guava fruit. *Sci. Hortic.* **2018**, *238*, 343–349. [CrossRef]
47. Zhu, X.; Wang, Q.; Cao, J.; Jiang, W. Effects of chitosan coating on postharvest quality of mango (*Mangifera indica* L. cv. Tainong) fruits. *J. Food Process. Preserv.* **2008**, *32*, 770–784. [CrossRef]
48. Yin, C.; Huang, C.; Wang, J.; Liu, Y.; Lu, P.; Huang, L. Effect of chitosan-and alginate-based coatings enriched with cinnamon essential oil microcapsules to improve the postharvest quality of mangoes. *Materials* **2019**, *12*, 2039. [CrossRef]
49. Yu, K.; Xu, J.; Zhou, L.; Zou, L.; Liu, W. Effect of Chitosan Coatings with Cinnamon Essential Oil on Postharvest Quality of Mangoes. *Foods* **2021**, *10*, 3003. [CrossRef]
50. Muy Rangel, D.; Espinoza Valenzuela, B.; Siller Cepeda, J.; Sañudo Barajas, J.A.; Valdez Torres, B.; Osuna Enciso, T. Efecto del 1-metilciclopropeno (1-MCP) y de una película comestible sobre la actividad enzimática y calidad poscosecha del mango 'Ataulfo'. *Fitotec. Mex.* **2009**, *32*, 53–60. [CrossRef]
51. Kerch, G.; Korkhov, V. Effect of storage time and temperature on structure, mechanical and barrier properties of chitosan-based films. *Eur. Food Res. Technol.* **2011**, *232*, 17–22. [CrossRef]

52. Qiuping, Z.; Wenshui, X. Effect of 1-methylcyclopropene and/or chitosan coating treatments on storage life and quality maintenance of Indian jujube fruit. *LWT-Food Sci. Technol.* **2007**, *40*, 404–411. [CrossRef]
53. Fawaz, S.A. Physiological studies on mango fruits handling. *FAO* **2000**, *33*, 137–149.
54. Fallik, E.; Lurie, S. Thermal control of fungi in the reduction of postharvest decay. In *Heat Treatment for Postharvest Pest Control: Theory and Practice*; CABI: Wallingford, UK, 2007. [CrossRef]
55. Kittur, F.; Saroja, N.; Tharanathan, R. Polysaccharide-based composite coating formulations for shelf-life extension of fresh banana and mango. *Eur. Food Res. Technol.* **2001**, *213*, 306–311. [CrossRef]
56. Hojo, R.H.; São José, A.R.; Hojo, E.T.D.; Alves, J.F.T.; Rebouças, T.N.H.; Dias, N.O. Quality of 'tommy atkins' mangoes in post-harvest with calcium choride spray use in the pre-harvest period. *Rev. Bras. Frutic.* **2009**, *31*, 62–70. [CrossRef]
57. Jongsri, P.; Wangsomboondee, T.; Rojsitthisak, P.; Seraypheap, K. Effect of molecular weights of chitosan coating on postharvest quality and physicochemical characteristics of mango fruit. *LWT-Food Sci. Technol.* **2016**, *73*, 28–36. [CrossRef]
58. Amin, U.; Khan, M.K.I.; Khan, M.U.; Ehtasham Akram, M.; Pateiro, M.; Lorenzo, J.M.; Maan, A.A. Improvement of the Performance of Chitosan—Aloe vera Coatings by Adding Beeswax on Postharvest Quality of Mango Fruit. *Foods* **2021**, *10*, 2240. [CrossRef] [PubMed]
59. Wang, C. Postharvest techniques for reducing low temperature injury in chilling sensitive commodities. In *Improving Postharvest Technologies of Fruits Vegetables and Ornamentals*; Artes, F., Gil, M.I., Conesa, M.A., Eds.; U.S. Department of Agriculture: Washington, DC, USA, 2000.
60. Wang, J.; Wang, B.; Jiang, W.; Zhao, Y. Quality and shelf life of mango (*Mangifera Indica* L. cv. 'Tainong') coated by using chitosan and polyphenols. *Food Sci. Technol. Int.* **2007**, *13*, 317–322. [CrossRef]
61. Medeiros, B.G.d.S.; Pinheiro, A.C.; Carneiro-da-Cunha, M.G.; Vicente, A.A. Development and characterization of a nanomultilayer coating of pectin and chitosan–Evaluation of its gas barrier properties and application on 'Tommy Atkins' mangoes. *J. Food Eng.* **2012**, *110*, 457–464. [CrossRef]
62. Cissé, M.; Polidori, J.; Montet, D.; Loiseau, G.; Ducamp-Collin, M.N. Preservation of mango quality by using functional chitosan-lactoperoxidase systems coatings. *Postharvest Biol. Technol.* **2015**, *101*, 10–14. [CrossRef]
63. Goulao, L.F.; Oliveira, C.M. Cell wall modifications during fruit ripening: When a fruit is not the fruit. *Trends Food Sci. Technol.* **2008**, *19*, 4–25. [CrossRef]
64. Lo'ay, A.A.; Doaa, M. The potential of vine rootstocks impacts on 'Flame Seedless' bunches behavior under cold storage and antioxidant enzyme activity performance. *Sci. Hortic.* **2020**, *260*, 108844. [CrossRef]
65. El-Banna, M.; Lo'ay, A. Evaluation berries shattering phenomena of 'Flame seedless' vines grafted on different rootstocks during shelf life. *Sci. Hortic.* **2019**, *246*, 51–56. [CrossRef]
66. Lo'ay, A.A.; Rabie, M.M.; Alhaithloul, H.A.S.; Alghanem, S.M.S.; Ibrahim, A.M.; Abdein, M.A.; Abdelgawad, Z.A. On the Biochemical and Physiological Responses of 'Crimson Seedless' Grapes Coated with an Edible Composite of Pectin, Polyphenylene Alcohol, and Salicylic Acid. *Horticulturae* **2021**, *7*, 498. [CrossRef]
67. Lo'ay, A.A.; EL-Ezz, S.F.A. Performance of 'Flame seedless' grapevines grown on different rootstocks in response to soil salinity stress. *Sci. Hortic.* **2021**, *275*, 109704. [CrossRef]
68. Lo'ay, A.A.; Ameer, N. Performance of calcium nanoparticles blending with ascorbic acid and alleviation internal browning of 'Hindi Be-Sennara' mango fruit at a low temperature. *Sci. Hortic.* **2019**, *254*, 199–207. [CrossRef]

Article

Application of *Auricularia cornea* as a Pork Fat Replacement in Cooked Sausage

Yuan Fu [1,2], Long Zhang [2], Mengdi Cong [2], Kang Wan [2], Guochuan Jiang [2], Siqi Dai [2], Liyan Wang [1,2,*] and Xuejun Liu [2,*]

[1] Engineering Research Center of Edible and Medicinal Fungi, Ministry of Education, Jilin Agricultural University, 2888 Xincheng Street, Changchun 130118, China; 20201617@mails.jlau.edu.cn

[2] College of Food Science and Engineering, Jilin Agricultural University, 2888 Xincheng Street, Changchun 130118, China; 20200704@mails.jlau.edu.cn (L.Z.); congmengdi@mails.jlau.edu.cn (M.C.); wankang@jlau.edu.cn (K.W.); jiangguochuan@mails.jlau.edu.cn (G.J.); daisiqi1232021@163.com (S.D.)

* Correspondence: wangliyan@jlau.edu.cn (L.W.); liuxuejun@jlau.edu.cn (X.L.); Tel.: +86-158-4301-5766 (L.W.); +86-153-0446-0733 (X.L.)

Abstract: The effect of *Auricularia cornea* (AC) as an alternative for pork fat on the physico-chemical properties and sensory characteristics of cooked sausage were evaluated. The results indicated that replacement of pork fat with AC led to a significant increase in the protein, ash, moisture, cooking loss, water holding capacity, springiness, and chewiness, especially isoleucine, leucine, proline, palmitic, palmitoleic, oleic, and arachidonic acids of the sausages. In contrast, AC reduced the level of fat (12.61%–87.56%) and energy (5.76%–56.40%) of the sausages. In addition, AC led to the mild lightness, yellowness, whiteness, and soft texture, while it did not affect the water activity of the sausages. From the sensory point of view, all sausages were judged acceptable, and the substitution of 75% of pork fat by AC exhibited best sensory characteristics. In a word, AC is a promising food to partially replace the pork fat in sausages.

Keywords: mushroom; fat replacement; meat product

1. Introduction

The World Health Organization (WHO) and Drug Administration (FDA) recommended a reduction in ingestion of total fat and saturated fatty acids (SFAs) [1], since recent reports have shown that high intake of SFAs will increase the risk of several chronic diseases, such as obesity, hypertension, colon cancer, cardiovascular diseases, and coronary heart diseases [2]. However, the popular sausage products have a high-fat content (20%–30%) and characterized by a high proportion of SFAs [3]. Thus, it is necessary to develop healthier nutritional sausages with lower animal fat.

Some healthy ingredients have been used to replace animal fat in meat products, resulting in better nutritional properties [3]. *Camellia* oil gel [4], interesterified palm kernel oil [5], hydroxypropyl methylcellulose oleogel [6], hazelnut [7], pre-emulsified perilla-canola oil [8], and different vegetable oils have been used to partially replace animal fat in meat products [9]. Certainly, two materials also can be used to replace the fat, such as whey proteins and sodium dodecylsulfate [2], konjac gel with vegetable powders [10], double emulsions with olive leaves extract [1], cellulose nanofibers and its palm oil pickering emulsion [11], canola oil hydrogels and organogels [12], inulin-based emulsion-filled gel [13], and carboxymethyl cellulose and inulin [14]. Quite a few studies had successfully achieved healthier products, while some fat replacer not only impaired the texture, but also accelerated lipid oxidation reactions with reduction of shelf life and loss of sensorial and nutritional values [6]. New fat substitutes should be developed.

Auricularia cornea (AC), also called "Yu Mu Er" in Chinese, is a pure white strain of the variant of *Auricularia cornea* species, which has been artificially cultivated by the team of Yu Li in Jilin Agricultural University and commercialized in China [15]. It is crystal clear appearance, tender, delicious taste, and rich in nutrients, including polysaccharides, dietary fiber, amino acids, and various trace elements [16]. The polysaccharides extracted from AC have antioxidant, hepatoprotective, and anti-tumor activities [17]. Thus, it exhibits potential to be a fat replacer.

The objective of the present study was to evaluate the effect of AC as an alternative for pork fat on the proximate composition, water activity, pH, color, cooking loss, water holding capacity, textural profile, free amino acids, free fatty acids, and sensory characteristics of cooked sausage.

2. Materials and Methods

2.1. Materials

The pork lean meat (protein: 20.16%; fat: 2.57%; moisture: 71.11%; ash: 1.18%) and back fat (protein: 0.54%; fat: 91.01%; moisture: 7.26%; ash: 0.17%) were obtained from Jilin Huazheng Agriculture and Animal Husbandry Development Co. Ltd. (Changchun, China) and stored at $-20\ °C$ until use. Dry *Auricularia cornea* was supplied by Engineering Research Center of Chinese Ministry of Education for Edible and Medicinal Fungi. All spices (salt; sugar; wheat; carrageen; isolated soy protein; dry starch; ice) were obtained from the local market. All the additives and chemical reagents were purchased from Sichuan Jinshan Pharmaceutical Co. Ltd. (Guangyuan, China) and Beijing Beihua Co. Ltd. (Beijing, China), respectively.

2.2. Sausages Formulation and Processing

The lean pork meat was minced and the back fat was cut into approximately 1.5 cm cubes. Dry *Auricularia cornea* was immersed in water at $40\ °C$ for 30 min, then cut into 0.5 cm slices. The sausages were prepared using the formulations that are shown in Table 1. Differently, the control sausage formulation was made with pork fat, and the other four formulations (AC25, AC50, AC75, and AC100) were prepared with substitution of 25%, 50%, 75%, and 100% of pork fat by AC, respectively. Each sausage (Control, AC25, AC50, AC75, and AC100) was added according to the formula and the ingredients were homogenized by the blender (German K + G wetter touch screen) for 140 s. The sausage was roasted at $68\ °C$ for 30 min, steamed at $80\ °C$ for 50 min, and smoked at $50\ °C$ for 150 min. The three processes were completed by an electric heating flue gas furnace (Zhucheng Yizhong Machinery Co., LTD., Weifang, China). After cooling, the samples were individually packaged in polyethylene bags using a vacuum packaging machine (Zhucheng Yizhong Machinery Co., LTD., Weifang, China) and stored at $4 \pm 1\ °C$ for subsequent analysis.

Table 1. Formulations of sausages with alternative of pork fat at *Auricularia cornea* (AC).

Ingredients (%)	Control	AC25	AC50	AC75	AC100
Pork lean meat	49	49	49	49	49
Pork back fat	21	15.75	10.5	5.25	0
AC	0	5.25	10.5	15.75	21
Salt	1.5	1.5	1.5	1.5	1.5
Sugar	1	1	1	1	1
White pepper	0.2	0.2	0.2	0.2	0.2
Carrageenan	0.3	0.3	0.3	0.3	0.3
Isolated soy protein	2.8	2.8	2.8	2.8	2.8
Dry starch	4.2	4.2	4.2	4.2	4.2
Ice	20	20	20	20	20
Total	100	100	100	100	100

Notes: Control (0%), AC25 (25%), AC50 (50%), AC75 (75%), AC100 (100%) were substitution of pork fat by AC, respectively.

2.3. Proximate Composition and Energy Value

The proximate composition of cooked sausages was detected foundation the Association of Official Analytical Chemists (AOAC, 2005). Energy value was calculated based on 9 kcal/g for fat, 4 kcal/g for protein and carbohydrate [18].

2.4. Water Activity and pH

The water activity of sausages was measured by a water activity meter (Rotronic, Bassersdorf, Switzerland). The pH of sausage was determined on pH meter (Mettler Toledo, Columbus, OH, USA).

2.5. Color

The color parameters of different sausages were determined by a previously described method [19].

2.6. Cooking Loss and Water Holding Capacity

Cooking loss of sausages was conducted according to [11]. Raw sausage samples (50 g) were cooked at 80 °C for 50 min, then calculated as follows:

$$\text{Cooking loss (\%)} = \left[\frac{m1 - m2}{m1}\right] \times 100\% \quad (1)$$

where m1 is the weight of raw sausages and m2 is the weight of cooked sausages.

The water holding capacity (WHC) was detected according to [20]. Sausage samples were respectively centrifuged for 30 min at 12,000× g centrifugal force, then calculated as follows:

$$\text{WHC} = (W2/W1) \times 100\% \quad (2)$$

where W1 is the weight of the sample before centrifugation and W2 is the weight of the sample after centrifugation.

2.7. Textural Profile Analysis

TPA was performed by using a texture analyzer (TMA-Pro, FTC, Washington, DC, USA) according to [11], with some slight modifications. Sausages were cut into 1.0 cm height and 1.5 cm diameter. The detection speed was 60 mm/min and minimum force 0.8 N. The characteristics of sausage were hardness, cohesiveness, springiness, gumminess, and chewiness. All samples were performed in triplicate at room temperature, and the average value was taken.

2.8. Free Amino Acids

The amino acids content was detected by the method according to [21], with slight modifications. The sample (0.2 g) was hydrolyzed by 6 mol/L HCl (10 mL) for 24 h at 110 °C. After cooling at room temperature, a hydrolyzed sample was volumed to 50 mL. 10 mL from the 50 mL hydrolysate was taken and dried, the dried sample adding 0.1 mol/L HCl solution to 10 mL. The solution was filtered through a 0.22 μm water membrane. After filtration, the filtrate (500 μL) and internal standard solution (50 μL) were mixed and derived. The derivative solution (2 μL) was injected into Liquid Chromatography (20AT-PDA (Diode Array Detector) Detector, Shimazu, Kyoto, Japan). The measurement conditions were as follows: chromatographic column (C18: Ajs-01 amino acid special analytical column, 3 μm, 4.6 mm × 150 mm), detection wavelength, 338 nm; column temperature, 50 °C. Elution gradient and flow rate are as follows: time: 0 s, 6 s, 8 s, 10 s, 23 s, 30 s,31 s, 34 s, 35 s, 35 s, 38 s; mobile phase B%: 5, 10, 10, 16, 40, 50, 100, 100, 55; flow rate (mL/min): 1.6, 1.6, 1.6, 1.3, 1.0, 1.6, 1.6, 1.6, 1.6.

2.9. Free Fatty Acids

Free fatty acid was detected by a method previously published according to [22], with slight modifications. Determination of fatty acid methyl esters (FAME) was performed using a GC/MS system equipped with a GC-7 Agilent HP-88 capillary column (60 m × 0.25 mm × 0.2 µm). The temperature profile of the oven was 100 °C for 13 min followed by increasing at 10 °C/min to 180 °C for 6 min, then increasing at 1 °C/min to 200 °C for 20 min, and then increasing at 4 °C/min to 230 °C for 10.5 min. Injector and detector temperatures were set to 270 and 280 °C, respectively. The conditions applied for gas chromatography were nitrogen as the carrier gas at a flow of 1.0 µL/min. Fatty acids were identified and quantified based on chromatographic retention times using reference standard Supelco 37 component FAME mix (Sigma Aldrich Chemical Co., St. Louis, MO, USA). Results obtained were presented as percentage of total fatty acids.

2.10. Sensory Evaluation

A thirty-member panel from the Departments of College of Food Science and Engineering at Jilin Agricultural University have evaluated the sensory attributes of samples. Sensorial analysis has used the method described by [23]. with some modifications. Unsalted crackers and water were offered to clean the palates and remove residual flavors between samples. The samples were reheated using a microwave oven, sliced to 2 cm thick, placed on plates and provided to panelists. The appearance, odor, taste, texture, and overall acceptability of the sausage samples were evaluated with a 10-hedonic scale from 1 (dislike extremely) to 10 (like extremely).

2.11. Statistical Analysis

One-way ANOVA was used to determine statistical significance in the data compared the sausages on the proximate composition, water activity, pH, color, cooking loss, water holding capacity, textural profile, free amino acids, free fatty acids, and sensory characteristics of cooked sausage. Duncan multiple range test was carried out to compare the mean between the two groups to determine which groups had significant differences compared with other groups ($p < 0.05$). All data were expressed as mean ± standard error.

3. Results

3.1. Proximate Composition and Energy Value

The results of proximate composition and energy value analysis of sausages are listed in Table 2. Significant differences ($p < 0.05$) were observed among the protein, fat, moisture, ash contents, carbohydrate, and energy value between control and fat replacement group (AC25, AC50, AC75, and AC100). The control shown the lowest protein content, close to the 12.73% protein content in model sausages as reported by [24], but lower than the protein content (13.77%) in frankfurters as described by [25]. As the proportion of AC increased, the protein content in the sausages increased by 1.42%–14.31%. These changes were attributed to the protein in AC.

The fat levels in the replacement sausages significantly ($p < 0.05$) decreased by 12.61%–87.56%, compared with those of the control. The fat content of AC100 was similar to those of the sausage with fried *Pleurotus eryngii* as fat replacements [11], but lower than those of frankfurters with porcine plasma protein hydrolysates and oxidized tannic acid to partially replace pork fat in [25]. It was because the replacement AC is extremely low in fat. The moisture levels of the sausages were significantly ($p < 0.05$) affected by the increase of AC concentration. Compared to the control, the moisture contents of AC25, AC50, AC75, and AC100 increased by 1.62%, 16.74%, 26.08%, and 31.64%, respectively. The reason was that the moisture content of the AC was higher after water treatment.

In terms of ash content, the replacement group was notably ($p < 0.05$) higher than the control. The ash content of sausages was from 3.12% to 3.57%. The consequences were approximate to the value in frankfurters with phenolic compounds in emulsion gel-based delivery systems as animal fat replacers, which was reported by [18], but higher than those in Bologna sausage and Toscana sausage with partial substitution of pork fat with canola oil, as determined by [3] and [26], respectively.

Table 2. Proximate composition (%), energy value (kcal/100 g of the product), water activity, pH, cooking loss and water holding capacity of sausages with replacement of pork fat by *Auricularia cornea* (AC).

Parameters	Control	AC25	AC50	AC75	AC100
Protein	12.65 ± 0.14 [a]	12.83 ± 0.07 [a]	13.16 ± 0.21 [b]	13.47 ± 0.02 [b]	14.46 ± 0.02 [c]
Fat	17.77 ± 0.65 [e]	15.53 ± 0.05 [d]	8.34 ± 0.05 [c]	5.47 ± 0.04 [b]	2.21 ± 0.27 [a]
Moisturer	53.69 ± 0.09 [a]	54.56 ± 0.06 [b]	62.68 ± 0.08 [c]	67.69 ± 0.04 [d]	70.68 ± 0.08 [e]
Ash	3.12 ± 0.01 [a]	3.22 ± 0.01 [b]	3.53 ± 0.03 [c]	3.55 ± 0.05 [c]	3.57 ± 0.06 [c]
Carbohydrate	12.77 ± 0.09 [b]	13.86 ± 0.05 [a]	12.29 ± 0.01 [c]	9.82 ± 0.01 [d]	9.08 ± 0.02 [e]
Energy value	261.61	246.53	176.86	142.39	114.05
Water activity	0.99 ± 0.01 [a]	0.98 ± 0.00 [a]	0.98 ± 0.00 [a]	0.98 ± 0.00 [a]	0.98 ± 0.00 [a]
pH	6.33 ± 0.03 [b]	6.30 ± 0.01 [b]	6.23 ± 0.03 [a]	6.21 ± 0.02 [a]	6.20 ± 0.01 [a]
Cooking loss	5.13 ± 0.12 [a]	11.73 ± 0.07 [b]	12.79 ± 0.08 [c]	13.69 ± 0.08 [d]	14.68 ± 0.09 [e]
Water holding capacity	82.08 ± 0.1 [a]	83.27 ± 0.01 [b]	86.45 ± 0.24 [c]	88.49 ± 0.36 [d]	91.16 ± 0.51 [e]

Notes: Values are given as mean ± standard error. Different letters in the same row indicate significant differences ($p < 0.05$). Control (0%), AC25 (25%), AC50 (50%), AC75 (75%), AC100 (100%) substitution of pork fat by AC, respectively.

Energy values of the control (261.61 kcal/100 g) was significantly ($p < 0.05$) higher than that of replacement groups, decreasing by 5.76%–56.40%. The energy value of the AC50 were similar with those obtained by [11]—that is 171.8 kcal/100 g of the pork sausages using deep-fried *Pleaurotus eryngii* as replacements for pork fat. The energy level in AC75 and AC100 sausages were lower than those found by [18]—that is 174–196 kcal/100 g of frankfurters with phenolic compounds in emulsion gel-based delivery systems as animal fat replacers, and [27]—that is 146.24–198.47 kcal/100 g of low-fat burger with high beta-glucans content and oat-hull-based ingredient as a fat replacer.

3.2. Water Activity and pH

The water activity and pH of cooked sausage samples are shown in Table 2. There were no significant differences ($p > 0.05$) in water activity levels among all the samples, suggesting that AC did not affect the bound water content in the sausages. The consequences were similar to the values in hot-dog style sausages with pork skin-based emulsion gels as animal fat replacers. The authors of [28] also reported that hydrolyzed collagen or pork skin/green banana flour gel used as a fat replacer in frankfurter-type sausages or Bologna sausages did not influence the water activity.

As the proportion of AC increased, the pH of sausages decreased by 0.47%–2.05%. These changes were put down to the lower pH of AC. The pH of the control was approximate to that obtained by [29], and no significant differences ($p > 0.05$) were found among the AC50, AC75, and AC100. This decline trend was approximate to the trend observed by [24] in the sausage with *Lentinula edodes* as the pork lean meat replacer, [11] in pork back sausages with fat replaced by fried *Pleaurotus eryngii*, and [29] in frankfurter-type sausages with hydrolyzed collagen as replacements for fat.

3.3. Color

The influence of AC on the color of sausages are shown in Table 3. L*, b*, ΔE*, and whiteness value of the sausages notably ($p < 0.05$) improved with an increase in AC. These changes exhibited that color of the sausages became slight lightness, yellowness, and whiteness, which was put down to the larger amounts of AC and smaller amount of pork fat in sausages. As the proportion of AC increased, a* value decreased significantly ($p < 0.05$), suggesting that the redness of the replacement group was less than control due to the AC with a white crystal-clear appearance. Similar observations in L* and b* were found by [11], who used boiled *Pleaurotus eryngii* to substitute the pork back fat in sausages.

Table 3. Color and textural profile of sausages with substitution of pork fat by *Auricularia cornea* (AC).

Parameters	Control	AC25	AC50	AC75	AC100
L*	54.37 ± 1.80 [a]	55.43 ± 1.13 [a,b]	56.44 ± 0.30 [b]	57.43 ± 0.67 [b]	59.68 ± 0.84 [c]
A*	16.06 ± 1.74 [b]	15.25 ± 0.37 [b]	14.28 ± 1.21 [a,b]	14.31 ± 0.40 [a,b]	13.23 ± 0.45 [a]
B*	15.08 ± 0.36 [a]	16.91 ± 0.62 [b]	17.33 ± 0.84 [b,c]	17.03 ± 0.26 [b,c]	17.98 ± 0.42 [c]
ΔE*	58.66 ± 2.24 [a]	59.92 ± 1.42 [a,b]	60.74 ± 0.68 [a,b]	61.58 ± 0.28 [b,c]	63.71 ± 0.64 [c]
Whiteness	49.33 ± 2.15 [a]	49.95 ± 1.47 [a,b]	50.99 ± 0.70 [a,b]	51.97 ± 0.31 [b,c]	53.91 ± 0.52 [c]
Hardness (N)	155.63 ± 11.14 [b]	141.17 ± 5.97 [b]	107.23 ± 8.15 [a]	102.70 ± 6.48 [a]	102.86 ± 4.36 [a]
Springiness (mm)	3.50 ± 0.22 [a]	3.58 ± 0.36 [a,b]	3.63 ± 0.59 [a,b]	3.88 ± 0.22 [a,b]	4.23 ± 0.42 [b]
Gumminess (N)	85.23 ± 13.31 [c]	54.33 ± 7.26 [b]	28.50 ± 8.94 [a]	25.67 ± 4.48 [a]	23.43 ± 5.87 [a]
Chewiness (N)	248.49 ± 8.60 [a]	253.13 ± 7.85 [a]	255.63 ± 7.09 [a]	262.12 ± 6.92 [b]	268.59 ± 8.94 [b]
Cohesiveness	0.55 ± 0.03 [c]	0.39 ± 0.07 [b]	0.27 ± 0.03 [a]	0.26 ± 0.01 [a]	0.23 ± 0.06 [a]

Notes: Values are given as mean ± standard error. Different letters in the same row indicate significant differences ($p < 0.05$). Control (0%), AC25 (25%), AC50 (50%), AC75 (75%), AC100 (100%) substitution of pork fat by AC, respectively.

3.4. Cooking Loss and Water Holding Capacity

Cooking loss is a crucial factor for juiciness and is associated with water or fat binding capacities during the heating process [25]. The cooking loss of sausages is presented in Table 2. The substitution of pork fat by AC exerted a significant effect ($p < 0.05$) on cooking loss (Figure 1). The sausages with low-fat replacer contents showed higher cooking loss. Cooking loss in of replacement group was 2.29–2.86 folds reference to that in the control. Because the AC found much more water than in pork fat. Another reason may be that the capacity of AC to retain water during cooking was decreased. AC100 showed the highest cooking loss, which was significantly higher than that of the control ($p < 0.05$). Similar observations were found by [11] who used boiled and deep-fried *Pleaurotus eryngii* instead of fat in pork sausage, and [24] who used *Lentinula edodes* as a pork lean meat replacer in sausage.

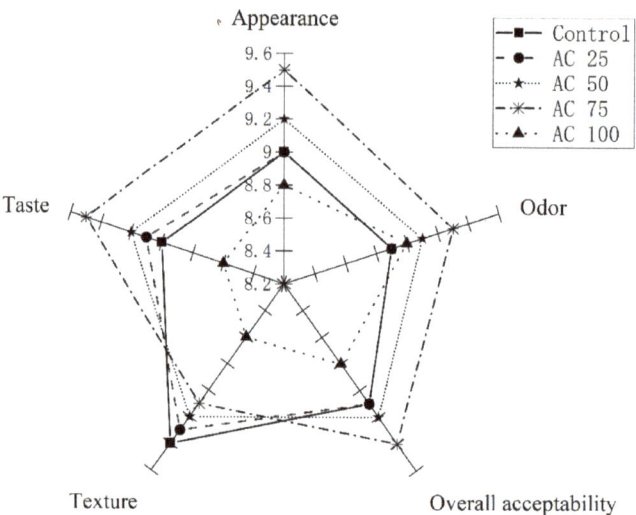

Figure 1. Sensory evaluation for the replacement of pork fat with AC in sausages.

The water holding capacity of the sausages is also shown in Table 4. The WHC increased as the percentage of AC increased. The WHC of the control was significant ($p < 0.05$) lower than that of all replacement groups. AC25, AC50, AC75, and AC100 were increased by 1.45%, 5.32%, 7.81%, and 11.06%, compared to the control, respectively. This phenomenon might be that the AC could help hold not only water but also fat to prevent their loss during cooking.

Table 4. Free amino acids (expressed as g/100 g of sausage) profile of sausages with substitution of pork fat by *Auricularia cornea* (AC).

Parameters	Control	AC25	AC50	AC75	AC100
		Essential			
Valine	1.61 ± 0.02 [e]	1.48 ± 0.05 [d]	1.43 ± 0.00 [c]	1.29 ± 0.01 [b]	0.98 ± 0.00 [a]
Threonine	0.67 ± 0.00 [e]	0.64 ± 0.00 [d]	0.60 ± 0.00 [c]	0.54 ± 0.01 [b]	0.47 ± 0.02 [a]
Lysine	1.41 ± 0.00 [d]	1.30 ± 0.06 [c]	1.30 ± 0.00 [c]	1.13 ± 0.00 [b]	0.95 ± 0.00 [a]
Methionine	0.16 ± 0.00 [c]	0.15 ± 0.00 [c]	0.12 ± 0.01 [b]	0.08 ± 0.01 [a]	0.10 ± 0.00 [a]
Isoleucine	0.53 ± 0.00 [a]	0.63 ± 0.01 [b]	0.70 ± 0.00 [c]	0.77 ± 0.01 [d]	0.79 ± 0.00 [e]
Leucine	0.93 ± 0.01 [a]	1.09 ± 0.00 [b]	1.24 ± 0.03 [c]	1.31 ± 0.01 [d]	1.38 ± 0.03 [e]
Phenylalanine	0.73 ± 0.01 [d]	0.73 ± 0.00 [d]	0.66 ± 0.00 [c]	0.60 ± 0.00 [b]	0.51 ± 0.03 [a]
Histidine	0.58 ± 0.02 [d]	0.53 ± 0.00 [cd]	0.51 ± 0.06 [bc]	0.46 ± 0.01 [ab]	0.45 ± 0.00 [a]
		Non-essential			
Serine	0.71 ± 0.00 [e]	0.64 ± 0.00 [d]	0.58 ± 0.00 [c]	0.56 ± 0.01 [b]	0.48 ± 0.01 [a]
Arginine	1.16 ± 0.04 [d]	1.03 ± 0.00 [c]	1.01 ± 0.03 [c]	0.89 ± 0.00 [b]	0.74 ± 0.00 [a]
Glycine	0.92 ± 0.00 [e]	0.84 ± 0.00 [d]	0.67 ± 0.00 [c]	0.66 ± 0.00 [b]	0.56 ± 0.00 [a]
Aspartic acid	1.64 ± 0.00 [d]	1.62 ± 0.01 [d]	1.53 ± 0.03 [c]	1.40 ± 0.06 [b]	1.15 ± 0.05 [a]
Glutamic acid	2.65 ± 0.03 [d]	2.59 ± 0.05 [d]	2.40 ± 0.05 [c]	2.17 ± 0.03 [b]	1.87 ± 0.01 [a]
Alanine	0.99 ± 0.00 [d]	0.91 ± 0.03 [c]	0.94 ± 0.03 [c]	0.77 ± 0.00 [b]	0.66 ± 0.00 [a]
Proline	0.88 ± 0.03 [a]	0.91 ± 0.03 [a]	0.95 ± 0.00 [b]	1.17 ± 0.02 [c]	1.26 ± 0.26 [d]
Cysteine	0.09 ± 0.00 [c]	0.08 ± 0.00 [b,c]	0.07 ± 0.01 [b]	0.07 ± 0.00 [b]	0.04 ± 0.01 [a]
Tyrosine	0.56 ± 0.01 [e]	0.49 ± 0.01 [c]	0.49 ± 0.00 [c]	0.42 ± 0.00 [b]	0.35 ± 0.00 [a]

Notes: Values are given as mean ± standard error. Different letters in the same row indicate significant differences ($p < 0.05$). Control (0%), AC25 (25%), AC50 (50%), AC75 (75%), AC100 (100%) substitution of pork fat by AC, respectively.

3.5. Textural Profile Analysis

TPA is an important indicator to evaluate the quality and acceptability of comminuted meat products, which is expressed as hardness, springiness, gumminess, chewiness, and cohesiveness [25]. The results of TPA of sausages is listed in Table 3, and significant differences ($p < 0.05$) were observed among the sausages. With the improve of AC content, the hardness, gumminess, and cohesiveness of sausages were decreased, but the springiness and chewiness were increased. Compared with the control, hardness, gumminess, and cohesiveness of replacement group decreased by 9.29%–33.91%, 36.25%–72.51%, and 29.09%–58.18%, indicating that these sausages had a softer texture. This downtrend was similar to low-fat burger with high beta-glucans and oat-hull-based ingredient as fat replacer [27]. By contrast, the springiness and chewiness of replacement group increased by 2.29%–20.86%, and 1.87%–8.09% in comparison with the control, suggesting that these sausages required more energy to be compressed and the work during chewing will be slightly greater. It could be attributed to the highest amount of AC. The moisture and fat content also could influence TPA parameters [30]. These results may be attributed to the fiber and the rich and elastic properties of AC. This result complies with the report by dos Santos Alves [31] where animal fat was replaced with pork skin-based emulsion gels. Similar observations were also made by [11] who reported that using *Pleaurotus eryngii* as replacements for pork back fat in sausages.

3.6. Free Amino Acids

Amino acids are important for the nutritional value and sensory properties of sausages [4]. The amino acids of sausages formulated with AC replaced pork fat are shown in Table 4. The main essential amino acids in the control group were valine, lysine, and leucine, and the main non-essential amino acids were glutamate, aspartic, and arginine. By contrast, the main essential amino acids in the replacement group were leucine, valine, and lysine, and the main non-essential amino acids were glutamate, proline, and aspartate. Compared with the control group, the contents of isoleucine, leucine, and proline in the replacement group increased by 18.87%–49.06%, 17.20%–48.39%, and 3.41%–43.18%, respectively. It may be due to the AC being abundant in these amino acids. Amino acids were linked with multiple health benefits. In the literature, isoleucine was a potent plasma glucose-lowering amino acid [32]. Zhang and Guo et al. [33] found that increasing dietary leucine intake could reduce diet-induced obesity and improve glucose and cholesterol metabolism in mice via multi mechanisms. Proline is critically essential for nutrition, antioxidative reactions, and immune responses [34]. In addition, the control was higher than other amino acids of replacement group. These results could be explained by the fact that the difference of amino acid content between AC and pork fat.

3.7. Free Fatty Acids

The results of free fatty acids in sausages with substitution of pork fat with AC are listed in Table 5. Significant differences ($p < 0.05$) were observed among the samples. The highest content of fatty acids in all the samples was oleic acid, followed by palmitic acid, stearic, and linoleic acid, indicating that replacement of pork back fat by AC did not change the major fatty acids. This aspect was similar to the previous study of [32] in Toscana sausage. With the increase of AC content, palmitic acid, palmitoleic acid, oleic acid, and arachidonic acid increased by 0.72%–5.10%, 10.71%–41.67%, 1.11%–20.04%, and 2.44%–75.61% in comparison with the control, respectively. It may be due to the AC being rich in these fatty acids. The palmitoleic acid could improve hyperglycemia and hypertriglyceridemia by increasing insulin sensitivity [35] showed that oleic acid plays a role in immunomodulation, and in treating and preventing cardiovascular or autoimmune diseases, metabolic disturbances, skin injury, and cancer. Arachidonic acid plays an important role in the brain, cognitive functions, skeletal muscle, and immune systems, which could also promotes and regulates type 2 immune responses against intestinal [36]. However, the control was higher than other fatty acids in the replacement group.

Table 5. Free fatty acids of sausages with substitution of pork fat by *Auricularia cornea* (AC).

Fatty Acid	Control	AC25	AC50	AC75	AC100
(C12:0)	0.08 ± 0.00 [b]	0.08 ± 0.00 [b]	0.08 ± 0.00 [b]	0.07 ± 0.00 [a,b]	0.06 ± 0.00 [a]
(C14:0)	1.37 ± 0.02 [d]	1.31 ± 0.01 [c]	1.26 ± 0.01 [b]	1.25 ± 0.00 [b]	1.23 ± 0.00 [a]
(C16:0)	23.72 ± 0.00 [a]	23.89 ± 0.01 [b]	24.15 ± 0.00 [c]	24.31 ± 0.00 [d]	24.93 ± 0.00 [e]
(C17:0)	0.32 ± 0.00 [d]	0.27 ± 0.00 [c]	0.28 ± 0.00 [c]	0.24 ± 0.01 [b]	0.20 ± 0.01 [a]
(C18:0)	15.73 ± 0.00 [e]	15.16 ± 0.00 [d]	14.69 ± 0.01 [c]	14.49 ± 0.00 [b]	11.81 ± 0.01 [a]
(C20:0)	0.26 ± 0.01 [d]	0.24 ± 0.00 [c]	0.23 ± 0.01 [b,c]	0.22 ± 0.00 [b]	0.16 ± 0.00 [a]
SFA	41.48 ± 0.01 [e]	40.95 ± 0.02 [d]	40.69 ± 0.01 [c]	40.58 ±0.01 [b]	38.39 ± 0.02 [a]
(C18:1)	38.78 ± 0.00 [a]	39.21 ± 0.01 [b]	39.46 ± 0.01 [c]	42.43 ± 0.00 [d]	46.55 ± 0.00 [e]
(C16:1)	1.68 ± 0.01 [a]	1.86 ± 0.00 [b]	1.90 ± 0.01 [c]	2.31 ± 0.00 [d]	2.38 ± 0.00 [e]
(C17:1)	0.25 ± 0.01 [d]	0.23 ± 0.01 [c]	0.23 ± 0.00 [c]	0.17 ± 0.00 [b]	0.15 ± 0.01 [a]
(C20:1)	1.71 ± 0.01 [c]	1.46 ± 0.01 [b]	1.46 ± 0.00 [b]	1.47 ± 0.00 [b]	0.92 ± 0.04 [a]
MUFA	42.42 ± 0.03 [a]	42.76 ± 0.03 [b]	43.05 ± 0.02 [c]	46.38 ± 0.00 [d]	50.00 ± 0.04 [e]
(C18:2)	15.60 ± 0.01 [e]	15.48 ± 0.01 [d]	14.39 ± 0.00 [c]	12.15 ± 0.00 [b]	9.76 ± 0.00 [a]
(C20:4)	0.41 ± 0.00 [a]	0.42 ± 0.01 [a]	0.47 ± 0.01 [b]	0.66 ± 0.00 [c]	0.72 ± 0.00 [d]
PUFA	16.01 ± 0.01 [e]	15.90 ± 0.00 [d]	14.86 ± 0.01 [c]	12.81 ± 0.00 [b]	10.48 ± 0.00 [a]

Notes: Values are given as mean ± standard error. Different letters in the same row indicate significant differences ($p < 0.05$). Control (0%), AC25 (25%), AC50 (50%), AC75 (75%), AC100 (100%) substitution of pork fat by AC, respectively. SFA: Saturated fatty acid; MUFA: Monounsaturated fatty acid; PUFA: Polyunsaturated fatty acid.

Excessive intake of Saturated fatty acid (SFA) will lead to the increase of fat content and cholesterol content in the body, which may be aggravate the risk of cardiovascular and cerebrovascular diseases [2]. The sausages incorporated with AC as a fat replacer showed less (1.28%–7.45%) SFA than the control. Additionally, Monounsaturated fatty acid (MUFA) of the replacement group increased by 0.80%–17.87% in comparison with the control. In general, AC is a healthier food that could reduce fat intake and be beneficial to human health.

3.8. Sensory Evaluation

The results of sensory evaluation for the replacement of pork fat with AC in sausages are shown in Figure 1. Replacing pork fat with 25% AC in sausages caused no significant ($p > 0.05$) change in sensory characteristics, indicating that a small amount of AC had no obvious effect on the sausage. With regard to appearance, odor, and taste, AC50 and AC75 showed a higher score, compared to the control. It may be due to AC having a unique flavor and reducing the greasy taste of sausages, which was very popular among the sensory evaluation team. This result is consistent with the report by [11] that the addition of fried *Pleaurotus eryngii* as a substitution for pork back fat-enriched these free amino acids and improved the taste and flavor properties of sausages. By contrast, AC100 got the lowest score in appearance, odor, and taste. Additionally, texture scores of replacement group were decreased significantly with the increase of AC, because AC made the sausage a soft texture. That may be characteristic of AC is different from pork fat. These results were in line with the TPA observed by instrumental measurements (Table 4). The highest score of overall acceptability in all the samples was AC75, followed by AC50, AC25, control, and AC100, indicating that excessive AC would reduce the sensory quality of the sausage. In general, all sausage groups were judged as acceptable (Overall acceptability > 7), and the best one was the replacement of pork fat with 75% AC.

4. Conclusions

The substitution of pork fat with *Auricularia cornea* (AC) in sausages was found to be a viable alternative in this study. AC improved the protein, moisture, and ash contents, especially isoleucine, leucine, proline, palmitic, palmitoleic, oleic, and arachidonic acids in the sausages. In addition, cooking loss and water holding capacity were also improved in comparison with the control. Meanwhile, AC significantly reduced the fat (12.61%–87.56%) and energy (5.76%–56.40%) levels of the sausages. Furthermore, the sausages demonstrated the yellowness, whiteness, lightness, and soft texture on account of AC. From the sensory opinior, the best sausage formulation was 75% of pork fat replacement with AC. The results shown that AC could be possibility used as healthier food to decrease the content of pork fat in sausages.

Author Contributions: Conceptualization, Y.F.; methodology and writing—original draft preparation, L.Z.; formal analysis, M.C.; data curation, K.W.; investigation, G.J.; writing—review and editing, X.L.; supervision, S.D.; project administration and funding acquisition, L.W. All authors have read and agreed to the published version of the manuscript.

Funding: The authors are grateful to the Plan of Science and Technology Development of Jilin Province of China (No. 20210202066NC) and the Opening Project of Engineering Research Center of Edible and Medicinal Fungi (Ministry of Education), Jilin Agricultural University (No. JJJW2021001).

Institutional Review Board Statement: Not applicable.

Informed Consent Statement: Not applicable.

Data Availability Statement: Data are contained within the current manuscript.

Conflicts of Interest: The authors declare no conflict of interest.

References

1. Robert, P.; Zamorano, M.; González, E.; Silva-Weiss, A.; Cofrades, S.; Giménez, B. Double emulsions with olive leaves extract as fat replacers in meat systems with high oxidative stability. *Food Res. Int.* **2019**, *120*, 904–912. [CrossRef]
2. Kwon, H.C.; Shin, D.M.; Yune, J.H.; Jeong, C.H.; Han, S.G. Evaluation of gels formulated with whey proteins and sodium dodecyl sulfate as a fat replacer in low-fat sausage. *Food Chem.* **2020**, *337*, 127682. [CrossRef] [PubMed]
3. de Souza Paglarini, C.; de Figueiredo Furtado, G.; Honorio, A.R.; Mokarzel, L.; da Silva Vidal, V.A.; Ribeiro, A.P.B.; Pollonio, M.A.R. Functional emulsion gels as pork back fat replacers in Bologna sausage. *Food Struct.* **2019**, *20*, 100105. [CrossRef]

4. Wang, X.; Xie, Y.; Li, X.; Liu, Y.; Yan, W. Effects of partial replacement of pork back fat by a camellia oil gel on certain quality characteristics of a cooked style Harbin sausage. *Meat Sci.* **2018**, *146*, 154–159. [CrossRef] [PubMed]
5. Kılıç, B.; Özer, C.O. Potential use of interesterified palm kernel oil to replace animal fat in frankfurters. *Meat Sci.* **2019**, *148*, 206–212. [CrossRef] [PubMed]
6. Oh, I.; Lee, J.; Lee, H.G.; Lee, S. Feasibility of hydroxypropyl methylcellulose oleogel as an animal fat replacer for meat patties. *Food Res. Int.* **2019**, *122*, 566–572. [CrossRef]
7. Saygi, D.; Ercoşkun, H.; Şahin, E. Hazelnut as functional food component and fat replacer in fermented sausage. *J. Food Sci. Technol.* **2018**, *55*, 3385–3390. [CrossRef] [PubMed]
8. Utama, D.T.; Jeong, H.S.; Kim, J.; Barido, F.H.; Lee, S.K. Fatty acid composition and quality properties of chicken sausage formulated with pre-emulsified perilla-canola oil as an animal fat replacer. *Poult. Sci.* **2019**, *98*, 3059–3066. [CrossRef]
9. de Carvalho, F.A.L.; Munekata, P.E.; Pateiro, M.; Campagnol, P.C.; Domínguez, R.; Trindade, M.A.; Lorenzo, J.M. Effect of replacing backfat with vegetable oils during the shelf-life of cooked lamb sausages. *LWT* **2020**, *122*, 109052. [CrossRef]
10. Kim, D.H.; Shin, D.M.; Seo, H.G.; Han, S.G. Effects of konjac gel with vegetable powders as fat replacers in frankfurter-type sausage. *Asian-Australas. J. Anim. Sci.* **2019**, *32*, 1195–1204. [CrossRef]
11. Wang, L.; Li, C.; Ren, L.; Guo, H.; Li, Y. Production of pork sausages using *Pleaurotus eryngii* with different treatments as replacements for pork back fat. *J. Food Sci.* **2019**, *84*, 3091–3098. [CrossRef] [PubMed]
12. Alejandre, M.; Astiasarán, I.; Ansorena, D.; Barbut, S. Using canola oil hydrogels and organogels to reduce saturated animal fat in meat batters. *Food Res. Int.* **2019**, *122*, 129–136. [CrossRef] [PubMed]
13. Glisic, M.; Baltic, M.; Glisic, M.; Trbovic, D.; Jokanovic, M.; Parunovic, N.; Vasilev, D. Inulin-based emulsion-filled gel as a fat replacer in prebiotic-and PUFA-enriched dry fermented sausages. *Int. J. Food Sci. Technol.* **2019**, *54*, 787–797. [CrossRef]
14. Guedes-Oliveira, J.M.; Costa-Lima, B.R.C.; Oliveira, D.; Neto, A.; Deliza, R.; Conte-Junior, C.A.; Guimarães, C.F.M. Mixture design approach for the development of reduced fat lamb patties with carboxymethyl cellulose and inulin. *Food Sci. Nutr.* **2019**, *7*, 1328–1336. [CrossRef] [PubMed]
15. Li, X.; Zhang, K. A review of the advances of the research on a white variant strain in Genus *Auricularia*. *Edible Med. Mushrooms* **2016**, *24*, 230–233. (In Chinese)
16. Si, F.; Liu, X.; Deng, J. Optimization of extraction process and physicochemical properties of dietary fiber from *Auricularia cornea* var. Li Root. *Food Ferment. Ind.* **2019**, *45*, 209–214.
17. Cao, Y.; Bao, H.; Li, X.; Bao, T.; Li, Y. Anti-tumor activities of *Auricularia cornea* fruiting body extract in H22 bearing mice. *Mycosystema* **2017**, *36*, 1289–1298. (In Chinese)
18. Pintado, T.; Muñoz-González, I.; Salvador, M.; Ruiz-Capillas, C.; Herrero, A.M. Phenolic compounds in emulsion gel-based delivery systems applied as animal fat replacers in frankfurters: Physico-chemical, structural and microbiological approach. *Food Chem.* **2020**, *340*, 128095. [CrossRef] [PubMed]
19. Wan, K.; Cong, M.; Teng, X.; Feng, M.; Ren, L.; Wang, L. Effects of Pine Bark Extract on Physicochemical Properties and Biological Activity of Active Chitosan Film by Bionic Structure of Dragonflfly Wing. *Coatings* **2021**, *11*, 1077. [CrossRef]
20. Jridi, M.; Abdelhedi, O.; Souissi, N.; Kammoun, M.; Nasri, M.; Ayadi, M.A. Improvement of the physicochemical, textural and sensory properties of meat sausage by edible cuttlefish gelatin addition. *Food Biosci.* **2015**, *12*, 67–72. [CrossRef]
21. Kumakura, K.; Kato, R.; Kobayashi, T.; Sekiguchi, A.; Kimura, N.; Takahashi, H.; Matsuoka, H. Nutritional content and health benefits of sun-dried and salt-aged radish (takuan-zuke). *Food Chem.* **2017**, *231*, 33–41. [CrossRef] [PubMed]
22. Laranjo, M.; Gomes, A.; Agulheiro-Santos, A.C.; Potes, M.E.; Cabrita, M.J.; Garcia, R.; Elias, M. Impact of salt reduction on biogenic amines, fatty acids, microbiota, texture and sensory profile in traditional blood dry-cured sausages. *Food Chem.* **2017**, *218*, 129–136. [CrossRef]
23. Liu, X.; Qu, H.; Gou, M.; Guo, H.; Wang, L.; Yan, X. Application of *Weissella cibaria* X31 or *Weissella confusa* L2 as a starter in low nitrite dry-fermented sausages. *Int. J. Food Eng.* **2020**, *16*, 20190344. [CrossRef]
24. Wang, M.; Guo, H.; Liu, X.; Jiang, G.; Li, C.; Li, Y. Roles of *Lentinula edodes* as the pork lean meat replacer in production of the sausage. *Meat Sci.* **2019**, *156*, 44–51. [CrossRef]
25. Chen, Y.; Jia, X.; Sun, F.; Jiang, S.; Liu, H.; Liu, Q.; Kong, B. Using a stable pre-emulsified canola oil system that includes porcine plasma protein hydrolysates and oxidized tannic acid to partially replace pork fat in frankfurters. *Meat Sci.* **2020**, *160*, 107968. [CrossRef]
26. Monteiro, G.; Souza, X.; Costa, D.; Faria, P.; Vicente, J. Partial substitution of pork fat with canola oil in Toscana sausage. *Innov. Food Sci. Emerg. Technol.* **2017**, *44*, 2–8. [CrossRef]
27. Summo, C.; De Angelis, D.; Difonzo, G.; Caponio, F.; Pasqualone, A. Effectiveness of oat-hull-based ingredient as fat replacer to produce low fat burger with high beta-glucans content. *Foods* **2020**, *9*, 1057. [CrossRef] [PubMed]
28. dos Santos, M.; Munekata, P.E.; Pateiro, M.; Magalhães, G.C.; Barretto, A.C.S.; Lorenzo, J.M.; Pollonio, M.A.R. Pork skin-based emulsion gels as animal fat replacers in hot-dog style sausages. *LWT* **2020**, *132*, 109845. [CrossRef]
29. Sousa, S.C.; Fragoso, S.P.; Penna, C.R.A.; Arcanjo, N.M.O.; Silva, F.A.P.; Ferreira, V.C.S.; Araújo, Í.B.S. Quality parameters of frankfurter-type sausages with partial replacement of fat by hydrolyzed collagen. *LWT Food Sci. Technol.* **2017**, *76*, 320–325. [CrossRef]
30. Salcedo-Sandoval, L.; Cofrades, S.; Ruiz-Capillas, C.; Solas, M.T.; Jiménez-Colmenero, F. Healthier oils stabilized in konjac matrix as fat replacers in n-3 PUFA enriched frankfurters. *Meat Sci.* **2013**, *93*, 757–766. [CrossRef] [PubMed]

31. dos Santos Alves, L.A.A.; Lorenzo, J.M.; Gonçalves, C.A.A.; Dos Santos, B.A.; Heck, R.T.; Cichoski, A.J.; Campagnol, P.C.B. Production of healthier bologna type sausages using pork skin and green banana flour as a fat replacers. *Meat Sci.* **2016**, *121*, 73–78. [CrossRef] [PubMed]
32. Doi, M.; Yamaoka, I.; Fukunaga, T.; Nakayama, M. Isoleucine, a potent plasma glucose-lowering amino acid, stimulates glucose uptake in C2C12 myotubes. *Biochem. Biophy. Res. Commun.* **2003**, *312*, 1111–1117. [CrossRef] [PubMed]
33. Zhang, Y.; Guo, K.; LeBlanc, R.E.; Loh, D.; Schwartz, G.J.; Yu, Y.H. Increasing dietary leucine intake reduces diet-induced obesity and improves glucose and cholesterol metabolism in mice via multimechanisms. *Diabetes* **2007**, *56*, 1647–1654. [CrossRef] [PubMed]
34. Wu, G.; Bazer, F.W.; Burghardt, R.C.; Johnson, G.A.; Kim, S.W.; Knabe, D.A.; Spencer, T.E. Proline and hydroxyproline metabolism: Implications for animal and human nutrition. *Amino Acids* **2011**, *40*, 1053–1063. [CrossRef] [PubMed]
35. Yang, Z.H.; Miyahara, H.; Hatanaka, A. Chronic administration of palmitoleic acid reduces insulin resistance and hepatic lipid accumulation in KK-A y Mice with genetic type 2 diabetes. *Lipids Health Dis.* **2011**, *10*, 120. [CrossRef] [PubMed]
36. Tallima, H.; El Ridi, R. Arachidonic acid: Physiological roles and potential health benefits—A review. *J. Adv. Res.* **2018**, *11*, 33–41. [CrossRef] [PubMed]

MDPI
St. Alban-Anlage 66
4052 Basel
Switzerland
www.mdpi.com

Coatings Editorial Office
E-mail: coatings@mdpi.com
www.mdpi.com/journal/coatings

Disclaimer/Publisher's Note: The statements, opinions and data contained in all publications are solely those of the individual author(s) and contributor(s) and not of MDPI and/or the editor(s). MDPI and/or the editor(s) disclaim responsibility for any injury to people or property resulting from any ideas, methods, instructions or products referred to in the content.

www.ingramcontent.com/pod-product-compliance
Lightning Source LLC
LaVergne TN
LVHW070648100526
838202LV00013B/907